WRITING ACROSS WORLDS

International migration has long been a dominant feature of world literature from both post-industrial and developing countries. The increasing demands of the global economic system and continued political instability in many of the world's regions have highlighted this shifting map of the world's peoples.

Yet, political concern for the larger-scale economic and social impact of migration has effectively obscured the nature of the migratory experience itself, the emotions and practicalities of departure, travel, arrival and the attempt to rebuild a home.

Writing Across Worlds explores an extraordinary range of migration literatures, from letters and diaries to journalistic articles, autobiographies and fiction in order to analyse the reality of the migrant's experience. The sheer range of writings – Irish, Friulian, Italian, Jewish and South Asian British, *Gastarbeiter* literature from Germany, *pied-noir*, French-Algerian and French West Indian writing, Caribbean novels, Slovene emigrant texts, Japanese-Canadian writing, migration in American novels, narratives from Australia, South Africa, Samoa and elsewhere – illustrate the diversity of global migratory experience and emphasise the social context of literature.

The geographic and literary range of *Writing Across Worlds* makes this collection an invaluable analysis of migration, giving voice to the hope, pain, nostalgia and triumph of lives lived in other places.

Russell King is Professor of Geography at the University of Sussex; **John Connell** is Associate Professor of Geography at the University of Sydney; **Paul White** is Reader in Geography at the University of Sheffield.

WRITING ACROSS WORLDS

Literature and migration

Edited by Russell King,
John Connell and
Paul White

London and New York

First published 1995
by Routledge
11 New Fetter Lane, London EC4P 4EE

Simultaneously published in the USA and Canada
by Routledge
29 West 35th Street, New York, NY 10001

Typeset in Garamond by
Ponting–Green Publishing Services, Chesham, Bucks
Printed and bound in Great Britain by
TJ Press (Padstow) Ltd, Padstow, Cornwall

British Library Cataloguing in Publication Data
A catalogue record for this book is available from the
British Library

Library of Congress Cataloging in Publication Data
A catalogue record for this book has been requested

ISBN 0–415–10529–3
0–415–10530–7 (pbk)

CONTENTS

CONTENTS

CONTRIBUTORS

Robert Aldrich is Associate Professor of Economic History at the University of Sydney.

Claire Alexander is British Academy Research Fellow in the Faculty of Social Sciences, Open University.

Suresht Renjen Bald is Professor of Politics at Willamette University, Oregon.

Arnold Cassola is Senior Lecturer in Maltese and Comparative Literature at the University of Malta.

John Connell is Associate Professor of Geography at the University of Sydney.

Jonathan Crush is Associate Professor of Geography at Queen's University, Kingston, Ontario.

Patrick Duffy is Associate Professor of Geography at St Patrick's College, Maynooth, Ireland.

Sabine Fischer is Lecturer in German for Social Scientists at the University of Sheffield.

Alec Hargreaves is Professor of French and Francophone Studies at Loughborough University.

Rosemarie Jones was Lecturer in French in the School of European Studies at the University of Sussex until her death in October 1992.

Roy Jones is Associate Professor of Geography, Curtin University of Technology, Perth, Western Australia.

Russell King is Professor of Geography, School of European Studies, University of Sussex.

Audrey Kobayashi is Professor of Geography and Director of the Institute of Women's Studies, Queen's University, Kingston, Ontario.

Moray McGowan is Professor of Germanic Studies at the University of Sheffield.

Jerneja Petrič is Associate Professor of American Literature in the Department of English, University of Ljubljana, Slovenia.

Federica Scarpa is Associate Professor of Translation Studies at the University of Trieste.

Marlena Schmool is Director of the Community Research Unit, Board of Deputies of British Jews, London.

Neil Larry Shumsky is Associate Professor of History at the Virginia Polytechnic Institute and State University, Blacksburg, USA.

Stanley Waterman is Associate Professor of Geography and Dean of the Faculty of Social Sciences and Mathematics, University of Haifa, Israel.

Paul White is Reader in Geography at the University of Sheffield.

PREFACE

This book had its origins in a specific place at a specific time, as does every migration event. The date was 28 March 1990 and the 'place' the Gozo ferry, en route to the subsidiary island of the Maltese archipelago. Gozo is a pretty, unpretentious Mediterranean island whose landscape is impregnated with evidence of migration. Although many Gozitans have migrated for good, it is the imprint of return migration that catches the eye: large American cars, out of place on an island with just a few country lanes to drive on; newly built houses crowned with stone eagles or kangaroos, votive symbols of successful emigrant missions to the United States or Australia. The name-plates of such houses – 'Maple Leaf House', 'Australia House', 'Tottenham Hotspur', 'God Bless America' – are the titles of the emigrant stories of those who dwell within.

It was on the Gozo ferry that two of the editors, Connell and King, hatched the plot for the book, an idea sparked off by a conference paper presented at the University of Malta the previous day by a local scholar, Arnold Cassola. This paper, slightly revised, appears as Chapter 11 of this book. A special note of thanks is due to Dr Cassola for his patience in waiting five years whilst the rest of the book's chapters were commissioned, written, revised and edited.

Why this book? For more than a century the study of human migration has formed an important part of the research agenda for social scientists – particularly geographers but also sociologists, anthropologists, social historians, statisticians, economists and others. There is a myriad of books and research papers recording the social, demographic, cultural and economic aspects of migration at scales ranging from the local through the regional to the global. The approaches adopted in such studies have often been aggregative, working from statistical data culled from censuses or other primary surveys: where more interpretative methods have been used, these have generally been developed on the basis of in-depth observational, participatory or ethnographic research. This social-scientific research, immensely rich and diversified in its own way, is often limited in its objectives, aiming to shed light on some single aspect of migration such as the decision to leave, residential location on arrival, or socio-linguistics. It fails to capture the essence of what it is like to *be* a migrant; and be, or not be, part of a community, a nation, a society – cut off from history and from a sense of place. It fails to portray nostalgia, anomie, exile, rootlessness, restlessness.

The humanities, and in particular the study of literature, have not concerned themselves directly with research questions dealing with migration, although changing residence, as a significant and life-transforming event, looms large in much literature. Yet a surprising amount of literature has been written by migrants themselves in novel, essay or other form. This book starts from the premise that 'non-academic' literature, written often (but by no means exclusively) by migrants, can offer powerful insights into the nature of the migration process and the experience of being a migrant. Literary accounts focus in a very direct and penetrating way on issues such as place perception, landscape symbolism, senses of displacement and transformation, communities lost and created anew, exploitation, nostalgia, attitudes towards return, family relationships, self-denial and self-discovery, and many more. Such insights are often infinitely more subtle and meaningful than studies of migrants which base themselves on cold statistics or on the depersonalised, aggregate responses to questionnaire surveys.

Thus the book seeks to explore, and to demonstrate the validity of, a cross-disciplinary approach which brings together the social scientist's concerns with explanation and the student of literature's expertise in the handling of text. It thus joins an established genre of work linking geography with literature but which so far has concentrated on exploring *place* rather than *movement*. Taking as read the legitimacy of humanistic and qualitative approaches to geography and social science, we would seek to justify the book on three grounds which are more specific. First, as already implied, we believe that in the field of migration studies cross-disciplinary collaboration between the humanities and the social sciences is a fertile way of extending the methodological range of research, of highlighting previously neglected aspects, and of identifying new questions for consideration. Extending the tools of analysis in this way allows the migration researcher to attain greater understanding of the subject-matter of migration. Second, the collaboration between social scientists and literature specialists can be of benefit to the latter through the development of wider contextual understanding of textual artefacts. Third, the cross-disciplinary collaboration here applied to a single topic has strong potential for application in other areas where interpretative study of human activity is involved.

Migrant literatures have evolved through various forms and a number of typologies can be suggested. A preliminary classification, suggested by Patrick Duffy in Chapter 2, draws a distinction between the largely autobiographical work of migrants themselves with their direct, personal accounts of migration, and general fiction by more (for want of a better term) 'professional' writers which reflects either directly or indirectly on migration. Duffy notes that many famous Irish writers were emigrants (Oscar Wilde,

Samuel Beckett, James Joyce, Sean O'Casey, Edna O'Brien ...) and whilst their own experiences of migration and exile were very different from those of the supposedly archetypal Famine migrant or Donegal bricklayer, their literary genius was often directed to themes of migration.

Some of the most telling accounts of the immigration experience are the work of authors who are not immigrants at all, but who are in some way the product of past migrations. Alec Hargreaves (Chapter 6) examines the exciting work of the young *Beurs*, the French-born sons and daughters of Algerian immigrants, stressing that it is only when they 'return' to Algeria that they experience the migrant condition in its most evident forms. Similarly Sabine Fischer and Moray McGowan (Chapter 3) point out that much of the post-1960s *Gastarbeiter* literature has been produced by writers whose ethnic identity may have exposed them to comparable discrimination but who are not, socially, *Gastarbeiter* themselves. In the work of Caribbean authors like V.S. Naipaul, David Dabydeen and the late Samuel Selvon we have expressions of migrant and ethnic identities that are more complex historically, reflecting mixtures of slavery, indentured migration and postwar migration to England; both Claire Alexander (Chapter 4) and Robert Aldrich (Chapter 7) address these issues.

An evolutionary series of forms of migrant literature can be suggested as follows. The model starts with 'pre-literatures' such as ethnic newsletters or community newspapers; these often represent an attempt to hold on to past identities. Other pre-literary forms include diaries, letters, songs and other oral narratives. Federica Scarpa makes use of migrants' letters as important documents of testament in her chapter on Friulian emigration, whilst Jonathan Crush (Chapter 15) explores the significance of *lifela*, a form of oral recitative poetry born out of the Basotho migrant experiences in the South African mines. Historically, of course, most migrant labourers were poorly educated, even illiterate. Their 'pre-literatures' have been condemned as rudimentary and naive but this is to miss their true value as evidence to reconstruct the experience of migration: as Crush observes, transcripts of *lifela* and other performance art speak tantalisingly of a world outside and beyond the intrusions of the interviewer and detached from the measured prose of scholarly analysis. Even popular music – Country and Western, blues, jazz, or the lyrics of U2 or The Pogues – is often about displacement, nostalgia for 'home', or the casualties of migration failure. Take me home, country roads!

The sequence continues with pre-literatures giving way to poems, short stories and *reportage* written and often published in the mother tongue, usually in newspapers and magazines produced from within the ethnic community. Fischer and McGowan (Chapter 3) regard this as a form of therapy by the victims of social processes, articulating their commonality of migrant experience by recording it in simple, conventional, often autobiographical forms. As with letters and oral histories, autobiographies and other first-hand accounts certainly have value, but not necessarily face value. Migrants rarely remember in neat and accurate chronology. More important, even the most prominent modes of remembering are subject to implicit social control or redirectioning based on the power exercised by existing dominant and popular cultural forms. The hegemony of the publishing industry imposes its own control over what is published or distributed widely – and therefore maybe only the accounts which conform to dominant stereotypes get printed, leaving out oppositional literature which more closely captures or mirrors the feelings and experiences of migrants. One of the aims of this book is to recover some of these oppositional voices, recognising that, even within the migrant experience of a particular group, there are likely to be wide differences of viewpoint and interpretation, not all of which will find outlets even within publishing channels controlled by members of the ethnic group.

The next stage of our simple model sees the emergence of fully-fledged creative literatures using a variety of more sophisticated narratives, including complex novels, plays, films and poetry. They are written in the languages of the dominant metropolitan powers which caused the migration to happen and received the migrants: English, French, German and so on. The authors, mostly of migrant ancestry, are often professional writers and their work ranges from portrayals of the direct migration encounters of primary migrants to more complicated issues of identity and ambivalence resulting from cultural confrontation and fusion. Some authors, like the Samoan Albert Wendt (Chapter 17), have used all these forms as their work has changed and evolved. Many of the literary products are postmodern explorations of the post-colonial, multicultural encounter. They also reflect the changed nature of migration itself in the late-twentieth-century world. The writings of Salman Rushdie for instance show, in the subtlest of ways, what it is like to be a post-colonial cosmopolitan, whilst not ignoring the wider histories and geographies of migration and diaspora. As well as demonstrating the global-isation of migration – in which a move from middle-class Bombay to London may be less of a cultural jolt than a migration from a remote Welsh hamlet to London – Rushdie also, tragically, provokes the global backlash of religious fundamentalism.

The final phase of the migration-literature model sees the linkage reversed. Literature no longer simply sheds light on migration processes, but migrants, or writers born out of the migrant experience, play an increasingly prominent role in shaping the development of erstwhile 'pure' national (and inter-national) literatures. Within the anglophone and francophone literary worlds, it is now clear that migration and colonialism lie at the very roots of certain literary structures. This is demonstrated for French West Indian and *pied-noir* writing by Robert Aldrich and Rosemarie Jones in Chapters 7 and 8 respectively. After the 1950s English literature became infused, and invigorated, with new writers from the newly independent English-speaking countries: initially North America and Australia, subsequently the Caribbean, South Asia and Africa. Such writers move easily amongst their various worlds: theirs are the literatures of post-migration.

The chapters of this book are contributed by colleagues based in universities and research institutes around the world working in a range of disciplines including geography, area studies, literary studies, politics, women's studies and history. Chapter 1 expands some of the themes touched on in this Preface and develops theoretical perspectives on the use of literature by social scientists. After a brief statement of the kinds of questions that social scientists seek to answer about migration, the introductory chapter sets out the underlying problematic: traditional social-scientific research (especially in geography) has felt uncomfortable in dealing with individuals; yet texts are intrinsically personal and individual-based artefacts. The chapter then goes on to review recent developments in ontology and method, particularly through concepts of structuration, that permit individual experience to be brought into a more general framework of understanding. The migrant is seen as the critical participant-observer into his/her own condition, enabling powerful insights to be made into the 'insider–outsider' dichotomy and the real 'lived experiences' of migration. But, of course, 'migrant texts' (of whatever type) must not be accepted uncritically, and Paul White's opening chapter also lists some of the questions to be asked of a text, and of its author, before the text's validity as evidence can be judged.

The rest of the book, Chapters 2–17, presents case-studies of migrant literatures and of themes explored in and through such writings. The chapters represent a wide geographical range: Europe, the Mediterranean, North and South Africa, North America, the Caribbean, Asia, the Pacific and Australia. They comprise a selection of what may be regarded as some of the major migrant literatures – the Irish, the Caribbean, *Gastarbeiter* writing from Germany – as well as some lesser-known but equally fascinating literatures such as the Slovene, Samoan and Japanese-Canadian. The variety of regional contexts analysed demonstrates that population dislocation and mobility give rise to a certain repeated set of themes and images whatever the spatial setting. Details may differ according to local economic and political situations, and the outcomes of constraining circumstances may also vary in the migratory (or non-migratory) strategies chosen by individuals or groups, but common elements emerge forcefully from the literatures.

The literatures analysed in the book deal with both individual circumstances and the wider embedding of migration as an aspect of societal make-up or even of a group's cultural identity. Certain societies exist – for example the Irish or the Caribbean – in which migration is (or was) not only the 'normal' life-path, conditioning the entire society, but also dominant in literature and other cultural artefacts such as folklore. In Chapter 9 by

Federica Scarpa, for example, a heroic vision of Friulian migrants is seen as contributing to the self-assertion of Friulian culture as a whole. Similarly in Chapter 8 by the late Rosemarie Jones the whole issue of the existence of a separate Algerian (under French colonialism) or *pied-noir* identity can be approached through the reading of a literature heavily saturated with metaphors related to various types of movement. And the foundation of certain aspects of Australian society on self-images of the country as offering opportunity, openness and welcome to migrants is examined, and challenged, in Chapter 16 by Roy Jones.

Several chapters of the book deal with the literatures of emigrant societies (Ireland, Malta, Friuli) and demonstrate the centrality of migration as an element in those societies. By definition, however, these origin-based literatures must also look outwards with the migrants; in doing so they often shed light on the 'other' as represented by the world of destinations, knowledge of which is part of everyday socialisation processes. But the emigration perspective is self-reflexive about the society of origin. It was James Joyce, who lived part of his writing life in Trieste, who stated that the route to Tara is through Holyhead: in other words, people left Ireland to understand it. On the other hand, once a migration takes place, the migrant may never be quite sure where home is, ever again. What ensues is a permanent mobility of the mind, if not the body, a constant dual or multiple perspective on place.

Most of the remaining chapters deal with literatures spawned by communities of migrants in their destination settings. A variety of types of literary output is dealt with and the chapter authors explore a diversity of migration-related themes. Some emphasise the more intensely personal representations of experience that spring from the pages of creative literature: an approach perhaps best exemplified in Claire Alexander's Chapter 4. Other contributions explicitly focus on how migrant experiences have differed according to the period or generation of the migrants. Suresht Renjen Bald's contribution (Chapter 5), for example, demonstrates the divergences in response between earlier and later generations of South Asian writers in Britain. The writing genre may also change, and perhaps 'evolve', with each 'generation' of migrants (by 'generation' we can mean either a later wave of primary migrants, or succeeding generations of offspring born of immigrant ancestors in the country of settlement). This complex sequencing of literary type, 'generation' of migration, and migration experience forms part of the structure of several chapters, including Sabine Fischer and Moray McGowan on *Gastarbeiter* literature, Alec Hargreaves on Algerian literature in France, Jerneja Petrič on Slovene literatures in North America and Australia, and Audrey Kobayashi on three generations of Japanese-Canadians. This serves to remind us once again that much 'migrant' literature is not, in fact, by migrants, but by writers who are labelled or racialised in some way by the societies in which they live, and in which they were actually born. The emergence of distinctive cultural voices amongst such writers reflects a growing series of issues concerned with group identity and the search for self-confidence, acceptance and legitimation in an environment of alienation and conflict.

Other chapters are more explicitly 'two-ended'. Movement between the two worlds, and two identities, of Samoa and New Zealand is the concern of Chapter 17 by John Connell, whilst in Chapter 13 Neil Larry Shumsky concentrates on the return migration theme as portrayed in a selection of early-twentieth-century American novels written by or about immigrants from Europe.

Taken together, the chapters of this book are a celebration of the literary approach to a key geographic phenomenon. The literatures of migrants, be they creative or *reportage*, offer an entrée into the interior world of migrant experience. Migrant literature is individual, subjective, diverse: it reflects but also may exaggerate or even invert the social experience that drives it. We may be surprised, and hence illuminated, by the migrant's feel for the quotidian and commonplace, and by migrant perceptions of the odd and the exotic. For some groups, migration is not a mere interval between fixed points of departure and arrival, but a mode of being in the world – 'migrancy'. The French West Indian concept of *errance*, wandering, which is present in all the literatures of the Caribbean, captures something of the historical depth and psycho-social significance of the emigration tradition in West Indian island societies. They were created by migration; their reproduction is through further migration.

The migrant voice tells us what it is like to feel a stranger and yet at home, to live simultaneously inside and outside one's immediate situation, to be permanently on the run, to think of returning but to realise at the same time the impossibility of doing so, since the past is not only another country but also another time, out of the present. It tells us what it is like to traverse borders like the Rio Grande or 'Fortress Europe', and by doing so suddenly became an illegal person, an 'other'; it tells us what it is like to live on a frontier that cuts through your language, your religion, your culture. It tells of long-distance journeys and relocations, of losses, changes, conflicts, powerlessness, and of infinite sadnesses that severely test the migrant's emotional resolve. It tells of new visions and experiences of the familiar and unfamiliar. For those who come from elsewhere, and cannot go back, perhaps writing becomes a place to live.

Appropriately for a project on migration, this book is a product of social networking on a global scale. From its birth on a 'voyage', it has evolved via bilateral and trilateral meetings between the editors in many places: Dublin, Sheffield, Brussels, Vienna and successive venues of the Annual Conference of the Institute of British Geographers over the past four years. Its chapters have been written in ten countries of the world. Many hands on many keyboards have contributed to editing the final product and in this connection we should like to thank several of the support staff of the School of European Studies at the University of Sussex: Vincenzo Raimo, Mary Clarke, Jean Crew, Louise Gerber and Jenny Money. Their help was vital.

Russell King
John Connell
Paul White

ACKNOWLEDGEMENTS

All translations of quotes from literature in foreign languages are by the authors of the individual chapters, unless otherwise stated.

The authors would like to acknowledge permission to reprint the following: 'Forest of Europe', from *Poems* by Derek Walcott, reprinted in Chapter 4 by permission of Random House UK and Farrar Strauss & Giroux, New York; excerpts from 'East Coker' and 'Little Gidding', from *Four Quartets* in *Collected Poems 1909–1962* by T.S. Eliot, reprinted in Chapter 1 by permission of Faber and Faber Ltd and Harcourt Brace & Company; extracts from *Sons for the Return Home* by Albert Wendt, reprinted in Chapter 17 by permission of the copyright owner c/o Curtis Brown (Aust) Pty Ltd; extracts from *The Limit of Love* by Frederic Raphael © Frederic Raphael, 1960, reproduced in Chapter 12 by permission of the author c/o Rogers, Coleridge & White Ltd, 20 Powis Mews, London W11 1JN; selection of poetry from *Brocade Pillow* by Ihei Ito, reprinted in Chapter 14 by permission of Weatherhill, New York.

Every effort has been made to obtain permission to reproduce copyright material. If any proper acknowledgement has not been made, we would invite copyright holders to inform us of the oversight.

1

GEOGRAPHY, LITERATURE AND MIGRATION

Paul White[1]

INTRODUCTION

We live in what has recently been termed 'The Age of Migration'.[2] Geographical movement can be seen as a crucial human experience. Such movement occurs within a striking concatenation of economic, political, social and cultural circumstances which provide both structural forces driving mobility and also the controlling mechanisms that limit and channel the selection of people and places involved. These circumstances lie within the range of traditional social-scientific concerns for the aggregate forces and developments occurring within human populations.

At the same time population movement transforms all the elements involved, not only in the structural circumstances that underpin migration systems, but also within the places and the people bound up in migratory experiences. Places of origin, of passage, and of destination of migrants are altered as a result of the flows of people that affect them. Transformations also occur in the lives of all those involved; not just the migrants themselves but also those who directly come into contact with them and those who, indirectly, are affected by social, political and economic changes induced by migration. The role of human agents is a determining one, but only within the structural context in which those agents are located.

Migration therefore changes people and mentalities. New experiences result from the coming together of multiple influences and peoples, and these new experiences lead to altered or evolving representations of experience and of self-identity. Such representations are then manifest in cultural artefacts of many kinds – new forms of dress, of food cultures and of consumerism, new styles of music and of poetry, new political ideologies, new forms of literary production. All of these can be seen to have their own claims as authentic materials. However this book, and this introductory chapter, explore just one aspect of these representations – that of literary output – without seeking to privilege or prioritise this over other artefacts.

Literary output can be considered at a variety of levels, of which two merit attention here. At one level we can consider individual works, but at another

1

we can consider a full body of literature that arguably hangs together through a relationship with a migratory record or history, often on a societal scale. At the first level, therefore, we may be dealing with individual authors and with the representation of the experience of particular people; at the second we may be concerned with responses in whole societies or nations that have been affected by population movement.[3] The remainder of this chapter will seek to develop a framework and to highlight a series of issues in which the concerns of the social scientist and of the student of literature can be seen to converge fruitfully in the analysis of texts relating to migratory experience.

MIGRATION AND IDENTITY SHIFT

A useful starting-point for the discussion of the representational outcomes of migration experiences lies in setting up a conceptual framework consisting of a series of possible shifts in identity that occur in relation to migration, both at the individual and at larger-group levels.[4] Such realignments of identity may both precede migration (and in a sense, therefore, 'cause' it), and they may also occur as a result of movement to a new location. Migration 'events' therefore occur within personal biographies that neither start nor end at those events, but which provide the context for them.[5] We may, perhaps, conceptualise a number of overlapping multiple identities which are the subject of constant renegotiation in the face of the conflicts and compromises of everyday life. At any point in our lives we can think of ourselves as relating to a number of identities – in gender terms (concerning gender roles and gendered behaviour: sexual identity may perhaps be better considered as a separate element), in terms of a stage in the life-course, in terms of age and family status, in terms of economic identity (related to occupational identity but also to attributes of consumption and savings propensities), in terms of linguistic, religious and other cultural identities and in terms of ethnic identity. In the analysis of identity shift through migration it can be argued that creative literature contains some of the most effective explorations of identity issues.

The act of migration often relates to the calling into question of many of these aspects of identity that make up the individual's personality and psychological self-image. This is not to say that migrants, before migration, have necessarily 'fitted in' to a homogeneous societal structure with no traces of discordance: indeed, sociological and anthropological studies have often suggested that migrants may be effectively 'lost' to their home communities long before they actually pack their bags and leave,[6] and of course not 'fitting in' may be a primary cause of migration. However, the words 'migration' and 'change' can almost be regarded as synonyms in this context – why migrate if such movement does not result in change, or does not accommodate an identity change that has already occurred?

Migrants, whether individually, in groups, or as whole displaced societies,

2

are open to new influences. Many of these provide a challenge to earlier self-perceptions and self-images, and through such challenges the compositional elements of multiple identities may be redefined. This is probably most significant in terms of ethnic identities, where many migrants may not have held a particularly strong view of their own ethnicity prior to movement, but where they may find themselves in situations where they are confronted by an alternative ethnic awareness that labels them and confines them to a stereotyped 'otherness' from which there appears little chance of escape, although a number of differently constrained responses are possible.

Other aspects of individual, group or societal identity are also open to transformation through migration. Much academic writing about migration has tended to ignore the significance of gender issues, but gender roles may be crucially affected by movement.[7] In other spheres, families are often broken up, temporarily or on a more long-term basis. Economic status is altered, with changed employment, changed income and wealth, and with changed patterns of consumption being very common. Secularisation may ensue, or alternatively there may be reassertions of cultural (religious) distinctiveness through a re-energising of attributes of distinction. Habitual language use may slide or be jolted from one tongue to another, with all that such a change implies about the means of representation in words – spoken or written. Projects, dreams and ultimate goals may be revised. All of these changes are discernible in creative writing about migration.

Shifts of identity are highly complex, sometimes unstable, and often have reversible elements built into them. The titles of various works on migration, produced by creative writers or by social-scientific researchers, suggest that migrants may live in a number of worlds, and move between them on a daily, annual or seasonal rhythm.[8] Other changes resulting from migration include attempts to re-create elements of former lives (possibly accentuating significant icons of that existence into quasi-talismans of high symbolic or ritual significance); attempts to integrate or assimilate completely (which may be blocked by a number of mechanisms within the 'host' society); or the creation of a new identity which is characterised by a feeling of independence from both the society of origin and the social structures of the destination. These changes in identity cannot be pinned down to a rigid linear continuum, for they represent the multiple and continually renegotiated outcomes of complex multifaceted phenomena operating both within individual biographies and for societies as a whole.[9]

A common feature of many migrants and migrant cultures is ambivalence. Ambivalence towards the past and the present: as to whether things were better 'then' or 'now'. Ambivalence towards the future: whether to retain a 'myth of return' or to design a new project without further expected movement built in. Ambivalence towards the 'host' society: feelings of respect, dislike or uncertainty. Ambivalence towards standards of behaviour: whether to cling to the old or to discard it, whether to compromise via symbolic events whilst adhering to the new on an everyday basis. The choices (or the paths taken, since in many cases 'choice' is not actually perceived to exist) depend not just on the individuals involved but also on the constraints of the situations in which migrants find themselves. And since these situations change on a variety of temporal scales, so the identities expressed through attitudes, behaviour and artefacts also change and may be marked by ambiguity.

The act of migration concerns people and places, but it also concerns time. The first movers settle down. More migrants follow the path of the earlier pioneers. The world these latter move to is not the same as for the first arrivals, since the existence of past movement will have in some way altered the conditions of reception, whether directly from people of similar origin, or in terms of the underlying social, economic and political conditions that will influence the experiences of the later arrivals. So, too, circumstances at the places of origin of the migrants change, in part because of earlier departures. Through time the identity of migrant groups and individuals changes, not simply because the people involved age, but because the experiences undergone progressively build up to influence the evolution of identities. What are at first immigrants, singly or in groups, progressively become something else. Here perceptions of what they become may actually be divergent. External labelling may see 'them' as 'communities' (often 'ethnic-minority communities') which may not, in fact, accord with the view from within. Externally driven categorisations can be over-rigid, with a great deal of over-generalisation so that, for example, all people of Afro-Caribbean origin in Britain are thought of as 'West Indians', ignoring the facts of individual island identity that are of great significance to the people concerned,[10] or in France labelling all North Africans as 'Arabs' when some are Jews and others are Berbers rather than of Arabic culture. Internally (and the very word is questionable since it implies a homogeneity of outlook amongst those involved) the ambivalence and ambiguity of identities already referred to often produces a much more circumspect approach to labelling, with divergences between generations, genders, and classes reflecting the different experiences undergone.

These features of migrant identities might be argued to lead towards concepts of pluralism or syncretism at a number of levels, both concerning migrant and non-migrant individuals and groups but also as describing the multiple experiences, reactions and self-identities of migrants themselves and of those of migrant origin. To seek to research pluralism is to seek to elucidate worlds of meaning and belief, of attitude, interpretation and behaviour, and to do so, as argued above, in a context of instability and change within these attributes. Among the representations of these worlds, those that occur through literature present exciting opportunities for analysis as inputs to research. Since the contexts for these worlds expressed in literature are so diverse, the overall project is one where social scientists and humanities scholars can fruitfully come together.

MIGRATION AS A MOTIF IN LITERATURE

The discussion that follows is confined to representations of migration defined from a geographical viewpoint as 'a change in the place of residence'.[11] No reference will be made to the very extensive genre of travel writing, part of which actually considers certain aspects of human character as discussed here. The equally extensive corpus of writing that has as a theme the relationship of people with place will also be put on one side, again despite the existence within it of certain elements common also to true writing about migration.[12] This is especially so of literature concerning the city – a significant proportion of which takes the viewpoint of the newly arrived migrant as one of its devices for exploring the human condition within the metropolis.[13]

Even leaving this body of literature on one side, the theme of migration *per se* is extremely common in writing produced over the last century – the period during which we have moved into the 'Age of Migration'. The frequency with which the theme appears is not simply a reflection of the realities of human existence, but also has an internal literary justification to it. Put simply, the theme of migration and its outcomes has been an inherently attractive one over recent decades to writers working in a number of different literary movements and traditions.

Accepting for one moment David Harvey's suggestion that the evolution from modernism to postmodernism reflects more elements of continuity than of change,[14] we can see that throughout the literary endeavours represented under these headings a number of themes relating to migration and its outcomes are of significance. Modernism takes the 'bewilderingly problematic' nature of human existence in the contemporary world and, emphasising the fragmentation of human experience, nevertheless still seeks to penetrate what is taken as a general, unifying, underlying reality, albeit through writing which makes use of multiple viewpoints and discontinuities.[15] Postmodernism is suspicious of such metanarratives, abandoning the belief in universals, and stressing the multiplicity and relativity of experiences while also regarding each strand as of potentially equal validity.

The relationships of both modernism and postmodernism to migration are strong. Fragmentation, dislocation and alienation are all very common themes in modernist writing: indeed as Hawthorne has pointed out, 'alienation becomes close to a cliché in modernist literature.'[16] This is often associated with pessimism about the individual condition and the outcome of individual projects, represented, for example, in the following extract from T.S. Eliot's 'Little Gidding', from *Four Quartets*:

5

And what you thought you came for
Is only a shell, a husk of meaning
From which the purpose breaks only when it is fulfilled
If at all. Either you had no purpose
Or the purpose is beyond the end you figured
And it is altered in fulfilment.[17]

Some of the associations of these negative emotions in modernism are with movement and change, and with the erosion of certainty that this entails.

The motif of alienation is continued in much postmodernist work, but is also accompanied by motifs of indeterminacy and of pluralism, which can be conceptualised as applying to cultures and societies as well as to literary style: indeed, postmodernism is as much a social ideology as a literary movement.[18] Certain of these motifs, however, lead into an assertion of the legitimacy of difference, and towards the celebration of such differences as a vibrant and enhancing aspect of contemporary life.

The act of migration accords closely with these wide issues of cultural evolution. Although for certain groups and societies (for example Western Ireland or some Caribbean islands) 'migrant cultures' exist where migration is accepted as the normal path for life, migration is generally about dislocation and the potential alienation of the individual from both old norms and new contexts. It is about change and, as argued earlier in this chapter, about identity. It is about movement.

In terms of the outcomes of migration there are also strong relationships between the establishment of migrant or ethnic communities and the legitimisation of 'otherness' in postmodernist discourse, with the search for an individual and group identity as a prime objective. In the Age of Migration these are themes that affect everyone, directly or indirectly, and put migration at the forefront of everyday influences in a role that is often disruptive. Social scientists have for some time seen migration as relating to marginality, both causing movement and resulting from it. A number of literary commentators are now beginning to argue that migration, dislocation and ensuing marginality are some of the most important influences subverting long-standing beliefs in the linearity of progress and the stability of cultural identity, and that these have been determining influences on the inner conditions of contemporary humanity.[19]

It is not, therefore, surprising that migration has been used as a topic through which writers have explored the human condition. We might also note that the ambiguities of language, including the use of metalanguages, metaphor and metonymy, can be constructively exploited in discussing what have been characterised earlier as the ambivalences and ambiguities inherent in many personal reactions to migration experiences.[20] One of the commonest uses of migration as a literary theme has actually been as a metaphor of death. Thus Stratis Haviaras, a Greek novelist who left his native country in 1967 and who now works as curator of the poetry collection in the Harvard University Library, uses emigration and death effectively as synonyms throughout his novel *When the Tree Sings*, about life in a small coastal town in southern Greece. At the end of the book the grandfather contrasts the habitual belief in the future, 'They'll make a fortune in exile, and they'll come back one day,' with the reality, 'No one ever returned. The ship came back empty, came back for more.'[21] This is a theme that recurs in other southern European literature.[22] The equation of migration with death is not, however, simply a literary device: around the turn of the century it was a common practice to provide an 'American Wake' for Irish emigrants about to set off across the Atlantic – not only a reflection of impending individual loss but also an indication of a whole community's attitude to migration and exile.[23] An alternative metaphor for migration, less often found but present nevertheless, is that of awakening or rebirth, relating to the migrant's transition to adulthood, to modernity, or to real self-discovery.[24]

It is useful to consider as examples two books that make full use of migration as a theme, but within a project that lies at a much deeper level of exploration: John Steinbeck's *The Grapes of Wrath* and Cesare Pavese's *The Moon and the Bonfires*. *The Grapes of Wrath*, which has sold more copies than any other of Steinbeck's novels, won the Pulitzer Prize for its author in 1940, and the award of the Nobel Prize for literature followed in 1962. The book tells the story of the uprooting of a poor family from the dust-bowl of Oklahoma during the Great Depression, and their migration westwards in search of a land flowing with milk and honey in California.[25] Steinbeck's motivations for writing the novel lay partly in a (successful) neo-realist endeavour to draw attention to a social problem, but in a pre-publication interview he also claimed that he had written a novel about 'desire', saying that 'this migration is the outward sign of want', and relating this to a wider human yearning: 'I have set down what a large section of our people are doing and wanting, and symbolically what all people of all time are doing and wanting.'[26]

But *The Grapes of Wrath* is about many other things too – the choice between home and the road, about loss, about matriarchy, about human roles in the natural environment (the novel both starts and ends with episodes of meteorological extreme events), about the need for an existential re-examination of the rules of living. These themes are all explored via the Joad family's journey, which consists of both a geographical relocation that calls into question the cosy significance of any concept of 'home', and of an inward journey into their own personality that each member of the family makes.[27]

Cesare Pavese's *The Moon and the Bonfires* makes equally complex use of a migration theme to produce one of the classics of postwar Italian fiction.[28] The protagonist is a man who returns to his native district (clearly identifiable in the novel as the area around Santo Stefano Belbo in the hill-country of southern Piedmont, south of Turin). The hero, Anguilla, returns from twenty years in America, seeking something to hold on to, and for an act of return that will be reassuring – the expected welcome, the glory of the returnee.[29]

> Once upon a time I'd had a longing within me (one morning in a bar in San Diego I nearly went mad with it) to come out on to the main road, to push open the iron gate between the pine and the lime trees at the corner, to hear the voices and the laughter and the hens and say 'Here I am, I've come back'.[30]

However, on his return Anguilla finds that his expectation that things will have remained the same as when he left is thrown back in his face: the war and the progress of time have transformed his old village and hardly anyone knows him or is interested in him.

> It was a longing I'd never get rid of now. I had come back. I had come out on to the road, I had made my fortune – I slept at the Albergo dell'Angelo and talked with the Cavaliere – but the faces, the voices, the hands which should have touched me and recognised me, were gone. They had been gone for a while. What was left was like a piazza the day after the fair, or a vineyard after the grape harvest, or going back to eat alone after someone has let you down.[31]

On leaving, the migrant so often expects to change and to run away from stability and an order which is felt to be unchanging. Yet, as Pavese realises, that world of the origin community may also be dynamic, so that the act of return cannot produce quite what the migrant expected. Pavese's Anguilla is searching not just for his geographical origins, but also for his own past and for an understanding of his own roots (he is, in fact, an orphan): he is partially disappointed in all these projects.[32] Thus once again, as in the case of *The Grapes of Wrath*, the author uses migration as a process occurring on two levels, both outward and inward. The re-exploration of place as well as personality through a return migration event is a common theme in much twentieth-century literature,[33] and it is a recurrent theme in some of the chapters that follow.

AUTHOR AND AUTHENTICITY

A striking fact about both Steinbeck and Pavese is that the powerful novels they wrote on a theme of migration were both written on the basis of research rather than drawing from autobiographical experience. Certainly Pavese had been born in the Langhe district in which he set *The Moon and the Bonfires* (and many other of his fictional writings), but he lived in Italy throughout his life. Steinbeck was a native of Salinas, in California, so that much of *The Grapes of Wrath* could, in fact, be seen as a polemic against his own community; but the basis for the story lay in research Steinbeck carried out for a series of articles commissioned by the *San Francisco News* in 1936, followed by more investigations both in California and along Route 66 back to Oklahoma in the following year.[34]

A very high proportion of creative writing relating to migration and its impacts is, however, strongly autobiographical. Motives for the production of such writing may be many and varied. Artistic or commercial considerations play a part, but there are also, in many cases, strongly personal motivations drawn from a possible need for catharsis, or to allow the act of writing to contribute to the re-definitions of identity alluded to earlier in the chapter.[35] In certain cases migrants have been deliberately encouraged to write by outside organisations, again for a number of reasons which often have a political objective of demonstrating cultural legitimacy.[36] Much of such group writing, as well as a proportion of the more individually motivated materials, can be seen as having a neo-realist purpose – of uncovering a less than optimal situation as a means of creating a policy climate for improvement. The material is directly based on lived experience, and although the styles and forms of language may be less complex than in the works discussed earlier, experiments are sometimes made with the blending of literary styles from different cultural origins.

Here, however, different readers need to ask different questions of the author. Whilst for the student of literature certain readings of literary criticism seek interpretation and understanding entirely through the text, for the social scientist interested in migration and the representation of migration experiences and outcomes there is more of a need to validate the credentials of the author. This relates strongly to the accepted need, within the more qualitative and interpretative strands of social-scientific research, to examine the positionality of sources in relation to social structures, power relationships and other influences: most often this concerns the researcher him- or herself in methodologies where the researcher is acting as a participant observer or as the motivator for in-depth interviews or group discussions that involve the double hermeneutic of exploration by both the researcher and the researched. This social-scientific need for more knowledge of the author should not, however, be taken as negating the use of certain texts: instead it produces the possibility of embedding the reading of texts by reference to the social, political and economic context in which they were written. These are not major differences between the approaches to literature adopted by the social scientist and the humanities specialist; they are instead differences of emphasis between text and author.

EMERGENT THEMES

Social scientists have traditionally tended to use literary sources as adjuncts to other materials – as commentaries, critiques, or as illustrations. In the words of Gerry Kearns, they have often been seen as 'tangential sources'.[37]

More recently, with the development of humanistic philosophies and approaches, particularly in geography, has come a wider engagement with literature for its own sake, but with the project still being generally aimed at uncovering empirical worlds of experience interpreted through literature taken as a secondary source.[38] The focus has generally been on 'worlds' as places,[39] but attempts have also been made to engage with the uses of literature in studies of situations defined by social context, and in broader work on societal structures themselves.[40] This is work where geographers and other social scientists come to have research objectives that are similar to those of scholars of cultural studies: thus Edward Said's work has proved a very fertile basis for exchange and discourse within a variety of disciplines.[41]

Research on migration from social-scientific perspectives has tended to focus on a certain range of issues concerning the characteristics of the migrants, the nature of the places of origin and of destination, and the underlying (political, social and above all economic) forces and structures that engender and condition movement. Much of this research has been operated at an 'objective' level with a concentration on statistical data sources: it is no coincidence that in the modern bureaucratic world data are generally more readily available on demographic events such as birth, death or (particularly international) migration than on many other aspects of human activities.[42] The corpus of established work has tended to operate on a limited set of scales: except in social anthropology and some social history, the role of the individual within migration has been poorly articulated, and at the same time the role of migration within a culture or society is often under-conceptualised and lacking in detailed study.[43] These are important gaps which can be constructively filled via the use of creative literature. The approach of the remainder of this chapter is to examine, albeit briefly, a number of texts to elucidate what such literature can contribute to the study of migration and its relations and effects at all levels. Throughout, however, the theoretical context of identity shift, outlined at the start of the chapter, is retained as a framework. Although the examples used here tend to highlight gender issues, this is not to say that the other elements of identity shift cannot be similarly considered through literature.

We might start with a recent English novel dealing with migration experiences that are as yet very poorly documented in academic studies – the experience of high-skilled 'transient migrants' from developed-world backgrounds, here seen through the eyes of women.[44] The author writes from a partly autobiographical perspective, having lived the life used as a background to her fiction. Hilary Mantel's *Eight Months on Ghazza Street*[45] concerns a woman who follows her civil engineer husband from Botswana to Jeddah where she confronts the compromises involved in seeking to retain her identity as an independent Western woman, with her attendant ideas on the morality of behaviour, in a context where all of these seeming fixities are daily challenged. In Jeddah it is policy, laid down by the company and by all the leaders of the British expatriate community, that the only viable response is to retreat into a state of mind in which the irritations and ambiguities of Saudi life are simply not thought about. Frances, the protagonist, finally manages to achieve this state but only after a series of traumatic and unexplained events which she observes and about which she asks too many questions. *Eight Months on Ghazza Street* brings out a number of themes relating to migration: the role of gender in migration decision-making and in adjustment (the steward on the flight from Heathrow to Jeddah points out that gender is crucial in Saudi Arabia – 'You're a woman, aren't you? You're not a person any more'); the roles of images and preconceptions (not so much for Frances but for those she meets); and the existence of defensively clustered social networks operating amongst Western expatriates, just as such networks evolve amongst ethnic-minority communities in Western cities.

Mantel's book is particularly interesting because it comments on the accepted Western view in which women from less-developed backgrounds follow their men to developed-world destinations and remain culturally and socially encapsulated in their new locations. Mantel shows Frances as more intelligent and aware than her husband, who is more willing to 'accept' the (albeit much more limited, since he is a man) restrictions on his thoughts and behaviour in Saudi Arabia. The writing of women migrants often challenges the hegemony of views of gender roles in migration, derived largely from male-dominated research. This is becoming increasingly the case in writing by migrant women from ethnic-minority backgrounds.

An example of such writing can be seen in the works of Buchi Emecheta. Emecheta was born in Nigeria and in 1962 joined her husband in England: however she was driven to leave him and to bring up her family alone, as well as making a living. Her breakthrough came through her stories for the *New Statesman* about life on a London council estate reserved for 'problem families' and from these came her output of novels and an autobiography.[46] Emecheta's works describe both the consciousness of migration and of an 'other' place in her childhood in village Nigeria, but also document the resultant struggles of migrant women, triply marginalised through their gender, their race and their poverty. There is a growing self-confidence among her women (reflecting her own rebellion at her husband's burning of her first writings), and the depiction of the growth of a community consciousness amongst black people in which women play a leading role. Many of these are themes common to much other migrant fiction.[47]

From a different literary tradition we may take *The Veil of Silence* written by the Algerian singer Djura.[48] Here again is an autobiographical story dealing in a complex fashion with a woman's identity between Algerian (more specifically Kabyle) and French cultural and social pressures, but ultimately celebrating the strength of an individual woman to survive. As the daughter of a migrant family in France, Djura was expected to conform to the traditional values of her parents' homeland and managed to complete an education only against the opposition of her male relatives. To escape these restraints she eventually fled to Algeria, paradoxically leaving France during the feminist revolution of the 1960s to seek freedom in Algiers, where her family instead arranged her imprisonment. On release by her French boy-friend (a major cause of family strife), Djura returned to France and became a film-maker and then a singer in an Algerian band, singing about the oppression of Algerian women and of their humiliation and the violence against them. But despite the fact that she had become the financial support of her family, it was her brother who attempted to murder her (and succeeded in killing her French husband) because of the stain she had brought on the family by marrying a Frenchman. *The Veil of Silence* depicts the complexities of a generation who move between identities, experiencing the exile's desire to retain cultural roots, whilst at the same time being drawn to assimilation and the abandonment of 'otherness'.

A common feature of many of these writings is therefore ambivalence. The migration event may seem clear-cut in the cold tables of statistical informa-tion, yet the event itself lies at the centre of a long-drawn-out (indeed, perhaps never completed) web of personal reflections, adjustments, reactions and repercussions that start in the individual's biography well before the move and which are played out for many years afterwards. The insights of literature provide a commentary on these processes.

Ambivalence and adjustment are themes treated elsewhere in migrant literature. In Prafulla Mohanti's book *Through Brown Eyes*, again an autobiographical account of the author's experiences, we see his early socialisation into the idea that Britain is a 'good' place and a natural desired destination for an Indian youth.[49] Following the path thus set before him, Mohanti moves to England and progressively finds himself becoming more English – a trait that is most apparent whenever he revisits India where he now sees the filth, malnutrition, petty officialdom and lack of efficiency to which he was previously oblivious. When in England he longs for India, and vice versa – a reflection of the arguments over movement between two identities made earlier in this chapter. Even where adjustment seems to be going well, as in some of the Sri Lankan poet Romesh Gunesekera's short stories, there are still cravings for the flavours of the life that has been left behind, and which draw ex-migrants together.[50]

These are all readings of literature which act to flesh out, often in great detail, many of the concerns of established work on migration, but which also raise a whole series of new questions which can not be approached through more aggregative approaches. Other types of literature may, however, not simply complement existing work but may stand alone. The example used here is that of writing about clandestine migration. Rey Ventura's auto-biographically-based book *Underground in Japan* has as its theme the experiences of illegal migrants, most of them from the Pacific-Asia region and especially from the Philippines, in a country that has officially set itself against any large-scale immigration, but which has a growing need for workers to take the jobs at the bottom of the occupational hierarchy. Ventura describes the labour-hiring arrangements in the informal economy:

> What everyone called the Centre was nothing more than an intersection on Kotobuki-cho where, at five in the morning, about a hundred workers were milling around, buying food from the noodle shops and stores, and eating their breakfast on the street. These were the *tachimbo*, the Standing Men, day-labourers dependent on the casual system of hire ... the Filipinos scattered around the four corners of the intersection. I discovered later that this technique of spreading out was part of a strategy in case of raids.[51]

As the French-Algerian *beur* writer, Mehdi Charef, has pointed out about dawn: 'It is the time of day possessed by the immigrant worker, after the milk delivery man and before the dustmen.'[52] Ventura has described, at one level, a relatively unknown situation in Japan,[53] but his book also reflects and comments on familiar migrant experiences in the Age of Migration in many other societies throughout the world.

A further theme approached only with difficulty through other research methods concerns the experience of flight, of exile as a refugee, of dislocation and of abandonment. Stories depicting such experiences are of value, drawn both from autobiography and from involvement as an outside observer. A recent example comes from Marion Molteno's *A Shield of Coolest Air*, in which two experiences of exile are simultaneously examined – that of a South African woman who finds herself alone in London after her husband has been deported from South Africa for radical journalism – and that of the Somali refugees whose cause she takes up.[54] Diverging interpretations of the flight of the oppressed are evoked through the terror and humiliations of those seeking sanctuary in Europe and the contrasting bureaucratic views of immigration officers: a pressing flood of emotional upheaval confronted with the decision that takes a moment to make but which has immeasurable consequences. Postwar population flows in Eastern Europe have also given rise to distinctive literature, and to a certain significance of a migrant culture, as when, in Ingeborg Drewitz's novel *Das Hochhaus* (The High-Rise), set in postwar Berlin, one of the commonest topics of conversation with relative strangers is not the weather but the ascertaining of common experiences as part of the 'refugee' movements out of Pomerania or Silesia to the east.[55]

Much of this body of literature serves to illuminate general aspects of human conditions through analysis of individual situations. Such an endeavour, accepted in certain academic disciplines, has traditionally made social scientists uneasy at attempting generalisations based on a sample of one. Yet even the sceptical may find consolation in the possibility of 'triangulation' – of comparing depictions and understandings between different sources and of bringing out a discourse between them, testing the plausibility of overall interpretations. Many bodies of literature are suffused with similar evocations of the experiences of migration, and can be taken as representing a certain 'mentality' within the societies depicted. The large body of Portuguese fiction from the 1950s to the 1970s depicting aspects of migration is a reflection of the omnipresence of this theme in Portuguese society during the same period.[56] At a more comparative level there are great similarities throughout Southern Europe in analyses of the rural worlds of out-migration during the twentieth century, and of the consequences for the communities left behind.[57]

Although couched in individual terms, such novels nevertheless add up to a very considerable body of evidence concerning the relationships of people and place, of the significance of territory, of the retention of multiple identities, and of the discordances that migration experiences bring to the lives of both individuals and societies at large. These were, indeed, the themes explored earlier in connection with Pavese's work, but they suffuse much other writing – sometimes fully articulated as the overall project of a piece, sometimes hidden but ever-present as the underlying structure that holds up the narrative.

But finally, it must be noted that amongst all the literature of migration the highest proportion deals in some way with ideas of return, whether actualised or remaining imaginary. To return may be to go back but it may equally be to start again: to seek but also to lose. Return has both a temporal and a spatial dimension. For the individual returning to their 'own' past and place it is rarely fully satisfying: circumstances change, borders in all senses are altered, and identities change too. But for many in the Age of Migration the time and place to be returned to are ill-defined. For those brought up in families with a background of migration, conceptualisations of 'here' and 'there', of 'home' and 'away' are confused. One of the novels by the Caribbean writer Joan Riley amply demonstrates this confusion. In *A Kindness to the Children* one of the two main characters is a British-born woman of Jamaican family origin who 'returns' to her cultural origins but finds herself alienated and unable to relate to Jamaican reality. The second main role is played by her cousin who has fled an oppressive childhood in Jamaica to survive in England in a form of self-exile from her past. Return to the Caribbean triggers a breakdown brought on by memories of the past.[58] 'In my beginning is my end.'[59]

CONCLUSION

Social-scientific study of migration has traditionally tended to emphasise the event and the aggregate: in dealing with the outcomes the focus has again been on groups. Recently there have been arguments calling for more biographical approaches, with an emphasis placed on individuals and their understanding of the 'realities' of their situation.[60] The world of creative literature opens up a whole new series of questions and issues that must rightly share prominence with more aggregate concerns, whilst also yielding new evidence on some

14

well-researched topics, as well as according with the new interest in individual experiences. Taken together, the contributions to this book consider, through their particular localised studies, all the aspects of multiple-identity reneg-otiation suggested at the start of this chapter. All illustrate the utility of literature as a representation of the migration experience, and of the embed-ding of movement within societal structures.

Creative or imaginative literature has a power to reflect complex and ambiguous realities that make it a far more plausible representation of human feelings and understandings than many of the artefacts used by academic researchers. In migration, above all topics, the levels of ambivalence, of plurality, of shifting identities and interpretations are perhaps greater than in many other aspects of life. The relationships between people and their contextual societies and places are intimate ones which are transformed by movement. Adjustment processes may never be fully completed: indeed, since we all continually refine our self-identities throughout our life-course it may be more truthful to say that migration intervenes in that process of renegotia-tion as a lasting force, rather than as a single event.

Migration has often been conceptualised as being an outcome of tensions between the individual's desires and opportunities – as a reflection of past circumstances and of expectations for the future. Others place it more firmly in a longer biographical context. Creative writing on migration often illumin-ates the processes of socialisation that occur to awaken an acceptance that migration is a viable, even sometimes the expected, means to achieve a goal (poorly articulated though it may often be). Migration also produces its own outcomes, for those directly and indirectly involved. Fictional and auto-biographical writing illuminates many of those outcomes in terms of both superficial happenings and deeper-seated attitudinal and behavioural changes. The study of such writings, alongside other sources, extends the range of understanding of the longer-term impacts of migration and thus our under-standing of the forces at work in modern society.

Finally, literature does not just reflect the circumstances that lead to its creation: a given corpus of writing also becomes a cultural force with the power to influence (and not just to reflect) societal mentalities. Questions of readership as well as of authorship certainly arise: multiple readings and interpretations may be possible, in which the author's 'meaning' is filtered through the experiences and conceptions of the audience. Meaning may thus be contested, but in many cases bodies of literature using similar metaphors and metanarratives reflect general societal norms and values. The myths set up or reflected by such writing may play powerful political roles. The writing of an individual may reflect the identity of that individual: the writing produced from within a particular society may become part of the mark of the identity of that society.

The postmodern world of greater diversity of experience is reflected in the growing variety of literary outputs, with migration and its outcomes as one of the dominant contemporary themes. 'Writers are citizens of many countries: the finite and frontiered country of observable reality and everyday life, the boundless kingdom of the imagination, the half-lost land of memory . . .'[61] In the field of migration studies, collaboration between social scientists and literary scholars studying such writers has the potential for increasing the understanding of an agenda that is wider than can be tackled by either group of scholars alone.

NOTES

1 The author wishes to acknowledge the help of comments received from Juliet Carpenter, Daniel Gutting, Elizabeth White and his two co-editors on an earlier draft of this chapter.

2 S. Castles and M.J. Miller, *The Age of Migration: International Population Movements in the Modern World*, London, Macmillan, 1993.

3 Both of these levels can themselves be analysed from both objectivist and subjectivist perspectives, as discussed by A. Giddens, 'Action, subjectivity and the constitution of meaning', *Social Research*, 1986, vol. 53, pp. 529–45.

4 The discussion which follows in this section owes much to the work of D. Gutting, 'Residential histories of the population of Turkish origin in Munich', Ph.D thesis in preparation, University of Sheffield.

5 K.H. Halfacree and P.J. Boyle, 'The challenge facing migration research: the case for a biographical approach', *Progress in Human Geography*, 1993, vol. 17, pp. 333–48.

6 See, for example, J. Galtung, *Members of Two Worlds: A Development Study of Three Villages in Western Sicily*, Oslo, Universitetsforlaget, 1971, ch. 4.

7 See A. Phizacklea (ed.), *One Way Ticket: Migration and Female Labour*, London, Routledge and Kegan Paul, 1983; E. Gordon and M. Jones, *Portable Roots: Voices of Expatriate Wives*, Maastricht, Presses Interuniversitaires Européennes, 1991.

8 For academic examples of this argument see J.L. Watson (ed.), *Between Two Cultures: Migrants and Minorities in Britain*, Oxford, Basil Blackwell, 1977; E. Lichtenberger, *Gastarbeiter: Leben in zwei Gesellschaften*, Vienna, Böhlau, 1984. An example from creative literature is Förderzentrum Jugend Schreibt (ed.), *Täglich eine Reise von der Türkei nach Deutschland*, Fischerhude, Atelier im Bauernhaus, 1980.

9 N. Hutnik, *Ethnic Minority Identity: A Social Psychological Perspective*, Oxford, Clarendon Press, 1991; K. Ljios (ed.), *Die psychosoziale Situation von Ausländern in der Bundesrepublik*, Opladen, Leske und Budrich, 1993.

10 C. Peach, 'The force of West Indian island identity in Britain', in C. Clarke, D. Ley and C. Peach (eds), *Geography and Ethnic Pluralism*, London, Allen and Unwin, 1984, pp. 23–42.

11 P. White and R. Woods, *The Geographical Impact of Migration*, London, Longman, 1980, p. 3.

12 This is a theme that has been extensively surveyed elsewhere: see G. Tindall, *Countries of the Mind: The Meaning of Place to Writers*, London, Hogarth Press, 1991; D.C.D. Pocock, *Humanistic Geography and Literature*, London, Croom Helm, 1981; W.E. Mallory and P. Simpson-Housley, *Geography and Literature: A Meeting of the Disciplines*, Syracuse, NY, Syracuse University Press, 1987. Geographers have also produced a large number of studies of individual authors and their sense of place or region: for some examples see B. Boulard, 'Paysages

normands dans l'espace de Flaubert et Maupasssant', *Etudes Normandes*, 1988, vol. 3, pp. 71–84; I.M. Matley, 'Literary geography and the writer's country (Britain)', *Scottish Geographical Magazine*, 1987, vol. 103, pp. 122–31; D.C.D. Pocock, 'Catherine Cookson country: tourist expectation and experience', *Geography*, 1992, vol. 77, pp. 236–43.

13 P. Keating, 'The metropolis in literature', in A. Sutcliffe (ed.), *Metropolis 1890–1940*, London, Mansell, 1984, pp. 66–80; M.A. Caws (ed.), *City Images: Perspectives from Literature, Philosophy and Film*, New York, Gordon and Breach, 1991; K.R. Scherpe (ed.), *Die Unwirklichkeit der Städte: Großstadtdarstellungen zwischen Moderne und Postmoderne*, Rowohlt, Reinbek bei Hamburg, 1988; E. Timms and D. Kelley (eds), *Unreal City: Urban Experience in Modern European Literature and Art*, Manchester, Manchester University Press, 1985.

14 D. Harvey, *The Condition of Postmodernity*, Oxford, Basil Blackwell, 1989, p. 116.

15 P. Faulkner, *Modernism*, London, Routledge, 1977, p. 75; Harvey, op. cit., p. 30.

16 J. Hawthorne, *A Concise Glossary of Contemporary Literary Theory*, London, Edward Arnold, 2nd edn, 1994, p. 122. Albert Camus is a paradigmatic modernist writer in this respect: see Chapter 8 of this book for discussion of his work.

17 T.S. Eliot, 'Little Gidding', from *Four Quartets*, London, Faber and Faber, 1944. Edition used here is that of 1959, p. 50.

18 I. Hassan, *The Postmodern Turn: Essays in Postmodern Theory and Culture*, Ohio, Ohio State University Press, 1987; J.-F. Lyotard, *The Postmodern Condition: A Report on Knowledge*, Minneapolis, University of Minnesota Press, 1984.

19 See, for example, I. Chambers, *Migrancy, Culture, Identity*, London, Routledge, 1993; N. Papastergiadis, *Modernity as Exile: The Stranger in John Berger's Writing*, Manchester, Manchester University Press, 1993.

20 W. Empson, *Seven Types of Ambiguity*, Harmondsworth, Peregrine, 3rd edn, 1961 (originally published 1930).

21 S. Haviaras, *When the Tree Sings*, London, Sidgwick and Jackson, 1979.

22 For example, A. Ribeiro, *When the Wolves Howl*, London, Jonathan Cape, 1963. Originally published as *Quando os Lobos Uivam*, Lisbon, 1958. The theme of migration as, or versus, death is further exemplified by Arnold Cassola in Chapter 11 of this book.

23 K.A. Miller, *Emigrants and Exiles: Ireland and the Irish Exodus to North America*, Oxford, Oxford University Press, 1986. See also the next chapter.

24 G. Lewis, *Day of Shining Red*, Cambridge, Cambridge University Press, 1980.

25 J. Steinbeck, *The Grapes of Wrath*, New York, Viking, 1939.

26 Quoted in D. Wyatt, 'Introduction', in D. Wyatt (ed.), *New Essays on the Grapes of Wrath*, Cambridge, Cambridge University Press, 1990, p. 6.

27 We might also note, relating to an earlier point, that Grandpa Joad dies on the day he leaves the old farmstead.

28 C. Pavese, *The Moon and the Bonfires*, London, Quartet, 1978. Originally published as *La Luna e i Falò*, Turin, 1950.

29 D. Thompson, *Cesare Pavese: A Study of the Major Novels and Poems*, Cambridge, Cambridge University Press, 1982, p. 226.

30 Pavese, op. cit., pp. 80–1.

31 ibid.

32 The passage from T.S. Eliot's Little Gidding, quoted above, is in fact used at the start of Thompson's discussion of *La Luna e i Falò*, see Thompson, op. cit., p. 223. See also A. Musumeci, *L'Impossibile Ritorno: La Fisiologia del Mito in Cesare Pavese*, Ravenna, Longo, 1980.

33 See, for example, many of Margaret Atwood's novels of return to Canadian settings: *Surfacing*, London, André Deutsch, 1973; *Cat's Eye*, London,

Bloomsbury, 1989. The contemporary French novelist Patrick Modiano explores similar themes with regard to return to Paris, for example in *Quartier Perdu*, Paris, Gallimard, 1984.

34 Wyatt, op. cit., pp. 13–14.

35 An interesting case here is that of Salman Rushdie: see J.P. Sharp, 'A topology of "post" nationality: (re)mapping identity in the *Satanic Verses*', *Ecumene*, 1994, vol. 1, pp. 65–76.

36 There is now a considerable outpouring of writings fostered by migrant or ethnic-minority writers' co-operatives, workshops and other organisations. A recent example is R. Ahmad and R. Gupta (eds), *Flaming Spirit*, London, Virago, 1994, produced by the Asian Women Writer's Collective in the United Kingdom. See also Chapter 3 of this book.

37 G. Kearns, 'Progress reports: historical geography', *Progress in Human Geography*, 1992, vol. 16, p. 407.

38 R.J. Johnston, *Geography and Geographers: Anglo-American Human Geography Since 1945*, London, Edward Arnold, 4th edn, 1991, p. 183.

39 J.D. Porteous, 'Literature and humanist geography', *Area*, 1985, vol. 17, pp. 117–22.

40 J. Silk, 'Beyond geography and literature', *Environment and Planning D: Society and Space*, 1984, vol. 2, pp. 151–78; P. White, 'On the use of creative literature in migration study', *Area*, 1985, vol. 17, pp. 277–83.

41 See especially Said's short article 'Narrative, geography and interpretation', *New Left Review*, 1990, no. 180, pp. 81–97.

42 This is not to say that the data are perfect, and it is arguable that the lacunae in migration data are currently increasing, necessitating new approaches to migration research: see P. White, 'Migration research', in P. Hooimeijer *et al.* (eds), *Population Dynamics in Europe: Current Issues in Population Geography*, Utrecht, Royal Netherlands Geographical Society, 1994, pp. 53–67.

43 A partial exception is the literature dealing with exile: a recent example is provided by H. Bernstein, *The Rift: The Exile Experience of South Africans*, London, Cape, 1994. Other exceptions include Stephen Fender's examination of the role of migration from Britain in creating America's literary culture and identity: S. Fender, *Sea Changes: British Emigration and American Literature*, Cambridge, Cambridge University Press, 1992. There are also several studies of the role of a 'myth' of America as a migration destination for Italians: see M. Chu, 'The popular myth of America in southern literature', *Journal, Association of Teachers of Italian*, 1983, no. 39, pp. 3–19; D. Heiney, *America in Modern Italian Literature*, New Brunswick, NJ, Rutgers University Press, 1964.

44 See Gordon and Jones, op. cit., for one of the few academic studies of expatriate wives; skilled transient migration in general is rapidly gaining interest, see A. Findlay and W. Gould, 'Skilled international migration: a research agenda', *Area*, 1989, vol. 21, pp. 3–11.

45 H. Mantel, *Eight Months on Ghazza Street*, London, Viking, 1988.

46 B. Emecheta, *In the Ditch*, London, Barrie and Jenkins, 1972; *Second-Class Citizen*, London, Allison and Busby, 1974: the autobiography is *Head Above Water*, London, Fontana, 1987. See also Emecheta's recent novel about return migration from London to Nigeria, *Kehinde*, Oxford, Heinemann, 1994.

47 An interesting collection of short stories reflecting this is K. Pullinger (ed.), *Border Lines: Stories of Exile and Home*, London, Serpent's Tail, 1993.

48 Djura, *The Veil of Silence*, London, Quartet, 1993.

49 P. Mohanti, *Through Brown Eyes*, Oxford, Oxford University Press, 1985.

50 R. Gunesekera, *Monkfish Moon*, Cambridge, Granta, 1992.

51 R. Ventura, *Underground in Japan*, London, Jonathan Cape, 1992, pp. 25–6.

52 M. Charef, *Le Thé au Harem d'Archi Ahmed*, Paris, Mercure de France, 1983, p. 44.
53 One of the few academic studies is J. Connell, *Kitanai, Kitsui and Kiken: The Slow Rise of Labour Migration in Japan*, University of Sydney, ERRRU Working Paper no. 11, 1993.
54 M. Molteno, *A Shield of Coolest Air*, London, Shola Books, 1992.
55 I. Drewitz, *Das Hochhaus*, Stuttgart, Werner Gebühr, 1975, p. 151. For a more recent period see also H. Müller, *Reisende auf einem Bein*, Berlin, Rotbuch, 1989, concerning the plight of those fleeing from Eastern Europe during the 1980s; and M. von der Grün, *Springflucht*, Darmstadt, Luchterhand, 1992, on the arrival of the *Aussiedler* in the Ruhr. See also the short stories of P. Huelle, *Moving House and other Stories*, London, Bloomsbury, 1994, on the interpretation of mass population movement in postwar Poland.
56 Ribeiro, op. cit.; O. Gonçalves, *A Floresta em Bremerhaven*, Lisbon, Seara Nova, 1975 (about migrant experience in Germany); L. Jorge, *O Cais das Merendas*, Lisbon, Europa-América, 1982 (a postmodernist novel bringing together emigration, tourism and the breakdown of community); F. Namora, *O Rio Triste*, Amadora, Bertrand, 1982 (on the Portuguese condition generally).
57 From Italian literature, C. Levi, *Christ Stopped at Eboli*, Harmondsworth, Penguin, 1982 (originally published as *Cristo si è Fermato a Eboli*, 1945); A. Giovene, *The Dilemma of Love*, London, Collins, 1973 (originally published as *L'Autobiografia di Giuliano Sansevero*, 1967); C. Levi, *Words are Stones*, London, Victor Gollancz, 1959 (originally published as *Le Parole sono Pietre*, 1955); from Spanish literature J. Goytisolo, *Campos de Níjar*, Barcelona, Seix Barral, 1959; I. Aldecoa, *Cuentos Completos*, Madrid, Alianza, 2 vols, 1973.
58 J. Riley, *A Kindness to the Children*, London, Women's Press, 1993. Another novel exploring similar themes of 'return' to an unexperienced ancestral homeland is that of Peter Handke, *Die Wiederholung*, Frankfurt, Suhrkamp, 1986, concerning a young Austrian of Yugoslav extraction.
59 T.S. Eliot, 'East Coker' from *Four Quartets*, London, Faber and Faber, 1944 (1959 edn, p. 23).
60 Halfacree and Boyle, op. cit.
61 Text by Salman Rushdie delivered to the 'International Parliament of Writers', Strasbourg, November 1993, reported in the *Times Literary Supplement*, 25 February 1994, p. 14.

2

LITERARY REFLECTIONS ON IRISH MIGRATION IN THE NINETEENTH AND TWENTIETH CENTURIES

Patrick Duffy[1]

INTRODUCTION

Emigration has been at the core of Irish consciousness for many generations. It is clearly evident in both census and song from the Great Famine of the mid-nineteenth century to the 1990s, and inevitably it has featured in varying degrees in the creative literature of the country. Indeed a great many of the writers of Ireland were themselves permanent or occasional emigrants and their writings often cast an interestingly intuitive light on the whole process of migration. While many writers like Oscar Wilde, Samuel Beckett and George Bernard Shaw quietly merged into their host society without a backward glance, others like James Joyce, Sean O'Casey or Edna O'Brien had notable reactions to the society from which they came.

Most creative writers, if they are not to wither away from the effects of provincialism, are by definition part of a wider intellectual and cultural world. For many Irish writers in the early years of the Irish Free State there was a constant tension between the isolationist conservatism of the state and the more liberal outward-looking world-view of the artist.[2] Many writers fled, either permanently or occasionally, from the cultural conservatism of Ireland in the same way that many ordinary emigrants have fled from the economic stagnation of the place. In any case, because emigration was so deeply ingrained in Irish society, one might expect that it would be reflected in some way in literary representations of life in Ireland, or in literary articulations of some of the social problems within Irish society.[3]

In what ways might one seek out perspectives on emigration in literature on Ireland? Two broad types of literature are worth examining. First there is the largely autobiographical work which contains direct, personal accounts of emigration. There are a number of important examples of this up to the 1950s – Patrick Gallagher and Mici MacGabhann have written about temporary and permanent migration from Donegal to Scotland and North America, for instance.[4] For later periods, Donal Foley, John Healy and Breandán

O hEithir offer differing reflections on the emigrants of the 1940s and 1950s.[5] Second, there is general fiction by writers at home and abroad which reflects either directly or indirectly on emigration as a reality of life in Ireland. These range from the novels and short stories of Patrick MacGill, Liam O'Flaherty, Sean O Faoláin and the plays of Brian Friel or J.B. Keane to the poetry of Patrick Kavanagh.[6] Because the process of emigration is such an emotional experience for most people, creative literature often captures and expresses the critical elements in what might be called a crisis for many individuals and families. Another collection of writing, not by Irish writers but by natives of the host society, also provides interesting perspectives on the Irish emigrant, particularly on the large-scale Irish emigration to America in the nineteenth century. In the final analysis, so extensive is the range of literature contained in short stories, novels, plays and poetry that it is impossible to cover it adequately in a single chapter such as this; the best that can be attempted is a fairly cursory review of trends evident in a limited sample.

In this restricted survey of Irish emigration as seen through the lens of literature, a chronological framework may be applied based on what appear from conventional migration studies to be significant phases of emigration from Ireland. First, for instance, one may focus on pre-Famine and post-Famine emigration in the nineteenth century, when millions moved to the New World. Second, one can take account of seasonal migration, principally to Scotland and England for the same period. Third, another group of writers reflected on twentieth-century emigration which was chiefly directed at England and peaked in the 1940s and 1950s. A final, and still developing literature relates to emigration in the past decade.

Irish society has experienced continuous out-migration for more than 150 years. The nature and dominant social impact of emigration from Ireland are reflected in a few broad statistics. There are estimated to be approximately 60 million people of Irish descent worldwide. The population of the island of Ireland fell from 8.5 million in 1841 to 4.3 million in 1921. During that period of time, 4.5 million emigrated overseas. In the years 1851–60, 1.2 million left the country; in 1861–70, 818,000 left: enormous numbers whose departure had a pervasive impact in every social and geographical niche on the island. The exodus continued on a slightly lower scale in the twentieth century: net emigration was 1.5 million up to 1971. Many rural districts, especially in the west of Ireland, demonstrate in their landscapes and community demography the consequences of persistent emigration over many generations. 'No wonder the country is full of ruins. You come on them in scores on scores, with maybe a tree growing out of the hearth, and the mark of the ridges they plowed still there, now smooth with grass.'[7]

Near Westport in 1925 a visiting American was struck by the looming presence of America in the lives of the people:

There were eleven people in the living room – an old woman, a young couple and their eight children, seven of whom were boys. . . . There were only three chairs in the room, but the young father immediately rose from his, made me sit down and gave me a cup of milk. Little boys of assorted sizes, resting themselves first on one leg and then on the other, stood against the walls. They seemed to be waiting for something. As I shared my raisin bread with them, there came to my mind the bizarre notion that they were waiting to grow up and go to America.[8]

It is no accident, therefore, that the theme of exile has tended to dominate the Irish psyche: it has become embedded in folk memory. The story of the 'Wild Geese' – the soldiery and native landowning class who were forced into exile in the seventeenth century – was reinterpreted in the emerging nationalism of the nineteenth century and recycled in popular attitudes to explain mass emigration. Exile and the wild geese analogy continued to be used in the twentieth century and John B. Keane's 1961 play, *Many Young Men of Twenty*, contains a line designed to strike appropriate chords in the audience: 'The Wild Geese are gone and now the goslings are flying.'[9] The historian Kerby Miller has comprehensively examined the exile theme in Irish emigration, which he has classed as migration from a pre-modern society.[10] According to Miller, a characteristic of such societies is strong attachment to homeland and home community; emigration is thus viewed as a reluctant but often necessary option. Because so much Irish emigration was involuntary flight from famine and severe economic oppression, it was easily viewed as reluctant exile. So poverty and exile are two major features of nineteenth-century emigrants – the one driving them out, the other maintaining emotional and sentimental links with home, which in turn led directly to a continuance of the process of emigration.

In introducing a discussion on literary reflections of Irish emigration, it is important to consider the social and economic setting of this emigration. Until the 1960s, it was predominantly rural in origin, principally from a demographic and settlement base of small and increasingly unviable farms: a society which was economically underdeveloped, with many archaic social structures. This increasingly maladjusted social and economic substructure shed its surplus population over a long period, its survival dependent on shedding further members. For much of the period in question, Irish emigration took people from a very rural, isolated and backward society directly into some of the most advanced and most urbanised economies and societies in the Western world. Irish migration represented rural–urban migration *par excellence*, an international emigration with all the stress and cultural and emotional trauma that this must have involved.

GOING INTO EXILE

The first stage in migration is the decision to go. Eventual departure obviously was an undertaking loaded with meaning and significance for the home community and as such it was the aspect of migration which is perhaps best represented in literature.

The 'American Wake', a farewell party for many emigrants on the night before departure, was full of symbolism. It frequently took on the ritualistic qualities of a wake on the death of someone, symbolising the enormity and finality of the departure. Ireland had the lowest rate of return migration of all European emigrations to America in the last century, so in most cases the departure was regarded as final. The wake allowed the emigrant to bid farewell to all in the close-knit community and presumably helped to maintain subsequent emotional connections and familial obligations. Mici MacGabhann described such a wake before he left Donegal for America in the 1880s:

> As night fell, most of the people of the district would gather into the 'convoy-house' – the old people in the first instance up to ten o'clock and then the young people. There'd be drinking and dancing then until morning. The person leaving would be keened [lamented] three times altogether during the night ... and the whole gathering would accompany him three or four miles along the beginning of the journey. Then they'd stand until the emigrant was well out of sight. It wasn't to be wondered at, of course, as often enough that would be the last sight of him a lot of them would ever have.[11]

Just before Liam O'Flaherty's emigrants left the house on the morning of departure, the mother

> burst into tears wailing: 'My children, oh, my children, far over the sea you will be carried from me, your mother'. And she began to rock herself and she threw her apron over her head. Immediately the cabin was full of the sound of bitter wailing. A dismal cry rose from the women gathered in the kitchen: 'Far over the sea they will be carried', began woman after woman, and they all rocked themselves and hid their heads in their aprons. [The mother stared firmly at her two children] as if by the intensity of her stare she hoped to keep a living photograph of them before her mind. The two children were sobbing freely. They left the house and the people filed out after them, down the yard and on to the road, like a funeral procession.[12]

Muiris O Súileabháin's departure from the declining island community of the Blaskets in the 1920s captures the experience of many emigrants from Ireland in the previous sixty or seventy years.[13] In his twenties, O Súileabháin had never been off the Blaskets. He describes the confusion and stress of the first-time rural emigrant, his embarrassed confusion at the railway stations, exacerbated by the hostility and incomprehension of people at his speaking of Irish, the fear of missing the train, the taunts about 'country cauboons' from the street urchins in Cork.[14] The traumas and tensions of the thousands who left the countryside remained private and personal for most of them, conveyed often only between the lines of letters written to loved ones at home. Only in some of the autobiographical writings is there an open description of the pressures of the journey in the transition from a quiet rural society to the bustle and impersonal world of the city.

> I looked down the street and up, people in plenty passing me in every direction and everyone seeming to have his eye on me. . . . I stopped, looking all round me. Oh Lord, where did the people come from? A man catching a bag, a woman running, another woman after her, chatter and confusion everywhere . . . I had only gone a few steps when an echo came back from the whole town of Dingle with the whistle the train threw out, and as for myself I was lifted clean from the ground. I looked around to see if anyone had noticed the start it took out of me, but nobody had . . .
>
> I had four hours to spend in the city [Tralee, population 10,000!], but if so, I had no intention of leaving the station, for I had no trust in the train but that it might go at any time. I put my bag against the wall and kept my eye on it always . . .
>
> Oh the confusion on the platform, my head split with the terrible roar throughout the place, boxes thrown out of the train without pity or tenderness, big cans, full of milk as I heard, hurled out on to the hard cement . . . And not a word of Irish to be heard! What would I do if there was not a word of English on my lips?[15]

In Dublin eventually he was

> blinded by the hundred thousand lights, lights on every side of me, lights before me, and lights above my head on the tops of poles. Soon I saw a small light coming towards me like a star through the mist. In half a minute it was gone by. It was another motor car . . . Trams and motors roaring and grating, newspaper-sellers at every corner shouting in the height of their heads, hundreds of people passing this way and that without stopping. . . . The trouble now was to cross the street . . .[16]

For multitudes of emigrants from Ireland this was the nature of their first trial on leaving their relatively sheltered rural backwaters. Even in the 1940s and 1950s, there were similar tensions. In 1944 Donal Foley left Waterford to emigrate to England. He described a scene which had been repeated throughout Ireland from the 1840s:

> The station at Waterford as I left was crowded with young women and men with white anxious faces and parents looking more anxious still. Most of the men were travelling on Wimpey's vouchers en route to the building sites of England, power stations, munitions factories, roads and bomb-damaged areas. Big strong men, pale and unsure, taking up their

first jobs, all of them fleeing from the demoralisation of unemployment, and ready to send money home to the helpless ones still left. . . . At Kilkenny the platform was crowded with young people, the mothers clinging on to them loath to let them go and anxious to get the last seconds of their companionship. So it was at all the little stations.

[Foley's] train stopped at Crewe first, a dark jungle of a junction with only the hissing of the steam trains, and the incomprehensible shoutings of the porters to be heard. Cups of tea could be got if one were quick and adventurous enough to dash across two platforms. I sat tight, afraid I would get lost if I ventured away from the London train . . .

That night I met my sister Cait in Mooney's Irish House in the Strand, then the longest bar in London. We clung to each other before leaving and cried profusely. She to make her way to New Eltham where she was a nurse and me to a vague address in Victoria, which I finally found after hours of feeling my way in the darkness [the war-time blackout]. My landlady was a kindly Waterford woman, so it wasn't a bad start.[17]

Most of the emigrants had strategies worked out to tide them over the problems encountered on the journey, the simplest of which was a linguistic euphemism aimed at playing down the significance of emigrating, which was described as 'taking the boat' or 'crossing the water'. Letters from former emigrants contained detailed instructions, based on first-hand experience, on each stage in the outward journey.

Seasonal migrants from the west of Ireland experienced less tension because they were to some extent accustomed to a gradual and repeated process of annual migration. On Mici MacGabhann's first trip to Scotland from Donegal in 1880:

we strode over to the quay of Derry. There were crowds of people getting on to the same boat – people from every corner of Donegal all bent on the same mission as ourselves. There they were, men and women, young and old, all off to Scotland looking for work to gather a bit of money to keep themselves and the families they were leaving behind them alive. You could see that the older ones, who had already tasted foreign parts, had no great wish to be on the move again. They knew they would have no ease nor peace until they returned . . .[18]

Seasonal migration was very much a normal pattern of rural life in nineteenth-century Donegal and most of the migrants travelled in groups with a couple of experienced senior members. Patrick Gallagher described his first trip to Scotland in 1889 with a degree of *sang froid*:

We walked from Cleendra to Letterkenny, a distance of thirty-six miles, and trained from Letterkenny to Derry on the Londonderry and Lough Swilly Railway built a few years previously. We arrived in Derry in good time for the Glasgow boat. We were in the same steerage compartment in which my father often travelled. The cattle and pigs were in the same flat. There was a bench or seat along the side of the ship, not enough sitting room for all the women and men in the compartment. There was no lavatory. If it came hard on you you had to go in among the cattle ... We landed at the Broomielaw at about 12 noon on the following day and broke up into squads. The squad I was with (six of us in all) went to Queen's Street Station, and took our tickets for Jadeborough, Roxboroughshire ... We reached Jadeborough late that night and got lodgings in Muldoon's Hotel, fourpence each for a bed. John Biddy and Hughie Mickey collected one shilling from each of us for the supper and breakfast. They went out to a small shop and got bread, bacon, tea, sugar and milk ...[19]

In looking at the light thrown on migration by Irish literature, it is useful to distinguish broadly between the type of emigrant who left Ireland – or the type of Ireland they left – in the years up to the Great War, and those who went to England in the years from the 1930s to the 1960s. The former came from what might be called 'truly rural' areas, communities that were fairly stable, isolated and integrated; the migrants were poor, naive, innocent and gauche. The second group came from a society that was more open, that indeed owed much of its modernisation to the influences of the earlier migrants, where there was increasing mobility and information feedback from the outside world.

Liam O'Flaherty's short story, 'Going into Exile', refers to a young brother and sister going to America from the west of Ireland at the end of the last century. It might be compared favourably for authenticity with the auto-biographies of MacGill, Gallagher or MacGabhann. There was an enormous chasm between the traditional peasant society which the young people were leaving and the urban world they were going to. Their thoughts reflected this divide.

Michael felt very strong and manly recounting what he was going to do when he got to Boston, Massachusetts. He told himself that with his great strength he would earn a great deal of money ... For the moment he forgot the ache in his heart that the thought of leaving his father inspired in him ... Each hungered to embrace the other, to cry, to beat the air, to scream with excess of sorrow. But they stood silent and sombre, like nature about them, hugging their woe ...

As she sat on the edge of the bed crushing her little handkerchief between her palms, [Mary] kept thinking feverishly of the United States, at one moment with fear and loathing, and the next with desire and longing. Unlike her brother she did not think of the work she was going to do or the money that she was going to earn. Other things troubled her, things of which she was half-ashamed, half-afraid, thoughts of love and of foreign men and of clothes and of houses where there were more than three rooms and where people ate meat every day . . .[20]

Although one should not overemphasise the contrast between rural peasant Ireland and the urban world, the literary evidence lends popular support to the anthropological notions of a stable, integrated and self-contained society contained, for example, in Arensberg and Kimball's studies in Clare in the 1930s.[21] The most woe-begotten and sorrowful literary representations of out-migration, however, contain inherent contradictions which are well represented in O'Flaherty's 'Going into Exile', where the weeping children are in effect being pushed into exile by their equally distressed parents. As Kevin Whelan has pointed out, Arensberg and Kimball's 'soft-focus on the family system blurs the crude realities of the inevitable break-up of the unit. It neglects to mention the cruelty of a system which efficiently dispersed family members . . . submitting them to a life lived far from family, friends, neighbours and relations.'[22] O'Flaherty's characters articulate an undercurrent of bitterness and dismay felt by these early rural migrants at the way their communities were rejecting them.

ESCAPING RURAL SOCIETY

For many of the later emigrants of the 1920s to the 1950s, there seems to have been a more tangible notion of 'escape' from the rural backwater into the more fast-moving mainstream of life in the city. In the works of MacGill, Kavanagh, O Faoláin, O'Connor, O'Brien and others, the main characters rationalise their emigration by expressing a disenchantment with Ireland and a wish to escape from the social and cultural claustrophobia of an inward-looking community.

In Patrick Kavanagh's *Tarry Flynn*, emigration comes at the end of a story of life in a rural parish in the 1930s. While there is little on the details of his leaving, there is a lot on the reason why many left and the inner tensions in the community. Tarry's much-travelled returned uncle undermined his attachment to the place by scorning the living conditions in such a backward spot, cynically observing 'What a life! How do you endure it? . . . there's no possibility to live in this sort of place. The best way to love a country like this is from a range of not less than three hundred miles.'[23] Much of Kavanagh's poetry also reflects on the narrowness and social impoverishment of Irish rural society in the 1930s and 1940s.

Edna O'Brien's *Country Girls*, published in 1960, unashamedly celebrates escape from the dead embrace and drudgery of the countryside to the

neon fairyland of Dublin . . . I loved it more than I had ever loved a summer's day in a hayfield. Lights, faces, traffic, the enormous vitality of people hurrying to somewhere . . . For evermore I would be restless for crowds and lights and noise. I had gone from the sad noises, the lonely rain pelting on the galvanised roof of the chicken-house; the moans of a cow at night when her calf was being born under a tree . . .[24]

This imaginary and exciting new world of a young rural migrant created by Edna O'Brien, perhaps out of her own early experience, is in sharp contrast to the lonely recollections of emigrants like Muiris O Súileabháin and Donal Foley described in the previous section.

Sean O Faoláin's characters had apparently rational attitudes to leaving. Well expressed by Frankie in *Come Back to Erin* (1940), they reflected the author's well-documented critical concern about Irish life in general: 'I must get out of here . . . out of Ireland. I've had enough of it . . . It's all down on top of you. Like a load of hay. There's no space here. No scope. It's too small.'[25]

Like O'Brien's country girls in Dublin, Frankie revelled in New York:

When he had found a cheap hotel, and began to walk the streets of the city, so immense as to excite, so untidy as to be comfortable, he almost hugged this sensation of release. Dusk gathered and he was sitting in the Park mesmerised by the vast cubes of buildings, re-cubed by their thousand windows, all now lit, and the brighter for the oncoming night. In each building, in each window, he found a picture-in-little of this place where people can be crammed together and yet never meet: every window, every man an individual pocket of energy; many, near, separate.[26]

Pat Lenihan, another of O Faoláin's characters, was filled with elation ever since the liner left Queenstown in Cork. Even the run-down New York streetscape was liberating:

he found Ninth Avenue beautiful. And yet, at any rate around Twenty-first Street, it is merely a dirty, paper-strewn street, darkened and made raucous by an overhead railway. There is the usual Greek fruit store, the usual wide-windowed restaurant and lunch counter, white-tiled like a public lavatory at home in Ireland, and with such names as Charlie's Lunch or The Coffee Pot; an old clothes shop, a cheap Sicilian haberdasher strayed up from Macdougal Street; there is a Palmist and Phrenologist, with big-breasted Polish gypsies always offering themselves in the doorway. The tramcars raced along the avenue under the thunder of the overhead railway. Only when the snow covers the dirt and smells and dulls the noise is the place really tolerable. Yet, to Lenihan, it had the charm of a foreign city.[27]

Richard Power expressed similar sentiments about Birmingham in the 1950s:

It was a wasteland, this black city under the glacial street lamps, a moonscape where you could hear the constant throbbing of machinery, with never a let-up. There was a curious vitality about it. In the pubs, as I passed, lights were gleaming, the clash of glasses, high-pitched conversation, the warmth of fellowship.[28]

Although Sean O'Casey's abandonment of Ireland in the 1930s was a highly personal and embittered decision, the rationale he gives to his main character in *Inishfallen Fare Thee Well* strikes many chords of disillusionment for emigrants from Ireland in the depressing interwar years:

the gay rustle of a girl's skirt will be hushed by the discreet rustle of the priest's cassock and the friar's gown. . . . Yes, London would mould him into a more fully-developed mind and man. The booming of Big Ben would deafen his new-listening ears to any echo from the bells of Shandon. Although he felt curious, and a little anxious, about meeting things he did not know, he felt relief at leaving behind the things he knew too well. . . . He was on the deck of the mail-boat, feeling her sway and shyly throb beneath his feet; watching the landing-stage drift afar away, getting his last glimpse of Eireann – separated for the first time from her, and never likely to stand settled on her soil again. It was bitterly cold, with a fierce, keen wind blowing, and soon it was sending sharp sleety hail and salty spray into his face, stinging it deeply – Ireland, spitting a last, venomous, contemptuous farewell to him.[29]

The migration decision-making process, however, is never clear-cut or black-and-white. Many migrants, if not most, are torn by conflicting emotions before they leave. There is the 'pull' of friends and family to remain, the 'push' of dullness and boredom; there is the 'pull' of possible excitement and freedom in the city, the 'push' of fear of the unknown. Is it possible that the rational and single-minded approach of O Faoláin's and O'Brien's migrants reflects more the contrived intellectual stance of their creators than the reality of indecision and contradiction? In Frank O'Connor's story 'Uprooted', there are hints of some of these conflicting and contradictory emotions in rural out-migration – the need to leave and the pain of loss.

He no longer knew why he had come to the city, but it was not for the sake of the bed-sitting room in Rathmines, the oblong of dusty garden outside the window, the trams clanging up and down, the shelf full of second-hand books, or the occasional visit to the pictures. . . . But no sooner did he set out for school next morning, striding slowly along the edge of the canal, watching the trees become green again and the tall claret-coloured houses painted on the quiet surface of the water, than all his fancies took flight.[30]

A visit home, however, reminded him why he had left. Nothing had changed, everything was the same:

The harness hung on the same place on the wall, the rosary on the same nail in the fireplace by the stool where their mother usually sat . . . it seemed to Ned that he was interrupting a conversation that had been going on since his last visit, and that the road outside and the sea beyond it, and every living thing that passed before them, formed a pantomime that was watched endlessly from the darkness of the little cottage.

This was the oppressive boredom that he had left behind. The rural idyll briefly beckoned, but he should not have returned, his roots had been severed:

He unbolted the half door and went through the garden and out to the road. There was a magical light on everything. A boy on a horse rose suddenly against the sky, a startling picture. Through the apple-green light over Carraganassa ran long streaks of crimson, so still they might have been enamelled. Magic, magic, magic. He saw it as in a children's picture-book with all its colours intolerably bright; something he had outgrown and would never return to . . .[31]

EMIGRANTS IN THEIR NEW WORLD

Literary evidence on the attitudes and aspirations of the emigrants away from home is more restricted. Probably the best material is provided by the autobiographies. These reminiscences contain a range of general themes which present a picture of a highly cohesive group of migrants who were seldom far from compatriots – not simply from Ireland, but from their own local districts back home. Most of the migrants had prior contacts in their destination areas. Like Foley finding lodgings with a Waterford woman in London in 1944, Mici MacGabhann, following his long journey from Cloghaneely parish in west Donegal to Bethlehem in Pennsylvania in 1886, obtained lodgings with a married couple from his home parish. Most of the emigrant characters in the stories of O Faoláin, O'Connor or O'Flaherty had siblings, other relatives or neighbours to meet them and help them on arrival at their destination.

The consequence of this was that the Irish emigrants were not readily assimilated into their host community. Harking back to Ireland is a constant theme of many of these displaced Irishmen. At the end of the last century, in MacGabhann's words, 'whatever we saw or heard or whoever we met in America, we had only one wish and that was to get back home to the old country.' In New York 'you'd think that there must be no one left in Cloghaneely – that they were all by now in New York.' Indeed such was the size and solidarity of the Donegal community in New York in the 1880s that there was an 'American wake' in reverse before they returned to Ireland, 'with a good bit of keening [lamenting] too'.[32] Donall MacAmhlaidh, who lived all his adult life in Northampton in England, never really accepted the fact that he had left home. Like many Irish emigrants, he always intended to return, and perhaps this prevented him making a full commitment to his host society. In Northampton and elsewhere, the Irish met together at Mass, in pubs and clubs, and in MacAmhlaidh's story Ireland is never far from his mind.[33] Likewise, Power's *Apple on the Treetop* is full of unintegrated Irish people in London in the 1950s, many of his friends and contacts being from the Aran Islands. On his first nights in London he stayed in lodgings owned by an Irish couple.

> There were up to thirty men staying there but I never saw half of them. Working the night shift most of them were ... There were four others in the bedroom, Cóilín and the old man in one bed, two Mayomen in another and I in the large double bed. I thought I'd have the bed to myself but I was awakened at two o'clock when somebody else pushed in under the same soiled quilt ... I found out the next morning that he came from Connemara.[34]

Power went on to refer to the animation 'that infused the lives of the expatriate Irish who lived in London, working, drinking, eating, condemned to relentless proximity to one another'; or, in Donal Foley's words, 'the welcoming pubs, the weekly wage packet and the comradeship of adversity'.[35]

Irish society at home continued to maintain a grip across the seas on the emigrant community abroad. The links with Ireland were functional as well as emotional, leading back into the familial system which had rejected so many of its offspring in order to survive. Religion, especially, was used as a potent welding force: from schooldays, young children were conditioned into an awareness of the 'dangers to faith' posed by emigration to the godless industrialised cities of England and the United States. The mother's farewell to her children in Keane's play is full of admonitions to say their prayers and go to Mass. This characteristic of Irish emigration was well established in nineteenth-century moralistic novels which were full of warnings about keeping the faith.[36] In the 1950s, there was a popular 'Prayer for Emigrants' recited in all churches in Ireland which had the effect of reminding families and future migrants about their duties as Irish Catholics.

In much the same farewell breath, many emigrants were also reminded of their duty to their parents. Patrick MacGill remembered that he was:

> born and bred merely to support my parents, and great care had been taken to drive this fact into my mind from infancy ... Often when my parents were speaking of such and such a young man I heard them say: 'He'll never have a day's luck in all his life. He didn't give every penny he earned to his father and mother.'[37]

These links with home were further consolidated, or expressed, in great reserves of nostalgia which often seem to have been close to the surface and which must have prevented many migrants from fully integrating into the host society. Recurring memories of Ireland abound in literary reflections on the emigrant experience. Even George Moore's character who re-emigrated in disillusionment to America, harked back to his 'first place' towards the end of his life: 'the things he saw most clearly were the green hillside, and the bog lake and the rushes about it, and the greater lake in the distance, and behind it the blue line of wandering hills.'[38]

O Faoláin's returned Yankee's somewhat maudlin sentiments on his homeland are fairly typical of the 'thatched cottage' nostalgia of the 1940s and 1950s, when thousands of young people in Ireland were scrambling to get out:

> My God, Frankie, to live here! To end my days here, in a cottage like that one down beyond the river, with a stack of turf against one gable to break the wind, and a donkey and butt all to myself . . . Imagine me on my own donkey and butt jolting down the road into the village to buy my groceries every week . . .[39]

Likewise, Healy's emigrants in America drank late-night toasts to Ireland: 'It's a great country, the old country. Here's to the old country. Here's to Charlestown [Co Mayo] and all the wunnerful guys in it . . . you're goin' back . . . you're lucky.'[40]

Songs and ballads formed a great body of what might be characterised as 'popular literature' which was readily accessible to the mass of the emigrants and which nurtured their links with home. Though the songs were often stereotyped, simplistic, even maudlin, they undoubtedly had wide appeal and touched many chords of emotion in the emigrant community. Feelings of exile, rejection and nostalgia are the foremost themes, reflecting in a simple manner many of the features of emigration which are evident in more orthodox literature. Titles include 'A lamentation for the loss of Ireland', 'The Donegal emigrant', 'The country I'm leaving behind', 'The emigrant's farewell', 'The exile's farewell', 'The exiles of Erin', 'I'll think of old Ireland wherever I'll go'. Here are just three typical verses.

> Farewell dear Erin, I must leave you,/And cross the seas to a foreign clime,
> Farewell to friends and to kind relations,/And my aged parents I left behind.
> My heart is breaking all for to leave you,/Where I've spent many a happy day,
> With lads and lasses and sparkling glasses,/But now I'm bound for Amerikay.
>
> ('The emigrant's farewell')

> I'm bidding you a long farewell/My Mary kind and true,
> But I'll not forget you darling/In the place I'm going to.
> They say there's bread and work for all/The sun shines always there,
> But I'll not forget old Ireland/Were it fifty times as fair.
>
> ('The Irish emigrant')

Just twenty years ago today,/I grasped my mother's hand,
She kissed and blessed her only boy/Going to a foreign land.
The neighbours took her from my breast/And told her I should go
But as they were taking her away/Her words came soft and low:
Goodbye Johnny dear/When you're far away
Don't forget your dear old mother/Far across the sea
Write a letter now and then/And send her all you can
And don't forget where e'er you roam/That you're an Irishman.

('Goodbye Johnny dear')

Songs like these were widely diffused from the 1920s when recording technology advanced in leaps and bounds in the United States, and as their sentiments indicate they were designed to keep alive memories of homeland and, in many cases, obligations to home communities.

Irish writers of fiction for the most part seem to have abandoned their emigrant characters at the boat. They disappear silently over the horizon, out of sight, out of mind and out of the story. Exceptions are writers like O'Brien, O Faoláin and O'Connor who themselves had experienced emigration and who make occasional excursions into the emigrants' new world in their works. In addition, for the nineteenth century, the literary evidence on the emigrant either at home or abroad is severely limited by the nature of the Anglo-Irish novel. Most writers in the nineteenth century were catering for an English market and they tried to represent Ireland as an exotic place. Emigrants, who in reality belonged to a large and nameless collectivity of peasants, had little role to play in the plots of these writers.[41] And so the irony is that in a century of enormous out-migration, these migrants made relatively little impression on the literary world. There were enough peasants left in the country to people the stories with minor characters.

One interesting literary source for the emigrants in their new world is contained in the work of American writers in the last century. In this work there are passing glimpses of the very large Irish community in America – the mostly poor and ill-educated masses who had poured out of post-Famine Ireland. They are everywhere in the streets, 'driving coaches or drays, pushing wheelbarrows, carrying hods, or peddling flowers or apples. They heckle street preachers, play the accordion in street cars, laugh and drink atop a cab.'[42] Apart from the obvious representations of labourers, policemen, and domestic servants – Eugene O'Neill's Long Day's Journey into Night has the obligatory garrulous but dense Irish maidservants, Cathleen and Bridget[43] – the Irish migrant emerges as part of a clearly identifiable self-conscious group. Much earlier, Sedgwick has the New York streets full of Irish peddlers and labourers, 'knots of Irishmen' with 'combustible passions'. One of the Irish characters is described as 'an ignorant, unprincipled foreigner, who had no name and no stake in society . . . There were thousands of such men in the city, they could be picked up anywhere, from the swarms about the Cathedral, to the dens of

Catherine-lane. . . .'[44] In general, these characterisations of the Irish immigrant are minor insertions of background colour: they are depicted as stock
types, incomplete characters with little evidence of their feelings, attitudes or
problems of assimilation and adjustment.

THE NEW IRISH EMIGRANTS OF THE 1980s AND 1990s

Following a lull for much of the 1960s and 1970s when, for the first time in
two centuries, there was a significant return migration to Ireland, emigration
resumed in the 1980s, when it is estimated more than 200,000 left Ireland.[45]
This new emigration has been quite distinct in many ways from earlier out-
movements. Reflecting fundamental changes in Irish society, the migrants
came to a large extent from a more 'urban' setting than formerly. Apart from
the pervasive 'urbanisation' of all of Irish society in the past thirty years, a
greater proportion of the population now lives in towns than ever before and
significant numbers of the new emigrants come from Dublin and the other
main towns of Ireland. Also important is the fact that the 'new emigrants' are
much more highly educated; a significant proportion have third-level education. A further distinction from the earlier emigrants is the fact that very large
proportions have entered the United States as illegal immigrants, due to
changes in official US attitudes to immigration since the 1960s. This illegal
status of many of the new Irish migrants has meant that they have had to
adopt many of the initial roles of their less privileged and less qualified
predecessors in the labour market; they are now as underprivileged by legal
status as the poor immigrants from the peasant communities at any time in
the past century. And at least part of this underprivilege is due to labour
exploitation of the newcomers by the established Irish-American community.
Mary Corcoran has comprehensively examined the ethnic identity of the new
Irish and has focused in particular on their relations with the home society
(resentful), the ethnic Irish-American community (instrumental) and the host
society in the US (from which they are legally excluded). She has noted that
the new Irish display the same problems of assimilation as their predecessors,
exacerbated in this case by their illegal status. The immigrants 'remain
strongly attached to the homeland, often idealising the family and community
left behind'. There are signs of assimilation, but these are countered to a great
extent by the continued identification with the homeland and the possibility
that at least some of them will return.[46]

In a great many ways, modern methods of communication have reduced
the impact of distance for the new emigrants which has meant that they have
been able to keep in touch with events in Ireland and their home areas to a
much greater extent than their predecessors. In the words of Dermot Bolger,
'Irish writers no longer go into exile, they simply commute.'[47] This has
continued to keep the image of Ireland and emigrants' awareness of their
ethnic identity alive. In terms of literary representations of this new migra-

tion, it may be a bit early to seek evidence. There is certainly a developing collection of songs which reflects the attitudes and concerns of the new emigrants in their new milieux. There are echoes of earlier migrations in this new emigration, but there are also important contrasts. One of the recurring themes among the new writers is the idea of dislocation in both Ireland and the host society:

> You're home in Ireland, but you're not home really. London is in your head, or New York, or Paris. But you're in *Ireland*. . . . You close your eyes and try to fight the overwhelming urge to be somewhere else. And you realise in that moment that you really are an emigrant now. And that being an emigrant isn't just an address. You realise that it's actually a way of thinking about Ireland.[48]

Like Edna O'Brien's country girls of the 1960s, for Joseph O'Connor's Evelyn,

> any time she did go home, she always wanted to come back to London after a week. It was the smallness of Ireland that got to her now, she said, the way everyone was the one colour, and knew everyone else.[49]

Many would argue that the new emigrants have been largely forgotten by the home society, which would not be unprecedented in Ireland: they have gone out of the story again. But in this case, in addition, the illegals are a potential embarrassment to the establishment in Ireland: the only advantage Emma Donoghue's disillusioned emigrant characters could think of for living in Ireland was its potato crisps![50] Claire Hendron kept away from the Irish pubs and cliques of Queen's, New York:

> Everything had become alien to her . . . her family, his family, New York; official America described her as an alien; she alienated herself and was in turn alienated by others. She would not participate in her new society; she'd never intended it to be her home.[51]

CONCLUSION

There is a range of literary sources which can be used to see how writers as artists, writing from their time and place, help us to understand the process of migration in Ireland. Autobiographical sources are, as expected, more factually descriptive. Impressionistic sources such as novels and short stories, which might be expected to reflect highly personal views of their authors, are also surprisingly representative of wider attitudes, supporting the notion that artists are valuable witnesses to broader changes in society. In general, however, the literary sources are limited in scope: for a country so steeped in emigration for so long, its depiction in creative literature is disproportionately small and confined mainly to the process and impact of leaving home. Recent writing, however, by a number of highly articulate and young Irish writers abroad is beginning to fill this gap for the current generation of emigrants.

In spite of the legacy of the Famine and nineteenth-century emigration as a poverty-stricken flight of peasantry, there is a continuity of behavioural attitudes running through the literature on emigration, even up to the 1990s. The exile motif continues; the attachment to home is still significant; the secret wish to come home persists. This must have been especially important in preventing full assimilation of migrants in the host society; in the case of American emigration today displacement and lack of integration is exacerbated by illegal status. There is also, however, evidence of continuing ambivalence in attitudes to emigration: exile is often juxtaposed with escape. The liberation and anonymity of the big city evident in the emigrant stories of the 1950s and 1960s also persists, although new 'places' become significant: in a recent short story, Heathrow Airport's arrivals building is characterised as the most favourite place on earth for emigrants who have just got away from Ireland.[52] For many young people the wider, bigger world continues to beckon.

NOTES

1 I wish to thank Seamus MacGabhann, Department of English, Maynooth College, and Professor Magda Stouthamer-Loeber, University of Pittsburgh, for suggestions on readings.

2 T. Brown, *Ireland, A Social and Cultural History, 1922–79*, London, Fontana, 1981, pp. 167–8.

3 See P.J. Duffy, 'Carleton, Kavanagh and the south Ulster landscape 1800–1950', *Irish Geography*, 1985, vol. 18, pp. 25–37.

4 P. Gallagher, *My Story*, London, Jonathan Cape, 1939; M. MacGabhann, *Rotha Mór an tSaoil*, Dublin, Foilseacháin Naisiúnta Teoranta, 1959, trans. as *The Hard Road to Klondyke*, London, Routledge and Kegan Paul, 1973.

5 D. Foley, *Three Villages*, Dublin, Egotist Press, 1977; J. Healy, *The Death of an Irish Town*, Cork, Mercier Press, 1968; B. O hEithir, *Lig Sinn i gCathú*, Dublin, Sáirséal agus Dill, 1976.

6 P. MacGill, *Children of the Dead End: The Autobiography of a Navvy*, New York, Herbert Jenkins, 1914; *Glenmoran*, New York, Herbert Jenkins, 1919; L. O'Flaherty, 'Going into exile', in *Short Stories of Liam O'Flaherty*, London, Jonathan Cape, 1937; S. O Faoláin, *Come Back to Erin*, London, Jonathan Cape, 1940; *The Stories of Sean O Faoláin*, London, Hart-Davis, 1958; B. Friel, *Philadelphia Here I Come*, New York, Noonday Press, 1965; J.B. Keane, *Many Young Men of Twenty*, Dublin, Progress House, 1961; P. Kavanagh, *The Complete Poems of Patrick Kavanagh*, New York, Peter Kavanagh Hand Press, 1972.

7 S. O Faoláin, 'A broken world', in *The Stories of Sean O Faoláin*, op. cit., p. 86.

8 H. Speakman, *Here's Ireland*, New York, Robert M. McBride, 1931, p. 163.

9 Keane, op. cit. The title-song in Keane's play commemorates the trauma of mass emigration: 'Oh, many young men of twenty said goodbye/All that long day/From break of dawn until the sun was high/Many young men of twenty said goodbye . . .'

10 K.A. Miller, *Emigrants and Exiles: Ireland and the Irish Exodus to North America*, Oxford, Oxford University Press, 1985.

11 MacGabhann, op. cit., p. 49.

12 O'Flaherty, op. cit., p. 148.

13 The Blasket Islands, which are now quite deserted, are located off the tip of the Dingle peninsula in the south-west of Ireland.

14 M. O Súileabháin, *Fiche Blian ag Fás*, Dublin, Clólucht an Talbóidigh, 1933, trans. as *Twenty Years A-Growing*, New York, Viking Press, 1933.

15 ibid., pp. 251–2, 254–8.

16 ibid., pp. 274, 279.

17 Foley, op. cit., pp. 52, 55.

18 MacGabhann, op. cit., p. 49.

19 Gallagher, op. cit., p. 26.

20 O'Flaherty, op. cit., pp. 139–41.

21 C.M. Arensberg and S.T. Kimball, *Family and Community in Ireland*, Cambridge, Mass., Harvard University Press, 1940.

22 K. Whelan, 'Aspects of society and culture in Ireland in the nineteenth and twentieth centuries', Dublin, National University of Ireland, unpublished PhD thesis, 1980, p. 188.

23 P. Kavanagh, *Tarry Flynn*, New York, Devin-Adair, 1949, p. 251.

24 E. O'Brien, *The Country Girls*, London, Hutchinson, 1960, p. 168.

25 O Faoláin, *Come Back to Erin*, op. cit., p. 122.

26 ibid., p. 127.

27 O Faoláin, 'A born genius', in *The Stories of Sean O Faoláin*, op. cit., p. 141.

28 R. Power, *Apple on the Treetop*, Dublin, Poolbeg Press, 1980, p. 197, trans. from R. de Paor, *Ull i mBarr an Ghéagáin*, Dublin, Sáirséal agus Dill, 1958.

29 S. O'Casey, *Inishfallen Fare Thee Well*, New York, Macmillan, 1949, pp. 379, 391, 395.

30 F. O'Connor, 'Uprooted', in *Short Stories of Frank O'Connor*, New York, Alfred A. Knopf, 1952, p. 196.

31 ibid., pp. 198, 199, 214.

32 MacGabhann, op. cit., pp. 138–9.

33 D. MacAmhlaidh, *An Irish Navvy: the Diary of an Exile*, London, Routledge and Kegan Paul, 1964. This book was originally published in Irish as *Dialann Deoraí*, Dublin, An Clóchomhar, 1960.

34 Power, op. cit., pp. 163, 168, 197.

35 Foley, op. cit., p. 54.

36 See, for instance, Mrs Sadlier, *Bessie Conway; Or the Irish Girl in America*, New York, Sadlier, 1861. Mrs Sadlier was an Irish Catholic who married an American publisher; the book warns of the dangers that can befall Irish girls in the New World.

37 MacGill, op. cit., p. 48.

38 G. Moore, *The Untilled Field*, London, Heinemann, 1914, p. 49.

39 O Faoláin, *Come Back to Erin*, op. cit., p. 121.

40 Healy, op. cit., p. 58.

41 See T. Leerssen, 'On the treatment of Irishness in romantic Anglo-Irish fiction', *Irish University Review*, 1968, vol. 20, pp. 251–63.

42 S.G. Bolger, *The Irish Character in American Fiction 1830–1860*, New York, Arno Press, 1976, p. 61.

43 E. O'Neill, *Long Day's Journey into Night*, London, Jonathan Cape, 1956.

44 C. M. Sedgwick, *Clarence, Or a Tale of Our Own Times*, Philadelphia, Carey and Lea, 1830, p. 96.

45 J.A. Walsh, 'Population change in the Republic of Ireland in the 1980s', *Geographical Viewpoint*, 1991, vol. 19, pp. 89–98.

46 M.P. Corcoran, *Irish Illegals – Transients Between Two Societies*, Westport (Conn.), Westport Press, 1993, p. 20.
47 D. Bolger (ed.), *Ireland in Exile*, Dublin, New Island Books, 1993, p. 7.
48 J. O'Connor, 'Introduction', in Bolger, op. cit., p. 14.
49 J. O'Connor, 'Four green fields', in Bolger, op. cit., p. 137.
50 E. Donaghue, 'Going back', in Bolger, op. cit., p. 161.
51 A. Hayes, 'The journey back', in Bolger, op. cit., p. 59.
52 O'Connor, 'Four green fields', op. cit., p. 137.

3

FROM *PAPPKOFFER* TO PLURALISM

Migrant writing in the German Federal Republic

Sabine Fischer and Moray McGowan

HISTORICAL CONTEXT

Labour migration and associated policies of implicit and explicit discrimination have been a feature of the German economy and society from the late nineteenth century through the forced-labour policy of National Socialist Germany to the present. During the postwar boom, the German Federal Republic's massive demand for labour soon could not be met even by the millions of refugees, displaced persons and returning soldiers, a situation exacerbated in 1961 when the construction of the Berlin Wall stopped the influx of workers from the German Democratic Republic. In these conditions of full employment the officially regulated recruitment of foreign workers was widely accepted. In 1955 the Federal Republic signed an agreement with Italy for the employment of Italian workers. Between 1960 and 1968, further agreements followed with Spain, Greece, Turkey, Morocco, Portugal, Tunisia and Yugoslavia. At first all signatories regarded these measures as temporary. The term *Gastarbeiter* (guestworker), which became established in the 1960s, also explicitly stressed that integration of foreign workers into German society was out of the question.

Their numbers grew rapidly, reaching a million in 1964. During the 1966–7 recession, 30 per cent returned to their countries of origin. But many were unable to resettle despite their savings and their new qualifications, and a massive new wave of labour migration began at the end of the recession. In 1973 the number of foreign workers in the Federal Republic reached a peak of 2.6 million.

The same year, another recession led the German government to decree the *Anwerbestopp* (1973), a halt on the recruitment of foreign workers. *Gastarbeiter* also bore the brunt of redundancy policies, but despite worsening circumstances the majority of migrants preferred to stay in Germany with its social-welfare benefits, instead of returning to a totally insecure future in their countries of origin. Their families began to follow them to Germany, initiating the period of settlement.

THE CONTEMPORARY SITUATION

By 1989 and the start of the process leading to German reunification (which brought added complexities to the situation of foreigners), the foreign resident population in the Federal Republic of Germany numbered 4.9 million and was still growing, whereas the number of foreigners in employment had declined from 2.6 million in 1973 to 1.7 million in 1989.[1] The majority of workers from the original catchment areas still held unpopular jobs with unpleasant working conditions and had fewer opportunities for advancement than their German colleagues. Nevertheless the *Gastarbeiter* population has undergone major changes in structure, attitudes and behaviour which clearly reflect underlying patterns of immigration in the Federal Republic.

For most *Gastarbeiter* Germany has become a country of permanent residence; by 1987 two-thirds had been living in the country for over ten years. Their population structure in terms of age and gender is now similar to that of the German population. Instead of hostels and similar low-grade temporary housing, most *Gastarbeiter* families today live in ethnic communities with certain similarities to those in classic countries of immigration such as the United States. Moreover their attitudes towards saving and consumption have changed markedly: more and more are abandoning plans for return and investing their capital in Germany, buying property and starting small businesses.

Migrant families in Germany display the generational conflicts symptomatic of immigration: tensions between the first generation of adult migrants with their close psychological relationship to their home countries and their intentions to return, and the 'second generation' who arrived as children and who have a more distant, critical view of these countries and are more orientated towards German society. Meanwhile a 'third generation', born in Germany, generally regards it as their home, their parents' countries of origin often as no more than a holiday destination.

Thus there is now a population of *de facto* immigrants who form an integral part of German society as ethnic minorities. The term *Gastarbeiter* is clearly no longer – if it ever was – suitable to describe this social group.

Despite this, the Federal Government still insists categorically that the Federal Republic is not a country of immigration.[2] German citizenship is based entirely on the Roman law concept of *ius sanguis*, the right of descent: persons born on German territory have the right of citizenship only if they are of German origin. Anyone of non-German origin is automatically regarded as 'foreign', and naturalisation depends on very strict conditions, such as renunciation of dual nationality. This makes it impossible for identity to develop in the same way as it does in, for example, the USA, where immigrants are able to be American and at the same time insist on their distinctive ethnicity.

This brings about a paradoxical situation for the *de facto* immigrants: they have become German residents with a foreign passport. But the attitude of the German state towards its ethnic minorities is itself a paradox. On the one hand it prevents foreign residents from integrating fully into German society, but on the other hand it demands at the same time their total assimilation to the 'legal, social and economic order of the Federal Republic, its cultural and political values' as specifically required of them by the *Ausländergesetz* (Foreigner Law) of 1991.[3]

This paradox, which reveals the Eurocentric idea of a German nation with a homogeneous culture superior to, and to be protected from, foreign influence, shapes the acculturation of ethnic minorities, and their strategies in response to exclusion on the one hand and the pressures of assimilation on the other.

The example of the Turks, the largest ethnic minority in Germany and one of those subject to most prejudice and discrimination, shows the increasing diversity of these strategies between integration, ethnic isolationism and denial of ethnic identity, reflecting the diversity within the ethnic minorities themselves.[4] Though there are socio-economic and cultural determinants in common, the acculturation process is an individual one, depending on each migrant's personal socialisation, opportunities, experiences, motives and perspectives. It can only be properly understood if these minorities are not treated as homogeneous groups and if the conflicts inherent in multi-culturalism are not papered over by assertions of a false harmony.[5] As Weigel puts it: 'A more careful, differentiated discourse is needed to analyze patterns of cultural and gender identities within the larger framework of Eurocentrism and German history.'[6]

THE EMERGENCE AND DEVELOPMENT OF MIGRANT LITERATURE

Every act of migration reflects wider social, political and economic developments, such as modernisation processes.[7] Thus, one may wish to draw on migrant literature as social evidence.[8] On the other hand, literature is individual, subjective and diverse. It may reflect, but may also exaggerate, challenge or invert the social experience that informs it. In finding a voice outside the established social discourse, and exploring the possible beyond the given, it may also resist functionalisation as social evidence. Interpretative practice needs to take account of both the individual and the social-historical significance of a text.

Just as with its overall subject matter, so the literary articulation of migration is not new in Germany, but the scale of demographic change through migration since the 1960s has generated literature of an entirely new quantity and quality. By the 1990s, it had attained a diversity contradicting all attempts to label it. *Gastarbeiterliteratur* (guestworker literature) *Migrantenliteratur* (migrant literature), *Ausländerliteratur* (foreigner literature), *Literatur deutschschreibender Ausländer* (the literature of foreigners writing in German): all are either too narrow,[9] potentially patronising or indeed racist (in implying that these texts are inferior appendages to some culturally homogeneous 'real' German literature), or so general that they erase crucial socio-economic, ethnic, cultural, gender or generational differences between the authors, between the patterns of experience their texts engage, and between the aesthetic possibilities their texts manifest.[10]

The development of this literature is linked to the phases of physical migration and psychological adjustment outlined above, though the relationship is not crudely synchronous. On the one hand, several factors initially inhibited the production of such a literature: the expectation of temporary sojourn, the lack of a written literary tradition in the social environments from which many *Gastarbeiter* were recruited, reinforced by the – more usually numbing than creatively stimulating – culture shock of transition to Western, urban-industrialised society. On the other hand, since its emergence in the 1960s much of the literature of *Gastarbeiter* experience has been by writers whose ethnic identity may have exposed them to comparable discrimination but who are not, socio-economically speaking, *Gastarbeiter* themselves.

The 1970s, however, did see a wider movement within which pre-literary forms – songs, diaries, letters, oral narrative – began to give way to poems, short stories and *reportage*, written and often published in the mother tongue (initially, predominantly Italian, since they were the first group of *Gastarbeiter*) in newspapers and magazines.[11] Mostly, they were written by men, reflecting both the demographic make-up of the first phase of labour migration and the additional barriers between women from these cultures and access to the means and traditions of literary expression.

The texts generally focused on immediate *Gastarbeiter* experience: workplace, hostel, station, government office, the annual journey home; the themes were the dreams of Germany as the promised land of material wealth; the reality of heavy, dirty, unhealthy work in poor conditions and the experience of prejudice, indifference and rejection; homesickness and dreams of return; life between two worlds and two languages.[12] It was seen, and to some extent correctly, as *Betroffenheitsliteratur*: therapeutic writing by victims of social processes, articulating, objectifying and establishing the commonality of experience by recording it in simple, conventional, usually autobiographical forms.

This literature developed in tandem with the growth of an infrastructure for its reception, which led in turn to its partial acceptance in the commercial literary market from the early 1980s.[13] German writing competitions for foreigners sponsored by the Institute for German as a Foreign Language at Munich University led to anthologies in a leading paperback imprint,[14] and to the founding in 1985 of an annual Adalbert von Chamisso Prize for such literature,[15] whilst also contributing to a widening awareness of this literature amongst German readers. However, the Institute also played a part in the stereotyping of this literature through a range of normative pronouncements.[16] This was still patronage by the dominant culture.

At the same time, a self-directed and more explicitly political trend was being instigated by writers themselves, some of whom had begun alongside German colleagues in the workers' literature group *Werkkreis Literatur der Arbeitswelt*. The *PoLi-Kunst-Verein* and the publishing collective *Südwind* were established in 1980 as a bridge between the numerous migrant groups in Germany. The goal was a 'polynational culture', the political theme the material reality of the *Gastarbeiter* and the furtherance of change. 'From tears to civil rights' was how Franco Biondi, one of its leading members, summarised this move from *Betroffenheit* (bewilderment) to political activism.[17] *Südwind* insisted on the terms *Gastarbeiter* and *Gastarbeiterliteratur* precisely because they name an exploitation process and an outsider status; the aim was to further a solidarity amongst the exploited which recognised the hierarchies *within* racial discrimination that help underpin socio-economic exploitation, but which sought to counteract attempts to play one migrant group off against another as a means of social control. In Biondi's *Aufstieg* (Climbing the ladder), a *Gastarbeiter* remarks: 'It's obvious, German: biggest fish. Italian, big fish. Turk, little fish. You [Pakistani], even smaller fish. African: all the worst jobs. Where there's rich and poor, always like that.'[18]

The affirmation of *Gastarbeiter* identity as a political act led, logically, to a policy of publishing in German as the potential common language of the migrant workers.[19] From 1980 on, a series of *Südwind-gastarbeiterdeutsch* anthologies appeared, whose shaping ideas were solidarity and the examination of the material conditions of immigrant experience. *Im neuen Land* (In the new land) focused on discrimination, lack of political rights, confrontation with an alien lifestyle, prejudices and fears; *Zwischen Fabrik und Bahnhof* (Between factory and railway station) on images of the home country and longings to return. *Annäherungen* (Approaches) examined the contacts between *Gastarbeiter* and the German population and between the various *Gastarbeiter* communities; *Zwischen zwei Giganten* (Between two giants) the specific experiences of the second and third generations.[20]

The *Südwind* group saw their work in the tradition of socialist workers' literature. With *Das Unsichtbare Sagen* (Speaking the invisible),[21] published in 1983, the subtitle of the anthologies changes from *Südwind Gastarbeiterdeutsch* to *Südwind-Literatur*. But this is not a depoliticisation: the goal is the articulation and dissemination of a *zweite Kultur*, a cultural experience excluded from the mainstream and common not just to migrant labourers but to the whole working class in Germany. In 1980, the Spaniard Antonio

Hernando responded to the question of his access, as a *Gastarbeiter*, to German culture: 'Not even the German workers have any access to the so-called "German culture". That is in fact a culture of privileged Germans for privileged Germans.'[22]

It was these and other anthologies with programmatic titles like *Sehnsucht im Koffer* (Longing in a suitcase)[23], rather than work by individual authors, which established this literature in the awareness of the German reading public, but in doing so perpetuated a normative perception of a homogeneous *Gastarbeiterliteratur*.

However, there were other currents. For example, Yüksel Pazarkaya and the publishers Ararat aimed to show German readers that Turkish culture was more than *Gastarbeiter* culture, by publishing translations both of modern Turkish literary classics and of Turkey's own *Deutschlandliteratur* – works on migrant experience in Germany published in Turkey itself.[24] Pazarkaya has written on *Gastarbeiter* experience since the 1960s, but rejects all categorisations: 'literature is literature'.[25] Highly educated, he views the German language not as a barrier to social participation, but as the path to the humanist tradition of Lessing and Heine, Schiller and Brecht, Leibniz and Feuerbach, Hegel and Marx.[26] The marked contrast to the position of Hernando demonstrates how class and educational factors may cut across the common experience of migration. Pazarkaya is an example of the danger of viewing the culture and experience of Turkish migrants (let alone all migrants) *en bloc*, since he belongs to a very specific group: middle-class intellectuals of a generation profoundly influenced by Atatürk's Westernising and Euro-peanising reforms.[27]

Some *Gastarbeiterliteratur* in the conventional sense of simple auto-biographical accounts of prejudice, identity conflict and broken dreams, continues to be written. But publication gave many authors confidence to reflect critically on the limitations of this approach. Moreover, the antho-logies, while promoting a misleading image of homogeneity, also established this literature as a market factor, which in turn widened access to publication for a much broader range of authors, many of whom never acquired the *Gastarbeiter* label, explicitly reject it or set out to deconstruct it. Authors of the second and third generations appeared, critical in new ways of the prejudice they encounter in Germany, but also of the self-pity, subservience, backwardness or greed of their parents' generation. Women writers became increasingly prominent. More single-author volumes were published, leading, in turn, to a recognition of the diversity and in some cases thematic, stylistic and linguistic sophistication of this literature, though its reception in the German media has remained dogged by stereotypes.

In the following section we examine a small selection of texts from the 1970s to the 1990s which are characteristic of the various phases of migrant literature in Germany whilst having qualities which take them beyond the merely typical. In focusing almost entirely on prose fiction, we necessarily neglect whole areas of a diverse picture, such as the vast output of poetry; the satires and cabaret programmes of Sinasi Dikmen;[28] the half-subversive, half-whimsical fairy tales of Rafik Schami;[29] or the development of ethnic theatre, such as the *Teatro Siciliano* in Frankfurt or the Turkish groups in several cities.[30]

FROM FIRST TO THIRD GENERATION

Franco Biondi's story 'Die Rückkehr von Passavanti' (Passavanti's return), first published in 1976 in the Italian emigrant newspaper *Corriere d'Italia*, has become a classic of the first phase of *Gastarbeiter* experience, showing the migrant labourer as disoriented, embittered, rootless, his dreams of return doomed to disillusion.[31] Passavanti is a bricklayer from a South Italian village. Recruited to Germany in 1960, he works as a labourer, in a fridge factory, in a tar works. In 1975 he returns to his village: for good, he says. He carries the same *Pappkoffer* (cardboard suitcase: an emblem of the first generation of *Gastarbeiter*) with which he set out, and little else to show for fifteen years in limbo. But he now finds the village *fremd* (alien), and can only squabble with family and friends. Eventually shunned and mocked as *der Deutsche*, he returns to Germany.

The text bitterly attacks the semi-feudal, subsidy-grabbing landlords who dominate the village. Thus unlike many early migrant texts it does not convert homesickness into an over-rosy portrayal of the homeland. However, it does not look beyond Passavanti's melancholy fatalism.

Güney Dal's novel *Wenn Ali die Glocken läuten hört* (When Ali hears the bells ringing), set amongst Turkish workers in Cologne (during an unofficial strike at Ford) and in Berlin, also has many typical themes and milieus of *Gastarbeiterliteratur*: ironic memories of the dreams of wealth that had motivated migration, hostel life, isolation, puzzlement at the alien language and culture, experience of everyday prejudice and racism, stress-related illness, corruption of traditional values, a sometimes crude location of the roots of migrant misery in capitalism.[32]

But the intercut scenes and multiple focalisers (the sociology student Ali, the Ford workers Schevket and Hamdi, the laboratory assistant Kadir) permit a complex and differentiated portrayal of how experiences before as well as after migration condition the very disparate attitudes of Turks in contemporary Germany. There are the politically active, the apathetic and the ignorant; the strike-breakers – dreams of wealth having lured most *Gastarbeiter* in the first place, some could always be bought off by management; the Turkish government agents hand-in-glove with the German employers to undermine the strike; finally the religious fundamentalists who despise the godless Germans, the communists and their own government – anybody and anything associated with modernity, secularisation and Westernisation.

Thus deep divisions run through the Turkish community itself. However, the novel also reflects the antagonism the Turks, above all, were already encountering in late 1970s Germany: they are seen by disadvantaged Germans not as class comrades but as ugly, loud and garishly dressed, dragging swarms of children, 'a rapidly spreading epidemic'.[33] The novel also shows the mutual process whereby each culture encounters the other with surprise, distance, suspicion, alienation and a sense of its own superiority, so creating a double ethnographic Other. It subverts German stereotypes of 'the Turks' and of the cowed, culturally inferior *Gastarbeiter* in general, but also perpetuates Turkish stereotypes of 'the Germans' as godless, heartless, bureaucratic and materialistic, generalisations which undermine the text's call for trans-ethnic class solidarity.

.However, with the Kadir figure, a parody of the ignorant, helpless, powerless, speechless Turkish migrant, Dal shows his literary self-awareness, even playfulness – traits more common in migrant literature than the stereotype would have it. Kadir is dependent on his elder son to interpret the German his younger son mumbles in his sleep, and trustingly takes pills his employer (director of a laboratory where Kadir cares for the animals) thoughtlessly gives him for his stomach pains. They are hormone pills; Kadir acquires breasts and, too ashamed to consult anyone, mutilates himself with a kitchen knife:[34] a grotesque metaphor for German treatment of *Gastarbeiter*, especially Turks, as subhuman beings, and for the helpless and self-destructive response of a man unprepared for the bewildering complexities of urban technological society.

Dal's *Europastraße 5*,[35] published in German in 1981, displays in its narrative form the multiple voices in the individual Turkish experience: the cultural and religious tradition, the family, the contemporary homeland, the dream of Germany and the reality, the nightmare experiences on the trans-Balkan route E5 that gives the novel its title: the consciousness of the central character, Salim, becomes the arena in which conflicting cultural forces fight their battles. The plot is gloriously grotesque and, like the story of Kadir's breasts, just believable (it is found, with variants, in other migrant texts). Salim's father dies in Berlin while staying there illegally. Salim tries to smuggle the body back to Turkey crammed in a television carton on the roofrack. Border delays, accidents, breakdowns, and alternate heat and heavy rain cause his plans to disintegrate along with the carton and the body in a nightmare of comic complications. *Europastraße 5* is a complex comic novel in which migrant experience is expressed not in autobiographical complaint, but in broken consciousness mediated by the narrative structure.

Though less sophisticated in form than Dal's work, Franco Biondi's *Abschied der zerschellten Jahre* (Farewell to the shattered years)[36] marks another historical stage of migrant experience: the third generation. Mamo, a young Italian, has received a deportation notice. He barricades himself in his flat with a Browning automatic, and reviews his life in a series of flashbacks.

Mamo cannot identify with the German society that oppresses him. Yet, born and brought up in Germany, he does not feel an *Ausländer*. He speaks a teenage working-class argot not specific to immigrants. 'He spoke like the natives. After all, he was one.'[37] Even the bitterness he feels for being treated as 'broken dolls in a lumber room'[38] is actually related to his generation and class as a whole, trained for useless jobs which will soon be automated. Like Hernando, Biondi sees class as well as ethnic identity at the root of exclusion and oppression. He dedicates his novella to 'all native foreigners of German and non-German origins'.

Mamo scorns the *Pappkoffer* images of immigrants in his well-meaning schoolbooks: 'miserable, moaning figures with cardboard suitcases . . . people choking on their homesickness and their tears'.[39] Those are his father's generation (that too of Biondi's own Passavanti). When the theme of *Gastarbeiter* is treated at school, he angrily resists the stares of fellow pupils. But gradually the experience of prejudice makes him into a *Gastarbeiter* by forcing him to develop a consciousness of his identity as the deviant, disliked Other. It is – like 'Jewishness' in Nazi Germany – an identity created by the oppressor. A fellow pupil taunts him with a Blutwurst which he tells him is a 'Turk's penis', indifferent to Mamo's actual ethnic identity: foreign is foreign. Mamo never felt much in common with Turks, but now begins to feel 'foreign'. He dreams of studying his face in a mirror: that is, he is forced to view his identity as an image of himself as transmitted to him in the gaze of the powerful, a theme which echoes women's writing of the period. His identity as a *Gastarbeiter* is part of a script written by the socio-economic forces which created the *Gastarbeiter* in the first place. This script now declares that he should be deported from the country of his birth: hence his act of rebellion.

This is a familiar theme of modern literature and film: the hero or heroine forced into an extreme decision, which is irrevocable, but which also engenders clarity, relief and a sense of power. Mamo compares it with the video game 'Player': there are no conjunctives, it is shoot or be shot. The myths and icons of his mental world and that of his class and his generation, German or migrant – films, rock music, the disco, the bowling alley and especially the video game – shape the metaphoric structures of his narrative. He is 'not a flying saucer on the screen, not a clay pigeon flashing by . . . not a skittle waiting for the ball, no, if I've got to topple over, if I'm going to be deported . . . then with dignity'.[40] Thus even the potentially authentic, identity-giving act, as a radical 'no' to socio-economic determination, is expressed in inauthentic metaphors from a sham world.

Ideological and social criticism of this kind links Biondi as much with German writers like Heinrich Böll or Günter Wallraff as with Italian-German or *Gastarbeiter*-German literature, a reminder that 'migrant' themes are often universal ones. But there is another crucial dimension. While the culture, values and language of his parents' south Italian world play no part in Mamo's discourse, it is intercut by narratives by Costas, Mamo's Greek neighbour.

In marked contrast to Mamo's colloquial tone, Costas relates folktales about human greed and folly, about the once-independent fishermen now in thrall to moneylenders and forced by tourist development to emigrate, and so invokes a rural idyll lost permanently through the migration process. By intercutting the present-day experiences of a young, urban, third-generation Italian who rejects his forebears' rural past with the reminiscences of a dying Greek ex-fisherman, Biondi points to the extreme diversity of migrant experience in Germany and the contrasting forms of its textual articulation, yet also asserts their common material causes.

Abschied der zerschellten Jahre is a first-generation writer's view of third-generation experience. Rather different is that generation's perspective as offered by Nevfel Cumart. Born in Germany in 1964 of Turkish parents but brought up by a German foster family, Cumart calls Turkey 'my homeland', but first visited it at the age of nine. He writes formally and linguistically confident, economical lyric poetry in his *German* 'mother tongue'.[41] It reveals a poetic subject troubled but enriched by its complex identity. Landing in Berlin from a visit to Turkey, his body stretches 'painfully / but also reassuringly / across Europe / across the bridge.'[42] Cumart offers his voice to 'those who sit between two stools / at home neither here nor there / who daily walk the narrow ridge / on the edge of two worlds', but this is not only an ethnic solidarity but also an existential one, with the lonely, the melancholic and the rebellious.[43] His work is not free of kitsch, and sometimes encourages a vicarious and socially unspecific identification with feelings of alienation. Other poems, though, provoke German readers to look behind their perception of *Gastarbeiter* identity as a one-dimensional set of economic problems and cultural pecularities, to see individuals with spiritual needs shaped by a complex dynamic of social forces.[44]

Cumart views Turkey with 'paralysing indecision / between magnetic longing / and repelling horror'.[45] The poem 'anatolische Frau' (Anatolian woman) evokes the crushing effects of rural poverty in a barren landscape: 'at sunrise / in the fields / with your hands / you break open / the furrowed scars / of the exhausted earth.'[46] Other poems attack the corruption of Turkish bureaucracy and the oppression of the military regime.[47] This refusal to idealise his 'homeland' makes more credible his view of the dehumanisation of the migrant labourer as a 'dark shameful chapter of German history'.[48]

Cumart's work demonstrates the complex interaction of the individual and the culturally typical, characteristic of the third generation. It does not ignore or repress his Turkish identity, but it also reflects his concerns as an inhabitant of an advanced industrial society, and explores poetic subjectivity and the travails of love in ways entirely familiar to the Western tradition. When one turns to a consideration of articulations of migrant experience by women writers, the complexities and diversity of Turkish identity, in particular, take on a further dimension.

WOMEN MIGRANTS: THE MINORITY IN THE MINORITY

Women migrants are confronted with further dimensions of discrimination. The patriarchy that defines their social identity in their countries of origin and which, as a result of internalisation, continues to condition their self-perception, is also found, in modified form, in the supposedly emancipated host country. This compounds the discrimination they experience as foreigners. The resulting search for a female identity in the face of the normative pressures of two cultures gives writing by migrant women its distinctive perspective.[49] However, the following examples also show how the diverse preconditions of the authors lead to diverse acculturation strategies.

There are virtually no literary texts in German by women from the first *Gastarbeiter* generation, whose educational background and social experience were characterised by deprivation and isolation. One of the few exceptions is the autobiography of the Yugoslav worker Vera Kamenko, first published in 1978.[50] Kamenko was recruited as a *Gastarbeiterin* in 1969, worked in factories for three years, spent three years in prison for murdering her child and was eventually deported. The first-person narrative reconstructs the circumstances and events which led her to murder, and records her prison experience.

However, writing is not purely therapeutic for Kamenko. Rather, a somewhat diffuse set of desires to address, to educate and to motivate other women in similar situations, prompt her to write in German, a language in which she is barely competent. The editor Marianne Herzog has sought to retain the text's authenticity, intervening only enough to ensure comprehensibility. The result, in the fractured and disharmonious quality of the language, is a very distinctive act of resistance against social and linguistic exclusion.

Beginning in the 1980s, Turkish women have come to predominate amongst contemporary female migrant writers. Their work manifests in most concentrated and dynamic form the multiple tensions of ethnic and sexual discrimination and of the perceived incompatibility between Islam and occidental Christian culture. Recurrent themes are the change of consciousness in succeeding generations of women – the recurrence of daughter figures in this writing stresses the dialectic of linkage to and liberation from a tradition of female Muslim identity – accentuated by experience of internal and external migration, and the confrontation with Western images of the Muslim woman in which these authors do not recognise themselves.

Especially in her early texts, Saliha Scheinhardt, who also needed linguistic help before publishing her first work in German, portrays women's lives very like Kamenko's. Her protagonists come from the underdeveloped regions of Turkey, from Anatolian villages or the slums of the big cities. Their fates lie completely in the hands of their fathers, brothers and husbands, who either force them to emigrate, subject them to extreme brutality or abandon them penniless at home when they themselves emigrate. Fear and dependency hinder many of the protagonists from developing constructive strategies for liberating themselves from this situation. Their suffering often expresses itself in desperate deeds. Thus Suna, the narrator of her first text, *Frauen, die sterben ohne daß sie gelebt hätten* (Women who die before they have lived) murders her husband, only to suffer such anxiety that even prison becomes a place of liberation. 'I ran back into the freedom of my cell. There I was at least protected; nothing else could happen to me there.'[51]

Whereas Kamenko goes little beyond the simple recording of events, Scheinhardt's protagonists recognise, in retrospect, some of the determinants of their lives. They see the principal causes of female oppression in the ignorance and fatalism of women, such as their own mothers, who accept unquestioningly their subjugation to an unbending patriarchal tradition. In *Drei Zypressen* (Three cypresses), the character Gülnaz portrays her mother as 'a quiet being, like the embodiment of patience, who carries out the heaviest jobs and makes no demands. She moves round the house, the barn, the field like a shadow.'[52]

With the critical capacity Scheinhardt gives her daughter figures comes an emancipatory duty. Zeynep Z., the second-generation migrant narrator of *Drei Zypressen*, appeals to her contemporaries: 'we, the daughters, bear the responsibility for the unfulfilled dreams of the women of the generations before us' and to ensure that 'those who follow us . . . will never have to go through what we have been through.'[53]

Scheinhardt has been motivated by her own unusual biography (her development from assembly-line worker to professional writer) to speak for those migrant women who cannot articulate their own experience. This prompts her choice of literary form: fictional biographies of underprivileged Turkish women, based on interviews, research and participant observation, but whose first-person narrative perspective invites emotional identification and solidarity from her German readership. However, both the content and the form of her work attract the derogatory *Betroffenheitsliteratur* label and tend if anything to confirm the widespread German cliché of the oppressed and backward Muslim woman, which until recently has tended to be perpetuated even by the sociological literature.[54]

By the mid-1980s, the origins and the lives of Turkish women in Germany had become as varied as the reasons for their original immigration. For example, many had sought asylum in the Federal Republic to escape the military government, amongst them students, academics and artists. For educated, big-city-bred Turkish women the German equation of Turkish with Anatolian peasant was a provocation, negating as it did all the emancipatory progress they had achieved in Turkey. For many Turkish women, this confrontation with prejudice represents a profound unsettlement of their sense of self.

This is especially evident in Aysel Özakin's texts. Before she chose exile in

Germany in 1981, she was already well known in Turkey as an author whose work had close links with Western feminism. However, the shock of exile seems to have led to a loss of creative autonomy and a return to auto-biographical, therapeutic modes of writing. All her texts written in Germany articulate not only the experience of everyday discrimination, but the way it is heightened by the anxieties of exile. Her main complaint is that she is not accepted as a writer with an individual artistic identity, but seen as a member of a faceless and inferior culture.

In Özakin's novel *Die Leidenschaft der Anderen* (The passions of others),[55] the narrator, an exiled Turkish writer, undertakes a reading tour through the Federal Republic, which becomes a symbol of her rootlessness. Her fear of any kind of discrimination prevents human contacts, even with audiences from the alternative culture which, like her, takes a stand against the 'official Germany'. She refuses to conform to the expectations even of these German audiences, who see in her only the 'atypical Turk' who has thrown off the shackles of Islam. Her feminist approach, advocating the assertion of female subjectivity against patriarchal structures, is often not understood by other exiled Turks. Like her author, she finds a sense of homeland and of identity only in her texts.

Confronted with the German stereotype, many Turkish women are brought face-to-face with a long-buried part of their own past. The image of the Muslim woman suffering under the patriarchal yoke reflects the real conditions under which their grandmothers, sometimes indeed their mothers lived, and forms the starting-point of their own personal development. In the introduction to her study of Muslim women between tradition and emancipation, the Turkish journalist Naila Minai writes:

> My Turkish-Tartar grandmother was educated at home, married a polygamous man and did not take off her veil even on journeys abroad. ... My mother never wore a veil, she went to local schools and eventually became a housewife in a monogamous marriage. I left my family to study in the United States and Europe [and] hitchhiked from country to country.[56]

The awareness of this transformation through three generations informs Minai's social research, in which she aims for a differentiated picture of changing gender roles in Arabic culture in order to call into question the typical Western view of the oriental woman.

In her novel *Das Leben ist eine Karawanserei* (Life is a caravanserai), published in 1992, Emine Sevgi Özdamar, in pursuit of a similar goal, also portrays a family over three generations.[57] War, political unrest and poverty are the motivating forces behind this odyssey from the Caucasus through a variety of Turkish landscapes and cities, ending with the narrator's emigration to Germany. The novel portrays the effect of political, social and cultural changes in Turkey on the family, especially on women's life conditions and on gender relationships. The story is narrated by the daughter. Out of the fairytales and stories of her grandfather, her grandmother's sayings and prayers from the Koran, and her own experiences and acute observations, she constitutes a personal historical identity, which at the novel's conclusion she takes with her to Germany as 'luggage'.

The women figures in the novel develop survival strategies in the face of various adversities and create freedom of action for themselves, though always within the limits of the given patriarchal structures. Özakin in contrast, in a poem dedicated to her grandmother, 'Zart erhob sie sich bis sie flog' (Gently she raised herself until she flew), develops a radical, if idealised, counter-image.[58] At moments of existential crisis, the memory of her grand-mother's liberation forms an inspirational model: she 'spent her childhood in an Ottoman palace as niece of the sultan, her youth as the wife of a powerful sheik and landowner and her old age as a divorced, free and happy woman in a poorhouse'.

During their exploration of their past these Turkish women writers repeatedly encounter a barrier. The break with the Arabic-Islamic culture of the Ottoman Empire which was instigated by Atatürk's revolutionary reforms brought women many advantages; but it also prevented full critical engagement with an important part of their history: 'I / The supporter / Of the reforms of the dress codes / And of the latin script / Learned as a small girl / To despise / The six-hundred-year rule / Of the Ottomans / I am trained for a future without a past.'[59] The massive challenge to their self-image which Turkish women experience in the foreign environment prompts many of them to attempt this retrospective critical engagement with Arabic culture.

In Özdamar's story 'Mutterzunge' (Mother tongue)[60] the protagonist has, with the loss of her native language in the foreign environment, also been alienated emotionally and physically from a part of herself. To regain her sense of self, she elects to learn the Arabic language which had once been forbidden by Atatürk. This is not the reactionary act it superficially appears, but one of refilling a space in her cultural-historical identity left vacant by the expulsion of Arabic and filled instead by Westernised values. The narrator deliberately undergoes forty days of ritualistic dependence on her teacher. Paradoxically, this subordination is her path to liberation: having entered and traversed the full historical experience of her culture, and been brought back to full contact with herself, she is able to adopt an active critical attitude to her new environment.

'Mutterzunge' repeatedly calls the assumed opposition between East and West into question. The protagonist continuously moves between opposites: between East and West Berlin, between Arabic Islam and Western-oriented Turkey. Languages intermingle. Borders are dissolved. However, this text, in which an emancipated Turkish intellectual voluntarily subordinates herself to an Islamic Koran teacher, has brought Özdamar repeated attacks from her German public, unable, it seems, to accept that a Turkish woman might choose a different path of self-discovery than the Western feminist route they Eurocentrically assume to be universally applicable.

Özdamar, who has trained and worked in theatre in Berlin, Bochum, Frankfurt and Paris since the late 1970s and is familiar with the Western as well as the Turkish cultural tradition, reportedly[61] deals with such criticism by quoting Hamlet: 'There are more things in heaven and earth, Horatio, than are dreamt of in your philosophy.'

The urbane confidence of this response is a reminder that 'migrant literature' in German has progressed from *Pappkoffer* to pluralism, and by now encompasses a huge range of writing from the naïvely autobiographical to highly self-conscious, sophisticated narratives by authors with an equally wide range of backgrounds and experience. Its analysis should always respect both specific factors of ethnic and cultural background, class, generation and gender, and the shaping force of the individual literary imagination. At the same time, the common, migration-related socio-economic and cultural experiences of exclusion and discrimination that inform it should never be forgotten or obscured.

NOTES

1 K. Bade (ed.), *Deutsche im Ausland. Fremde in Deutschland. Migration in Geschichte und Gegenwart*, Munich, Beck, 1992, p. 396.

2 ibid., p. 398.

3 H. Heldmann, 'Zum Ausländergesetz 1991', *Vorgänge*, 1991, vol. 1, p. 63.

4 E. Uzun, 'Gastarbeiter – Immigranten – Minderheit. Vom Identitäatswandel der Türken in Deutschland', in C. Leggewie and Z. Senocak (eds), *Deutsche Türken. Türk Almanar*, Reinbek bei Hamburg, Rowohlt, 1993, pp. 49–66.

5 It is necessary to see the groups dealt with here as part of a wider, highly diverse, complex and dynamic pattern of migrant and minority experience and its cultural articulation in Germany: other groups not considered in this chapter include German Jews (see H. Broder and M. Lang (eds), *Fremd in eigenen Land. Juden in der Bundesrepublik*, Frankfurt am Main, Fischer, 1979), Afro-Germans (see K. Oguntoye *et al.* (eds), *Farbe bekennen. Afrodeutsche Frauen auf den Spuren ihrer Geschichte*, Berlin, Orlanda Frauenverlag, 1986; K. Obermeier, 'Afro-German women: recording their own history', *New German Critique*, 1989, vol. 46, pp. 172–80; T. Campt, 'Afro-German cultural identity and the politics of positionality: contests and contexts in the formation of a German ethnic identity', *New German Critique*, 1993, vol. 58, pp. 109–26), the Sinti and Roma communities (G. Fienbork *et al.* (eds), *Die Roma – Hoffen auf ein Leben ohne Angst*, Reinbek bei Hamburg, Rowohlt Taschenbuch, 1992), asylum-seekers (R. Ludwig *et al.*, *Fluchtpunkt Deutschland*, Marburg, Schuren, 1992), and the *Volksdeutsche* or ethnic Germans from Eastern Europe (L. Ferstl and H. Hetzel, *'Wir sind immer die Fremden'. Aussiedler in Deutschland*, Bonn, Dietz, 1990; B. Malchow *et al.*, *Die fremden Deutschen. Aussiedler in der Bundesrepublik*, Reinbek bei Hamburg, Rowohlt Taschenbuch, 1990). There is also the historically distinct experience of migrant workers in the German Democratic Republic (M. Krüger-Potratz *et al.*, *Anderssein gab es nicht. Ausländer und Minderheiten in der DDR*, Münster and New York, Waxmann, 1991).

6   S. Weigel, 'Notes on the constellation of gender and cultural identity in contemporary Germany', *New German Critique*, 1992, vol. 55, p. 49.

7   S. Castles and M.J. Miller, *The Age of Migration. International Population Movements in the Modern World*, London, Macmillan, 1993, p. 18.

8   P. White, 'On the use of creative literature in migration study', *Area*, 1985, vol. 17, pp. 277–83.

9   The last term would, for example, exclude the work of Güney Dal or Aras Ören who write in Turkish, though they live in Germany, and who write extensively about migrant experience. They are also published in German. By Ören see *Was will Niyazi in der Naunystrasse*, Berlin, Rotbuch, 1973; *Deutschland. Ein türkisches Märchen*, Düsseldorf, Claassen, 1978; *Die Fremde ist auch ein Haus. Berlin-Poem*, Berlin, Rotbuch, 1980; *Mitten in der Odyssee*, Düsseldorf, Claassen, 1980.

10 See H. Kreuzer, 'Gastarbeiter-Literatur, Ausländer-Literatur, Migranten-Literatur? Zur Einführung,' *Zeitschrift für Literatur und Linguistik*, 1984, vol. 56, pp. 7–11; A. Teraoka, 'Gastarbeiterliteratur: the Other speaks back', *Cultural Critique*, 1987, vol. 7, pp. 77–101; H. Suhr, '"Ausländerliteratur": minority literature in the FRG', *New German Critique*, 1989, vol. 46, pp. 71–103; L. Adelson, 'Migrantenliteratur oder deutsche Literatur? Torkans *Tufan: Brief an einen islamischen Bruder*', in P.M. Lützeler, (ed.), *Spätmoderne und Postmoderne. Beiträge zur deutschsprachigen Gegenwartsliteratur*, Frankfurt am Main, Fischer, 1991, pp. 67–72.

11 See U. Reeg, *Schreiben in der Fremde. Literatur nationaler Minderheiten in der Bundesrepublik Deutschland*, Essen, Klartext Verlag, 1988, pp. 17–79; G. Chiellino, *Literatur und Identität in der Fremde. Zur Literatur italienischer Autoren in der Bundesrepublik*, Kiel, Neuer Malik Verlag, 1989, pp. 19–27.

12 See H. Schierloh, *Das Alles für ein Stück Brot. Migrantenliteratur als Objektivierung des 'Gastarbeiterdaseins'*, Frankfurt am Main, P. Lang, 1984; R. Dove, 'Writing in the margin: social meaning in "Gastarbeiterliteratur"', *Quinquereme*, 1986, vol. 9, pp. 16–31. H. Heinze, *Migrantenliteratur in der Bundesrepublik Deutschland*, Berlin, Express Edition, 1986; H. Hamm, *Fremdgegangen – Freigeschrieben. Eine Einführung in die deutschsprachige Gastarbeiterliteratur*, Würzburg, Königshausen und Neumann, 1988.

13 M. Frederking, *Schreiben gegen Vorurteile. Literatur türkischer Migranten in der Bundesrepublik Deutschland*, Berlin, Express Edition, 1985.

14 I. Ackermann (ed.), *Als Fremder in Deutschland: Berichte, Erzählungen, Gedichte von Ausländern*, Munich, Deutscher Taschenbuch Verlag, 1982; *In zwei Sprachen Leben: Berichte, Erzählungen, Gedichte von Ausländern*, Munich, Deutscher Taschenbuch Verlag, 1983; *Türken deutscher Sprache: Berichte, Erzählungen, Gedichte*, Munich, Deutscher Taschenbuch Verlag, 1984; K. Esselborn (ed.), *Über Grenzen*, Munich, Deutscher Taschenbuch Verlag, 1987.

15 H. Friedrich (ed.), *Chamissos Enkel: Zur Literatur von Ausländern in Deutschland*, Munich, Deutscher Taschenbuch Verlag, 1986.

16 H. Weinrich, 'Um eine deutsche Literatur von außen bittend', *Merkur*, 1983, vol. 37, pp. 911–20; 'Gastarbeiterliteratur in der Bundesrepublik Deutschland', *Zeitschrift für Literaturwissenschaft und Linguistik*, 1984, vol. 56, pp. 12–22. See also Teraoka, op. cit.

17 F. Biondi, 'Von den Tränen zu den Bürgerrechten: ein Einblick in die italienische Emigrantenliteratur', *Zeitschrift für Literaturwissenschaft und Linguistik*, 1984, vol. 56, pp. 75–100.

18 C. Schaffernicht (ed.), *Zuhause in der Fremde. Ein bundesdeutsches Ausländer-Lesebuch*, Reinbek bei Hamburg, Rowohlt Taschenbuch, 1984, p. 56 (first published 1981).

19 Later, seeing how little German some migrants could understand even after years in the country, *Südwind* experimented with parallel texts: for example, G. Giambusso, *Jenseits des Horizonts. Al di là dell'Orrizonte*, Bremen, Edition CON, 1985.

20 F. Biondi *et al.* (eds), *Im neuen Land*, Bremen, Edition CON, 1980; *Zwischen Fabrik und Bahnhof*, Bremen, Edition CON, 1981; *Annäherungen*, Bremen, Edition CON, 1982; *Zwischen zwei Giganten: Prosa, Lyrik und Grafik aus dem Gastarbeiteralltag*, Bremen, Edition CON, 1983.

21 H. Bektas *et al.* (eds), *Das Unsichtbare Sagen! Prosa und Lyrik aus dem Alltag des Gastarbeiters*, Kiel, Neuer Malik Verlag, 1983.

22 Schaffernicht, op. cit., p. 136.

23 F. Biondi *et al.* (eds), *Sehnsucht im Koffer*, Frankfurt am Main, Fischer, 1981. See also Schaffernicht, op. cit.; Förderzentrum Jugend Schreibt (ed.), *Täglich eine Reise von der Türkei nach Deutschland*, Fischerhude, Atelier im Bauernhaus, 1980; G. Giambusso (ed.), *Wurzeln hier. Le Radici Qui. Gedichte Italienischer Emigranten*, Bremen, Edition CON, 1982; E. Kroupi and W. Neumann (eds), *Das Fremde und das Andere. Verständigungstexte*, Munich, Lauer und Richter, 1983; S. Taufiq (ed.), *Dies ist nicht die Welt, die wir suchen. Ausländer in Deutschland*, Essen, Klartext, 1983.

24 See W. Riemann, *Das Deutschlandbild in der modernen türkischen Literatur*, Wiesbaden, Harrassowitz, 1983.

25 Y. Pazarkaya, 'Literatur ist Literatur', in I. Ackermann and H. Weinrich (eds), *Eine nicht nur deutsche Literatur. Zur Standortbestimmung der Ausländerliteratur*, Munich, Piper, pp. 59–64.

26 Schaffernicht, op. cit., p. 135.

27 Teraoka, op. cit., p. 87; H. Scheuer, 'Das Eigene ist das Fremde', *Der Deutschunterricht*, 1989, vol. 41, pp. 96–104.

28 S. Dikmen, *Wir werden das Knoblauchkind schon schaukeln*, Berlin, Express Edition, 1983; *Der andere Türke*, Berlin, Express Edition, 1985.

29 R. Schami, *Das letzte Wort der Wanderratte*, Kiel, Deutscher Taschenbuch Verlag, 1984; *Der erste Ritt durchs Nadelohr*, Kiel, Deutscher Taschenbuch Verlag, 1985; *Erzähler der Nacht*, Weinheim, Beltz, 1989.

30 G. Stenzaly, 'Ausländertheater in der Bundesrepublik Deutschland und West-Berlin am Beispiel der türkischen Theatergruppen', *Zeitschrift für Literatur und Linguistik*, 1984, vol. 56, pp. 125–41.

31 The edition used here is F. Biondi, *Passavantis Rückkehr. Erzählungen*, Munich, Deutscher Taschenbuch Verlag, 1985, pp. 39–61.

32 G. Dal, *Wenn Ali die Glocken läuten hört*, Berlin, Edition der 2, 1979.

33 ibid., p. 84.

34 ibid., pp. 151–2.

35 The edition used here is G. Dal, *Europastraße 5*, Munich, Piper, 1990.

36 F. Biondi, *Abschied der zerschellten Jahre*, Kiel, Neuer Malik Verlag, 1984.

37 ibid., p. 82.

38 ibid., pp. 24–5.

39 ibid., pp. 40–1.

40 ibid., p. 97.

41 N. Cumart, *Ein Schmelztiegel im Flammenmeer*, Frankfurt am Main, Dagyeli, 1988, p. 107.

42 ibid., p. 42.

43 ibid., p. 61.

44 ibid., pp. 67–8.

45 ibid., p. 108.

46 ibid., p. 105.

47 N. Cumart, *Das ewige Wasser*, Düsseldorf, Grupello Verlag, 1990.
48 N. Cumart, *Ein Schmelztiegel im Flammenmeer*, op. cit., p. 66.
49 S. Weigel, 'Eine andere Migrantenliteratur oder eine andere Frauenliteratur', in K. Brieglieb and S. Weigel (eds), *Gegenwartsliteratur seit 1968*, Hansers Sozialgeschichte der deutschen Literatur, Band 12, 1992, Munich, Carl Hanser, pp. 222–6.
50 The edition used here is V. Kamenko, *Unter uns war Krieg*, Berlin, Rotbuch Verlag, 1982.
51 S. Scheinhardt, *Frauen, die sterben ohne daß sie gelebt hätten*, Berlin, Express Edition, 1983, p. 46
52 S. Scheinhardt, *Drei Zypressen*, Frankfurt am Main, Dagyeli, 1990. First published 1984.
53 ibid., p. 139.
54 U. Boos-Nunning, 'Frauen und Migration – Kultur im Wandel', *Informationsdienst zur Ausländerarbeit*, 1989, vol. 4, pp. 12–13; H. Lutz, 'Orientalische Weiblichkeit. Das Bild der Türkin in der Literatur konfrontiert mit Selbstbildern', *Informationsdienst zur Ausländerarbeit*, 1989, vol. 4, pp. 32–8.
55 A. Özakin, *Die Leidenschaft der Anderen*, Hamburg and Zürich, Luchterhand, 1992. First published in 1983.
56 N. Minai, *Schwestern unterm Halbmond. Muslimische Frauen zwischen Tradition und Emanzipation*, Munich, Deutscher Taschenbuch Verlag, 1991, pp. 9–10.
57 E.S. Özdamar, *Das Leben ist eine Karawanserei. Hat zwei Türen. Aus einer kam ich rein. Aus der anderen ging ich raus*, Cologne, Kiepenheuer und Witsch, 1992.
58 A. Özakin, 'Zart erhob sie sich bis sie flog', Hamburg, Galgenberg, 1986, p. 7.
59 ibid., p. 26.
60 E.S. Özdamar, 'Mutterzunge', Berlin, Rotbuch Verlag, 1991.
61 A. Burkhard, 'Vom Verlust der Zunge. Annäherung an das Fremde: Emine Sevgi Özdamar im Literaturhaus', *Frankfurter Rundschau*, 46, 23 February 1991.

4

'RIVERS TO CROSS'

Exile and transformation in the Caribbean migration novels of George Lamming

Claire Alexander

Who is that dark child on the parapets
of Europe, watching the river mint
its sovereigns stamped with power, not with poets,
the Thames and the Neva rustling like banknotes,
then, black on gold, the Hudson's silhouettes?
From frozen Neva to the Hudson pours,
under the airport domes, the echoing stations,
the tributary of emigrants whom exile
has made as classless as the common cold,
citizens of a language that is now yours . . .
Derek Walcott, 'Forest of Europe'[1]

INTRODUCTION

In 'Middle Passage Anancy', Andrew Salkey places Anancy the spiderman on a rock in the mid-Atlantic to witness the Dance of the Souls of the Dead Slaves. Before him rise the victims of the Middle Passage, dancing their fate to the beat of tribal drums, gunshots and the voices of Europe. Containing and releasing the souls of the ancestors, the ocean becomes a repository of both continuity and change, of history and transformation, at once eternal and infinitely mutable:

> boiling up and bursting and letting go all the stories of pain and suffering and brutality and horrors it been hiding quiet mongst the stillness of shipwreck and planewreck and sargasso grass and submerge mountain and earthquake whispers.[2]

As he watches the spectacle, Anancy knows 'this history have teeth fasten into people memory, all across the land, two side of the Atlantic' and feels 'a roots change deep down inside him spiderman self'; yet he is also aware of the flickering passage of time, of 'Sizzling history words and phrases splashing themself with foam, water and air,' and – as his resting place starts to submerge – of the need to move on.[3] Anancy's vision is that reserved for the Traveller, one whose knowledge of time and place leads inexorably to movement and transformation. Thus, the woman of family wisdom exhorts:

> Nothing on the land and nothing under the water ever stay steady too long, you know. Everything moving up and down and sideways, according as to how life and death always going. So hold on tight and watch yourself Anancy! A travelling spiderman got to read the natural moves he living with, light and dark, or else he going to turn into desert pinckney, just circling round on himself, sun and moon making fool out of him, constant . . .[4]

The tension between inertia and movement, between tradition and its reinvention, what Paul Gilroy has termed elsewhere the dynamic of 'roots' and 'routes',[5] is explored by Salkey in the ongoing dialogue between Anancy the spiderman and Caribbea, the sea and source of constancy. Caribbea, significantly imaged as female, is thus positioned in opposition to the male traveller;[6] as she tells Anancy, 'Plain, straight and deep. I don't travel go nowhere. I not like you. I stay and stay and stay.' Anancy, by contrast, is the embodiment of exploration and migration, 'Travelling. Always going some-where. Always looking it outside. Even when I not on the road, in the air, I travelling in me head while I look like I staying steady on the land.'[7]

The image of Anancy, spiderman, traveller and trickster shadows the themes of movement and transformation which form central motifs in the migration literature of George Lamming. More than this, Anancy signifies the struggle for an artistic form and style which gives voice to the Caribbean experience of dislocation and emergence – of exile and arrival. He is both agent and interpreter; transformer and translator.

It can be argued that the trope of migration is central to an understanding of post-colonial literature and experience. Thus, Homi Bhabha writes of the 'middle passage' of contemporary culture, in which *fin de siècle* images of transition and transgression erode and subvert the traditional certainties of Western modernity in the 'third space'. Increasingly, an exploration of 'in-between spaces' negotiates and contests absolute and essentialist notions of origin and nation, belonging and citizenship, through a performative articula-tion of reimagined perceptions of temporality and spatiality.[8] Moreover, as Gilroy has claimed, the image of the Atlantic and its slave ships provides emotive and suggestive mediatory figures in which the borders of the modern world are confronted and translated. He writes:

> The history of the Black Atlantic . . . continually criss-crossed by the movements of black people – not only as commodities but engaged in various struggles towards emancipation, autonomy and citizenship – provides a means to reexamine the problems of nationality, location, identity and historical memory.[9]

Upon the surface of this connecting non-place move the ships, 'mobile elements that stood for the shifting spaces in between the fixed places that they connected'; traversing imagined cultural and political boundaries and testifying to the transformative power of those they carry. This 'mood of restlessness' can be seen to dominate black expressive cultures, proclaiming the affirmation of diaspora and exile which critiques the grand narratives of Enlightenment modernity.[10] More than this, black diaspora cultures reveal the artificiality of the nation as a bounded and holistic cultural entity; disrupting the historical certainties of the centre with the voices of the margins.

In narrating the nation from the margins, Lamming's migration literature intervenes in what Bhabha has termed the 'double time of national identity'.[11] Rather than the transgression of absolute historical and cultural boundaries, in which the migrants are placed outside and in opposition to the wider imagined nation,[12] migration is viewed as a process in which such boundaries are at once revealed as artificial and reconstructed. It thus explores and exploits the tension between the linear historical construction of the nation and its repetitive reimagination; the division of the pedagogical and the performative. Similarly, it articulates the ambivalences of permanence and temporality – of roots and routes, of Caribbea and Anancy – inhabiting the creative space between 'a politics of fulfillment and a politics of transfiguration'.[13] If, as Bhabha has argued, black people are transfixed in the dominant English imagination as peripheral, as always emerging,[14] the counter-narratives of much Caribbean migration literature vision this act of emergence as a transformative sign: it thus constitutes rather a movement towards insertion and translation of the cultural and artistic imagination, than positing an alternative to it.

This chapter aims to explore some of the ways in which the themes of migration and exile are articulated in the 'England' novels of George Lamming: *The Emigrants* and *Water with Berries*.[15] It argues that in both novels Lamming is engaged in the struggle over the re-narration of British national identity and the artistic imagination. Through both narrative content and structure, the works seek to intervene in and subvert traditional certainties and to create imaginative space wherein black people act and think and feel, displacing their subject position and transforming their environment. While acknowledging their relation to a larger, more complete body of literature by both the author and a wider Caribbean creativity,[16] the works considered here are significant in their confrontation with themes of exile, 'otherness' and arrival. At this level, the novels considered do more than record the subjective experiences of the 'facts' of postwar migration to Britain – they translate experience in the 'in-between spaces' where difference is imagined and contested.

STATES OF EXILE: *THE EMIGRANTS* AND *WATER WITH BERRIES*

In *The Pleasures of Exile*, Lamming writes, 'To be colonial is to be in a state of exile. And the exile is always colonial by circumstances.'[17] It is this state of exile – of dislocation and alienation – which forms the central theme of Lamming's two novels about migration to England, *The Emigrants* (1954), the semi-autobiographical sequel to *In the Castle of my Skin* (1953),[18] and the later, more allegorical, *Water with Berries* (1971). For Lamming, the experience of exile is an essential and inescapable element in the histories of colonial peoples; it is, moreover, not merely a state of being but of incessant movement, of repetition and of transformation. It thus lies closer to what Bhabha[19] has termed a sense of 'unhomeliness', than migration across absolute borders between fixed imagined cultural locations; it is a permanent state of becoming, which mediates and transmutes the perception of place and time, of origin and belonging. Lamming writes:

> The pleasure and paradox of my own exile is that I belong wherever I am. My role, it seems, has rather to do with time and change than with the geography of circumstances; and yet there is always an acre of ground in the New World which keeps growing echoes in my head.[20]

The tension between exile and belonging is figured by Lamming as the struggle between Prospero and Caliban, master and slave, a theme explored more fully in *Water with Berries*. The site of struggle is, however, no longer Prospero's island of exile, but his homeland, where Caliban intrudes and corrupts through his celebration of impurity. This sense of the impure, the hybrid, marks out Lamming's creative dilemma as a Caribbean author, a descendant of Prospero and Caliban. He writes:

> For Caliban is man and other than man. Caliban is his convert, colonised by language, and excluded by language. It is precisely this gift of language, this attempt at transformation which has brought about the pleasure and paradox of Caliban's exile. Exiled from his gods, exiled from his nature, exiled from his own name! [21]

In both *The Emigrants* and *Water with Berries*, Lamming is concerned with the intervention of Caliban into both Prospero's land and narratives. He thus enters into the doubled-time and space of the narration of nationhood to place Caribbean migrants within Britain as at once transformable and transformative agents, challenging the dominant imagination through the translation of its language and forms. In particular, Lamming visions the role of the writer as transcending the worlds of coloniser and colonised, creating a new imaginative vision of community, in which 'To be an exile is to be alive.'[22]

The equation of migration with existence is common to both *The Emigrants* and *Water with Berries*. In the former, movement is figured both in the symbolism of the ship which carries the migrants to England, and in the development of narrative structure, which is at once linear and circular. Similarly, the perception of time is imaged as repetitive as well as uni-directional; both processual and inevitable. The voyage itself is portrayed as the suspension of both time and place, a movement into the 'beyond', which is both unknowable and transformative. Thus, Lamming writes of the non-time of the journey, in which the suspension of process is echoed in the repetitive strain 'We were all waiting for something to happen.'[23]

The sense of unknowing, of change masked through the endless repetition of custom, is played out through the ambiguous spatiality of the ship, which provides both the boundaries of activity and the medium for crossing. Lamming notes its symbolic transformation from absence to presence in the minds of its passengers, highlighting its transitional and mediatory power: 'The ship was merely the vehicle that had taken them from one experience to another.'[24] Within its confines and motions, the passengers are marked by both linear and circular actions. Collis, the writer-narrator figure, reflects that:

> Everybody is in flight and no one knows what he's fleeing to. . . . This is a kind of sudden big push from behind; something that happened when you weren't looking. Perhaps we were all living without looking. *And now here in mid-ocean when decisions don't mean a damn because we've got no reality to test their efficacy;* only here and now we realise telling ourselves with an obvious conviction, we want a better break.[25]

In contrast to this linear push into the 'beyond', the awakening of consciousness that the journey symbolises and evokes in the minds of the emigrants is visioned as an encircling and centripetal movement; a coming-together of individuals in the fusion of a collective West Indian identity. This is especially apparent in the almost dramatic spatial interplay of characters on the open deck of the ship, which contrasts significantly with the later emphasis on interiors and closure. Lamming writes:

> The ship's deck seemed to urge an unusual exposure. People coming together in hurried relationships seemed determined to make use of their time, and talked about themselves with almost complete freedom. In the groups it seemed that they had found their right level. The Jamaican and the Barbadian who sat with Lilian and Tornado belonged to the same world. Those who didn't belong to that world had to remain on the fringe.[26]

It is significant that a number of the characters are identifiable primarily – and sometimes solely – as synonymous with their countries of origin, a feature also found in *Water with Berries*. Yet Lamming emphasises the potential for change within even this apparent continuity, in which landscape and history are transfigured; of the sleeping bodies on the ship's deck, he notes:

> It seemed possible that the habit which informed a man of the objects he had been trained to encounter might be replaced by some other habit new and different in its nature, and therefore creating a new and

different meaning and function for those objects. It seemed that this could happen even in a man's waking life: that change which deprived the object of its history, making it a new thing, almost unknown . . . a kind of blank.[27]

The moment of arrival in England constitutes a moment of just such enigmatic transformation; a realisation of the 'beyond' in which selfhood is revealed as transitory and elusive. Lamming writes: 'Beyond their enclosure was no-THING . . . So they remained within the cage unaware of what was beyond, without a trace of desire to inhabit what was beyond. It was unnatural and impossible to escape into something that didn't matter.'[28]

It is the drive to 'make things matter' which informs the experiences of the emigrants in England; the need for incessant movement which alone testifies to existence and meaning. Within the claustrophobic interiors of England, the life of the emigrants has been seen as leading simply to disintegration and madness, a descent mirrored by the looser, episodic narrative structure of the second half of the novel. Collis' assertion that 'I have no people' can be seen as a testament to the loss of community in the face of societal rejection from the motherland, upon which all identity has been based.[29] However, while at the level of linear narrative development, the experience of the migrants can indeed be seen as negating, Lamming also constructs a circular plot formation, which reasserts the role of community and its transformative potential. Each of the characters thus undergoes transformation within Britain, as Higgins descends from cook to drug dealer to madman; Dickson from schoolteacher to drunkard; Collis from writer to factory worker; each testifying to Tornado's belief that 'a man ain't anything in particular . . . He move wid de current.'[30]

This disintegration is subverted, however, by the circularity of the narrative, through which the characters continually meet and reassert a newly formed communal identity, partially mirrored by Lamming's use of more standardised English in dialogue. As the African, Azi, proclaims, 'Nowadays it seems we will all soon come to an almost perfect unity and brotherhood.'[31] Indeed, it is only in this repetition, the encountering and re-encountering of the other migrants at different times, in new and strange places, that existence is assured. Thus, Lamming writes of Dickson:

> He had to be assured that he was still there under his clothes, inside his skin and *these were possibly the only people who could probably restore the life, the identity*, which the eyes of others had drained away.[32]

Within this repetition comes transformation, the most dramatic example of which is the story of Miss Bis, a middle-class mulatto woman who flees the Caribbean after a romantic entanglement. In England, Miss Bis is 'made' into Una, a prostitute, and in her new persona encounters and marries Frederick, the cause of her original flight, now her unwitting redemption.

Collis reflects, 'It was now that Frederick and Miss Bis in new roles would begin again.'[33] This circularity is present also in the disjunction of time and narrative development throughout the novel; thus, the suspension of time during the voyage is reflected in, and subverted by, the time lag between sections, and by the non-sequential development of events in the last chapter, significantly entitled 'Another time'. The confusion of non-event and event, continuity and change are reiterated by the writer's meeting with Lilian: '"Nothing ever seem to happen", Lilian said, "then when it happen, it all happen at once." She seemed to be fighting with the order of the incidents which she wanted to relate.'[34]

The writer himself constitutes part of this circularity: as the novel begins with the persona of a first-person narrator, who later merges into the figure of the writer Collis, so in the final section, he re-emerges as a separate voice, testifying to the passage of time and the transformation of space:

> I had walked this street for more than 2 years, at first curious, with a sense of adventure which offered me the details of the houses and the fences. *Now it was my street. It seemed I had always walked it.*[35]

It is significant that it is through the act of walking, of movement, that the writer is able to observe and, ultimately, possess the space of the streets; a possession which, he seems to state implicitly, is reliant on the repetition of the act, the constant, performative reassertion of belonging and ownership.

More dramatically, the final dispersal of the characters into the night, disappearing into the invisible landscape of which they form a part and have created, signifies at once the alienation from and acceptance of a future which is infinitely experienced and strange. This confrontation with the beyond is articulated as an ambiguous suspension of time and place which echoes that of the voyage, 'Something was bound to happen.'[36]

In *The Emigrants*, Frederick attacks Collis, the writer-figure: 'People like you just look in . . . We all move about you. You encounter us, use us, put us aside as it suits you but you never really participate.'[37] For Lamming, the creativity of the Caribbean author lies in the encounter between Prospero and Caliban, in which the theme of exile remains central. It is this artistic tension – between coloniser and colonised, stasis and migration, death and creativity – which Lamming explores in *Water with Berries*. More tightly and oppressively structured than *The Emigrants*, and framed in an allegory of colonialism which draws on *The Tempest*, *Water with Berries* confronts the role of the artist as participant and transformative agent; themes shadowed in the earlier novel. Here, as previously, movement is equated with life; stasis can only result in death, both metaphorical and literal. Colonial power throughout the novel becomes synonymous with death, figured primarily in the representation of the Old Dowager, with her necrophiliac husband and her daughter, Myra, whose barren sexuality is inseparable from disease and destruction.[38] The ability of this power to deform is recognised by the pilot, who murdered his brother and is in turn killed by the Old Dowager:

> That experiment in ruling over your kind. It was a curse. The wealth it fetched was a curse. The power it brought was a curse. That's why my brother found it to his liking. He knew it could deform whatever nature it touched.[39]

Colonial power is enacted throughout in opposition to movement, and in particular to the creativity of the artists who are the central characters. This is symbolised by the actor, Derek's enforced casting as a corpse, 'Derek, who is still learning how to die'.[40] Derek's final act of rape, significantly performed on stage, can be seen as the violent assertion of an existence which confronts the colonial gaze and disrupts the endless cycle of death which threatens to annihilate him. 'Whatever his role . . . he saw his life assume the mantle of a corpse.'[41] The Old Dowager's obsessive control of Teeton mirrors this enforced and repetitious stasis and is excised through violent transformation, symbolised by fire. Teeton's burning of the Old Dowager's body is thus a reflection of Roger's arson, enacting an ambiguous purification which answers and equates the notion of impurity and sterility, symbolised by the *whiteness* of ash. As Munro has argued, however:

> If Teeton's, Roger's and Derek's final acts of violence proceed from a will to freedom and a desire to escape from the false 'safety' of their refuge in London, they also renew the pattern of violence begun by Prospero's conquest and enslavement of Caliban.[42]

The pattern of renewal – of struggle and enslavement – is explored throughout the novel through the use of 'doubled-time', in which, as in *The Emigrants*, the linear narrative development is cross-cut through images of circularity, bringing together past, present and future in endless and in-escapable repetition. This repetition is imaged in Teeton's evocation of the unknown slave: 'Slave was the name of unique predicament. *Like time, it signified every kind of moment*; grew echoes in every corner of his own history.'[43] In *The Pleasures of Exile*, Lamming characterised this conjunction of moments as the confrontation of the living with the dead, through which alone reconciliation and transformation could be achieved. Thus he writes of the Ceremony of Souls in Haiti:

> It is the duty of the Dead to return and offer, on this momentous night, a full and honest report on their past relations with the living. . . . The Dead need to speak if they are to enter that eternity which will be their last and permanent Future.[44]

The equation of Death and the past with Prospero and with the colonial experience is drawn explicitly in *Water with Berries*, where Teeton recounts the story of the Ceremony of Souls to Myra in the non-time and non-space of the darkened heath. Myra herself is both victim and perpetrator of colonialism, a source of disease and death, who finds peace – if not escape – in the *narration* of her ordeal to the unseen, unknown figure of Teeton, the descendant of slaves.

The implication of the past and the future in actions of the present is played out through the narrative circularity of time and space. Teeton's actions, in particular, are constantly intruded upon and transformed by a consciousness of temporal disjuncture; he is described as 'yielding to the sudden intrusions of the past'.[45] His interaction with the Old Dowager is dominated by cyclical time; of their first meeting, Teeton reflects: 'It seemed to contain all the elements of initiation: the disciplined gifts of secrecy which had transformed his error into a friendship that would remain a permanent part of his future.'[46] Isolated with the Old Dowager on the island, Teeton becomes aware of both the oppression and transformative potential of the temporal 'beyond': 'The minutest change in the order of incidents, some slight reshuffle in the phases of an intention that had no specific future, may start a terrible dislocation.'[47] Death, which is closely associated with women throughout the novel, signifies the closure of potential, in which all meaning becomes unknowable and unchangeable, even for the living; thus Teeton realises of his wife's suicide: 'her death had now deprived him of any chance to know what their future might have been; to know what their past could really have meant.'[48]

Opposed to the theme of death and sterility is the celebration of artistic creativity. For Lamming, however, as for his central characters, art proves an ambiguous medium, dominated by the colonial encounter, and struggling for its own expression. If Teeton abandons his art for violent revolution, Lamming uses the language of *Water with Berries* to explore the tensions of Caliban's new exile. Unlike Myra's, Teeton's use of language becomes a prison rather than a release; he thus asserts both freedom, 'Every day I do something, I say something, and what I do or say belongs to me,' and its denial, 'The moment the words fell from his mouth it seemed they had become exhausted by their service. He didn't feel what was inside them.'[49] Language becomes inseparable from history, both bound by its past and the means of its transcendence; it becomes inescapably implicated in the temporal cycle of events. The Old Dowager's control over Teeton is thus linguistic as well as emotional; he becomes imprisoned by the repetition of words. She tells him:

> When we talk, like we are talking now, it never gets lost. . . . Some time in the future, heaven knows where you and I might be, this conversation, word for word, will travel back to you. And every sound, every note of the voice, yours and mine, intact.[50]

This temporal and linguistic closure is subverted by the desire for movement which frames the novel's events. It is through completing the spatial cycle – by return to the island of origin and reconciliation with its history –

that spiritual/artistic transformation is deemed possible. It is only through a confrontation with the past that the future can be secured; only by completing the journey of exile, transformation and return, that history can be narrated and thus transcended. Teeton reflects:

> This was the last chance to return to his own roots. He knew that this need would never bless him with its passion again. And if he failed it now, there could be no remnant of hope in any future which remained to him.[51]

Teeton's actions, culminating in his murder of the Old Dowager, contest temporal and spatial closure, leading to the final transcendence and trans-formation through fire, 'He had burnt the Old Dowager out of his future.'[52] Yet his ambitions are simultaneously thwarted by his deeds; his return is impossible and resolution unattainable. It is significant that the novel ends with the same suspension of time that marks out the earlier novel, 'They were all waiting for the trials to begin.' As Potaro, leader of the revolutionaries, asserts, Teeton – like Anancy – must continue to travel: 'Alive or dead. Near or far. It can make no difference. No grave will hold us in this place. His heart will never reconcile to waiting here.'[53]

CONCLUSION

In *Moses Ascending* (1975), Galahad confronts Samuel Selvon's eponymous writer-narrator:

> 'In any case, who tell you you could write? . . . You think writing book is like kissing hand? You should leave that to people like Lamming and Salkey.'
> 'Who?'
> Galahad burst out laughing. Derisively, too. 'You never heard of them?'
> 'I know of Accles and Pollock, but not Lamming and Salkey'.
> 'You see what I mean? Man Moses, you are still living in the Dark Ages! You don't even know that we have created a Black Literature, that it have writers who write some powerful books what making the whole world realize our existence and our struggle.'[54]

The struggle over artistic form and expression in Lamming's migration novels finds echoes in the re-creation and re-narration of post-colonial experiences of exile and 'otherness' in a range of Caribbean writers from McKay to Naipaul, Walcott and, of course, Selvon himself. In a recent article, Gabrielle Watling writes of the 'mimic men' of the new Caribbean literature in which the construction of 'otherness' interrogates and disrupts the unified narrative of colonial discourse, 'and reveals a fractured dominant culture which is incapable of supporting either its own fantasy of identity or the fantasies of its colonial subjects'.[55] In seeking to 'realise the existence' of the margin, Lamming and his contemporaries challenge the dominant and transfixing artistic gaze of the colonial centre, and intervene in the imagination of a homogeneous and exclusive nationhood. Descendants of both Prospero and Caliban, these writers explore and celebrate artistic hybridity and syncretism, asserting the presence of difference, the ambiguities of selfhood and the potentialities of transformation.

This chapter has sought to explore some of the ways in which Lamming has reimagined the experience of migration from the Caribbean to England. In 'writing across two worlds', Lamming's work reveals the constructedness of both and creates imaginative space in which nation and nationhood, identity and belonging, are narrated, interrogated and transformed. In this, it could be argued that it lies closer to the 'reality' of migration as a process of becoming[56] than has been formerly acknowledged:

> The interpretation me give hist'ry is people the world over always searchin' and feelin', from time immemorial, them keep searchin' and feelin'.[57]

NOTES

1 D. Walcott, *Poems, 1965–1980*, London, Jonathan Cape, 1992.
2 A. Salkey, *Anancy, Traveller*, London, Bogle L'Ouverture, 1992, pp. 13–14.
3 ibid., pp. 12, 14.
4 ibid., p. 15.
5 P. Gilroy, *The Black Atlantic*, London, Verso, 1993.
6 This chapter is concerned primarily with a male view of the experiences of migration, in which movement is seen as masculine, and constancy as feminine. This relationship is, however, ambivalent: Salkey notes elsewhere that Anancy is both male and female, and that Caribbea's source of constancy is crucial to Anancy's sense of identity.
7 Salkey, op. cit., p. 27.
8 H. Bhabha, *The Location of Culture*, London, Routledge, 1994, pp. 1, 5.
9 Gilroy, op. cit., p. 16.
10 ibid., p. 111.
11 Bhabha, op. cit., p. 145.
12 The ways in which black migrants to Britain have been created as marginal and 'alien' in the dominant British imagination is argued in the CCCS (Centre for Contemporary Cultural Studies) Collective, *The Empire Strikes Back*, London, Hutchinson, 1982, and in Paul Gilroy's *There Ain't No Black in the Union Jack*, London, Hutchinson, 1987.
13 Gilroy, *The Black Atlantic*, op. cit., p. 112.
14 Bhabha, op. cit., p. 157.
15 G. Lamming, *The Emigrants*, London, Michael Joseph, 1954; *Water with Berries*, London, Longman, 1971.
16 The novels considered here could be usefully compared with the treatment of similar themes in V.S. Naipaul's *The Mimic Men* (1967), Samuel Selvon's *The Lonely Londoners* (1956), *Moses Ascending* (1975), and Caryl Phillips' *The Final Passage* (1985), amongst others.

17 G. Lamming, *The Pleasures of Exile*, London, Michael Joseph, 1960, p. 229.
18 G. Lamming, *In the Castle of my Skin*, London, Michael Joseph, 1953.
19 Bhabha, op. cit., p. 9.
20 Lamming, *The Pleasures of Exile*, op. cit., p. 50.
21 ibid., p. 15.
22 ibid., p. 24. See also I. Munro, 'George Lamming', in B. King (ed.), *West Indian Literature*, London, Macmillan, 1979, pp. 126–46.
23 Lamming, *The Emigrants*, op. cit., p. 10.
24 ibid., p. 36.
25 ibid., pp. 52–3, my emphasis.
26 ibid., pp. 58–9.
27 ibid., p. 83.
28 ibid., p. 105.
29 Munro, op. cit.
30 Lamming, *The Emigrants*, op. cit., p. 185.
31 ibid., p. 129. The masculine nature of the experience of Lamming's novels has already been commented upon.
32 ibid., p. 258, my emphasis.
33 ibid., p. 270.
34 ibid., p. 227.
35 ibid., p. 223, my emphasis.
36 ibid., p. 271.
37 ibid., p. 246.
38 Throughout the novel, Lamming equates female sexuality with colonialism and with stasis and closure; this is, however, riven with ambiguity – Myra and the Old Dowager are both victims as well as perpetrators of colonial power. This ambiguity is extended in Teeton's reaction to his wife's death, which suggests a difference in the forms of closure represented by 'Caribbea', the wife, and the English colonial women. This can be compared with the later novels of Samuel Selvon, in which black women are seen as bringing stability and a sense of community to the migrants, see *The Housing Lark* (1965) and *Moses Ascending* (1975). The complex interaction between gender and 'race' in these novels needs and deserves further more detailed consideration, which is beyond the scope of the present chapter.
39 Lamming, *Water with Berries*, op. cit., p. 229.
40 ibid., p. 93.
41 ibid., p. 219.
42 Munro, op. cit., pp. 142–3.
43 Lamming, *Water with Berries*, op. cit., p. 93, my emphasis.
44 Lamming, *The Pleasures of Exile*, op. cit., p. 9.
45 Lamming, *Water with Berries*, op. cit., p. 34.
46 ibid., p. 38.
47 ibid., p. 205.
48 ibid., p. 117.
49 ibid., p. 115.
50 ibid., p. 29.
51 ibid., p. 181.
52 ibid., p. 247.
53 ibid., pp. 246, 249.
54 S. Selvon, *Moses Ascending*, London, Davis-Poynter, 1975, pp. 42–3.
55 G. Watling, 'Embarrassing origins: colonial mimeticism and the metropolis in V.S.Naipaul's *The Mimic Men* and Sam Selvon's *Moses Ascending*', LINQ, 1985, vol. 20, pp. 68–77.

56 Compare the experiences of migration and nationhood in J. Western, *A Passage to England: Barbadian Londoners Speak of Home*, Minneapolis, University of Minneapolis Press, 1992, and C.E.Alexander, *'The Art of Being Black': the Creation of Black British Youth Identities*, Oxford, Oxford University Press, forthcoming.
57 Lamming, *The Emigrants*, op. cit., p. 68.

5

NEGOTIATING IDENTITY IN THE METROPOLIS

Generational differences in South Asian British fiction

Suresht Renjen Bald[1]

INTRODUCTION

All migrants leave their past behind . . . it is the fate of migrants to be stripped of history, to stand naked amidst the scorns of strangers upon whom they see the rich clothing, the brocades of continuity and the eyebrows of belonging.[2]

The notion of origin is as broad and robust and full of effect as it is imprecise. . . . History slouches in it, ready to comfort *and* kill. . . . But the only way to argue for origin is to look for institutions, inscriptions and then to surmise the mechanics by which such institutions and inscriptions can stage such a particular style of performance.[3]

Stripped of their belonging(s), and denied their history, all migrants occupy a vulnerable position. But when the migrant is a member of a formerly colonised people and the border s/he crosses marks the land of the former coloniser – as is the case of migrants of South Asian origin in Britain – the narrative of immigration comes to include not only the loss of 'continuity' and the search for 'belonging', but also the experience and negotiation of racism and colonialism. Relationships between immigrants from South Asia and white Britons are mediated by traces, memories, and history of two hundred years of British imperialism and South Asian resistance (and complicity). At a time when Britain's economy and international power are in decline, white Britons' categorisation of South Asians (and Afro-Caribbeans) as the essentially inferior and culpable 'other', helps maintain the myth of a superior white race. Thus categorised, migrants of South Asian origin and their British-born children struggle in different ways to counter their hosts' constructions of them; but these struggles are limited and structured by the discourses that define them.

The diverse realities of migrants' lives are represented in the works of

British writers of South Asian origin. In the last two decades, the fiction authored by these writers has expanded our understanding of the complex negotiations of identity that South Asian migrants and their children engage in every day in a dominantly white Britain. First, these works have revealed how the discourse of racism and colonialism changes with the shifts in the British economy and polity; and second, this literature shows how migrants' reactions to their marginalisation vary according to the generation, gender, and class of the author's fictional characters and the particular era in which the story is set. In the face of British efforts to erase differences among migrants by making people of South Asian origin fit into one depersonalised, marginalised category of the inferior 'other', South Asian British writers re-present migrants' stories in voices that affirm and validate their multiplicity and individuality, and problematise 'otherness'. By contrast, in the fiction of Anglo-British writers South Asian Britons are generally voiceless or faceless.[4]

This chapter will examine contemporary South Asian British fiction to explore the worlds of first-generation South Asian migrants and their British-born children in order to gain a better understanding of the workings of the politics of inclusion and exclusion that erase and inscribe difference.[5] Focusing mainly on the writings of Kamala Markandaya, V.S. Naipaul, Hanif Kureishi, Salman Rushdie and Farrukh Dhondy, this study seeks to show how the responses of migrants and their children to their experiences are constituted as much by the discourse of racism and colonialism as by their location in the discourses of the lands of their origin.

CONTEXT

People of South Asian origin have been living in Britain since the early eighteenth century. Rozina Visram's *Ayahs, Lascars and Princes* gives a fascinating account of the Indian domestic workers, retired seamen and sepoys (soldiers), radical freedom fighters, cricketers, and an occasional member of the princely families, who, for different reasons, chose to make Britain their home. Rather small in number, these migrants tended to be a source of curiosity and sometimes entertainment to the indigenous popu-lation.[6] But after the Second World War the situation changed dramatically. The number of South Asian migrants grew, eventually comprising 3.4 per cent of the population; instead of curiosity and amusement they now evoked primarily resentment and fear, and became targets of abuse.

Several factors contributed to this change. Following the Second World War Britain needed cheap labour to help rebuild the ravaged economy. It promised jobs with good pay for those in the South Asian sub-continent and the Caribbean who were willing to work in its factories, and in general fill the labour gap created by war deaths. The sub-continent, at that time, had gone through a traumatic partition which had displaced almost ten million people. After being uprooted from their homes, some of the victims of partition were receptive to starting a new life in Britain. Indians, and Pakistanis from both its Western and Eastern wings,[7] therefore began to arrive in Britain in relatively large numbers in the 1950s. A smaller number of migrants of South Asian origin came to Britain from the Caribbean. By the early 1960s the South Asian presence in Britain had created enough concern for the Conservative Government to propose controls on immigration from the 'coloured' Commonwealth. Passage of the 1962 Immigration Act with its strict immigration controls prompted South Asian workers to get their families, whom they had left in their homelands, to join them in Britain before the Act's provisions were implemented.[8] White Britons' perception of an increase rather than a decrease in the number of dark-skinned migrants contributed to further deterioration of relations between the indigenous and the migrant populations, and immigration became a hot issue during the 1964 and 1968 elections.

The late 1960s and early 1970s saw another wave of South Asian migrants. Most of these came from East Africa where the implementation of Africanisation policies by the independent states of Uganda, Tanzania and Kenya led to the exodus and sometimes expulsion of people of South Asian origin. In addition, during this period, the number of South Asians and Caribbeans who had come to Britain in search of an education or professional training grew; some of them opted to stay on in Britain rather than return home.

In the 1980s and 1990s, therefore, it was not uncommon to find South Asian families with children, and sometimes grandchildren, who were born and brought up in Britain. Indeed, 45 per cent of Britons of South Asian origin are now UK-born; the majority of them have never visited their ancestral 'homelands'. Educated and raised in Britain, their values, language and concerns are often closer to their Anglo and Afro-Caribbean peers than their parents. Yet, because of the colour of their skin, and in some instances, their mode of dress, they are perceived as 'immigrants' and often referred to derogatively as 'Pakis' by white Britons. By conferring migrant status on these British-born brown-skinned citizens, white Britons question their 'British-ness'. As the provocative title of Paul Gilroy's book reminds us, *There Ain't No Black in the Union Jack*.[9] South Asian writers' fictional accounts of migrant experiences in particular eras re-present migrants' and their children's 'construction' and 'reconstruction' in Britain. Collectively the works of these writers explore the complex discourses of racism and colonialism within which members of the South Asian diaspora must shape and define themselves; but, while the works illustrate the power of discourses that seek to define the migrants, these works also articulate alternative discourses within which first- and second-generation South Asian Britons shape their strategies of resistance and self-empowerment.

EVOLVING DISCOURSES AND
NEGOTIATION OF IDENTITY

One important writer who takes us into the world of first-generation South Asian Britons is Kamala Markandaya. Markandaya came to England as a professional reporter in the 1940s with no intentions of settling in Britain. Her novel *The Nowhere Man*, set in the late 1960s, recounts how the changing discourses of racism and colonialism limit the choices of the protagonist Srinivas, and define and shape his struggles. The novel underscores the frightening power of the definer.[10] Srinivas and his wife Vasantha are both Brahmins from a long lineage of scholars and landed gentry. However, when they arrive in Britain in the mid-1920s Srinivas is unable to pursue a career in keeping with what he calls 'the ambition sown in ancient nurseries' which made him remind himself, 'I am a Brahmin, a scholar, I intend to take my rightful place in the ranks of the learned.'[11] Every job interview in which he offers his learning ends in derision toward his abilities and training. Srinivas' experience illustrates the persistence of racial arrogance that made Macaulay in 1835 place a higher value on a shelf of Western literature than a library full of Arabic and Sanskrit works.[12] Only now, this arrogance has essentialised the Indian's inferiority.

To the white interviewers, Srinivas' inadequacies stem not from what he knows or is capable of knowing, but in his being a colonised Indian; by virtue of his colonial status and his brown skin he is automatically perceived as being inferior to *any* white Briton.

> One day he knew. Quite suddenly. One moment he was preparing for yet another interview, arranging dog-eared and much thumbed certificates neatly in a folder; the next, he knew that he would not go. Not now, not ever. . . . Shortly afterward he became a trader.[13]

But, though the British reduce Brahmin Srinivas to a job generally associated with people two castes below him in the Indian social hierarchy, they cannot change the temperament inscribed by centuries of discursive practices. His 'Brahmin' training prevents Srinivas from thinking and acting like a trader, and hence make it difficult for him to be successful in the occupation he is allowed to follow in his adopted land.

British-born Kureishi, like his protagonist in *The Buddha of Suburbia*, is of South Asian and English descent. Though he is a generation younger than Markandaya, his portrayal of the older immigrants in *The Buddha of Suburbia* nevertheless echoes Srinivas' experience of racism. Indeed, *The Buddha of Suburbia* picks up where *The Nowhere Man* ends. The protagonist's father Haroon, born into a well-established affluent Bombay family, comes to London in the 1950s to study law. He marries a working-class English woman and as a clerk in the Civil Service he earns £3 a week. 'His life, once a cool river of balmy distraction, of beaches and cricket, of mocking the British . . . was now a cage of umbrellas and steely regularity.' At work Haroon is denied promotions; he claims, 'the whites will never promote us. . . . Not an Indian while there is a white man left on earth . . . they still think they have an Empire when they don't have two pennies to rub together.'[14]

Another character in the novel, Jeeta, a princess in Pakistan, 'whose family came on horseback'[15] when she wed her husband Anwar, helps her husband sell groceries in London. Like Srinivas, they can only aspire to the rank of tradespeople. The discourse of white racism subverts the discourses of caste and class that ascribed Srinivas' legitimate role, and defined Haroon's and Jeeta's position in the lands of their origin.

Though Srinivas is acutely aware of his 'loss of status' at the hands of white racism, the tragedy of his situation lies as much in his need to 'remember' his past as his loss. He manoeuvres between the dominant Hindu discourse of the land of his origin and white British discourses of racism and colonialism to negotiate identity. But these negotiations are complicated by the changing face of British racism. The racism Srinivas encountered in pre-war years was different from what he experiences in the 1960s. Though in the pre-war and war years Srinivas was not considered good enough for the job for which he applied, his neighbours, secure in their superior imperial position, were not hostile towards him. They saw him and his wife as 'Oriental', different from themselves but harmless. And during the war the bombing 'ripped away veils, not to say whole walls, revealing weeping surfaces and intimate interiors, and making it difficult for conventions to rule with their previous inflexible rod.' The neighbours sought shelter in Srinivas' basement, with the neighbourhood behaving 'like one big family. It [war] seemed to draw everyone closer.'[16] But the camaraderie of the war years disappeared in times of peace. In the 1960s when the dust had settled to reveal a declining economy and a fading international presence, Britain needed an 'other' to bolster the British sense of superiority. Migrants from the 'coloured' Commonwealth filled the bill. As Roxanne Doty argues, now

> Nation and race were made the framework for a discourse of order and security. Insecurity and disorder, including national economic decline and decline as a world power, became linked with a dilution of Britishness which was associated with Commonwealth migration. [17]

Avowedly racist speeches by public figures like Enoch Powell blamed the migrants for unemployment, housing shortages and inadequate public services. The anti-immigrant discourse, according to Doty,

> focused on three aspects of immigration; the numbers involved, health, and crime. . . . Numbers were used as scare tactics. They served to signify the breakdown of unity and identity rather than any concrete situation. . . . There were also extensive references to the health situ-

ation, carrying with them implications of dirt, contamination, and even moral decay. Suggestions were made that Commonwealth immigrants did 'not always conform to our ideas of sanitation'. Similarly, the theme of crime was prevalent. . . .[18]

Phrases used to describe Commonwealth immigrants – 'hordes', 'terrifying', 'swamped', 'breed' – were often drawn from the vocabulary generally reserved for animal migrations.[19] Fred, the white supremacist son of Srinivas' neighbour, claims in *The Nowhere Man*, 'they [coloured immigrants] came in hordes, occupied all the houses, filled up the hospital beds and their offspring took all the places in schools.' The reader is not surprised when Fred and his friends first call Srinivas an 'ape', and then, almost a hundred pages later, beat him and tar and feather him.[20] By turning Srinivas into an animal, Fred and his friends define their superiority to and distance from Srinivas. Indeed, defined thus, the distance between Srinivas and Fred and his friends becomes unbridgeable.

While Fred and his friends animalise Srinivas, their parents ostracise and isolate their neighbour as if he were a leper. Srinivas thus falls victim to another constitutive element of the British discourse of racism. Treated as if he is diseased, Srinivas develops symptoms of a disease associated in the popular mind with the 'colonies'. He *becomes* a leper.

British discourses are not the only ones within which Srinivas is defined and shaped. He is a Brahmin in whom 'the early years [spent in his "homeland"] are most deeply etched . . . and the memories persist'.[21] Faced with white racism, he draws on his Brahmin training with its emphasis on non-violence, self-control and detachment to construct his responses. He prefers to bear pain rather than inflict it. But when Fred sets fire to Srinivas' home, the burden of pain (that eventually consumes him) becomes too much to carry even for Srinivas' Brahminically inscribed self. In the end, Markandaya's Srinivas does not die from the racially fuelled fire that burns his house and Fred but does not even singe Srinivas; he dies from the pain and shock of seeing his neighbours of thirty years turn against him.[22] Unable to hate or express anger, he has no place in the racially polarised Britain of the 1960s and early 1970s. He can only die.

The metamorphosis of post-colonial migrants in Britain is also the central theme of Salman Rushdie's *The Satanic Verses*. However, the transformations experienced by Rushdie's 'Brown Uncle Tom' British citizen, Saladin Chamcha, and the post-colonial Indian movie-star of B-class Hindi films, Gibreel Farishta, evoke in each of them responses quite different from those displayed by Srinivas. Neither Saladin Chamcha nor Gibreel Farishta belongs to the dominant Hindu majority in India. Farishta, like his creator Rushdie, is of Muslim heritage, and Chamcha is a Parsi; thus neither is inscribed with specifically Brahminical notions of non-violence, self-control, and passive resistance.

According to the story Rushdie tells, Saladin and Gibreel are ejected from a jumbo jet which has been blown up over the ocean en route to London. They arrive in Thatcher's England with just the wet clothes on their backs. Their landing without any papers or possessions increases the power of the police to define who they are. To do so, the police, one of the more racist constituents of the British state, search for 'identifying' marks. Saladin who has 'foul breath' and is unshaven with bumps on his forehead, is promptly classified as an illegal alien. His pleas that he is a British citizen are met with 'uproarious guffaws'. Faced by such a reaction Saladin starts to grow Satanic horns and the hoofs of a goat. 'Not much of a price to pay for survival, for being reborn, for becoming new.'[23] Gibreel on the other hand acquires a 'halo'. Seeing him dressed in a smoking jacket and jodhpurs, with 'a pale, golden light ... streaming softly outwards from ... behind his head' the police wish to 'reveal' to him 'their secrets. ... A more reputable looking gentleman you couldn't wish to see, in his smoking jacket ...'[24]

Clearly, it is not difference *per se* to which the police react, for both Saladin and Gibreel display marks that distinguish them from white mainstream Britons; it is *particular* marks of difference that concern them. Given the discourse of racism within which the police interpret difference, Saladin's foul breath and animal-like demeanour translate into 'coloured immigrant', but Gibreel's smoking jacket and jodhpurs, marks of class, make the police colour-blind. The discourse of class erases that of racism. As the police handcuff Saladin and lead him to the 'Black Maria', the police van, Gibreel does not intervene. Complicity with the 'enemy' is synonymous with his survival.

Abandoned by the post-colonial Indian (or the modern Indian State), and repudiated by the whites he admired and emulated for thirty years, Saladin is brought by the police to the hospital at the Detention Centre for illegal immigrants. Here Saladin wakes up to

> animal noises: the snorting of bulls, the chattering of monkeys, even the pretty-polly mimic-squawks of parrots or talking budgerigars. ... His nose informed him that the sanatorium, or whatever the place called itself, was also beginning to stink ... jungle and farmyard odours mingled with a rich aroma similar to that of exotic spices sizzling in clarified butter.[25]

The two smells, animal and South Asian (curry), become one, echoing Fred's descriptions of Srinivas. At the sanatorium Saladin learns that, unlike Srinivas, he is not alone; there are other 'mutated' beings like him at the hospital. A manticore, a creature with 'an entirely human body', but the 'head ... of a ferocious tiger', who was once a top-paid male model in Bombay, introduces Saladin to the other inmates:

> The manticore ground its three rows of teeth in evident frustration. 'There's a woman over that way,' it said, 'who is now mostly water-buffalo. There are businessmen from Nigeria who have grown sturdy

76

tails. There is a group of holiday-makers from Senegal who were doing no more than changing planes when they were turned into slippery snakes ... The point is ... some of us aren't going to stand for it ... Every night I feel a different piece of me beginning to change ... ' 'But how do they do it?' Chamcha wanted to know. . . . 'They *describe* us ... That's all. They have the power of description, and we succumb to the pictures they construct.'[26]

Animalisation and demonisation of South Asian Britons is paralleled by their 'exotic' or 'quixotic' representations by Western media and popular culture. In the mid-to-late 1960s when the Beatles and the rest of Western 'counter-culture' were romancing India and things Indian, and when the Nehru jacket had hit the fashion ramp, chic Londoners saw India as synonymous with wisdom and spirituality. Myths of India's 'non-materialist' culture served to highlight the West's culture of consumerism. Transcendental meditation or TM entered the vocabulary of those who considered themselves 'hip'.

This 'exoticisation' of South Asians by some whites is one of the stories told in Kureishi's *The Buddha of Suburbia*. In this coming-of-age novel, the narrator's father, Haroon, does not turn into a 'leper' or a 'goat/devil', but he becomes the 'Buddha of Suburbia' – an Indian wise man who is perceived by his white followers to have all the answers to the problems of the materialistic West. With the help of his white woman friend and lover, Eva, Haroon becomes the local 'guru' for North London. But his newly acquired 'spirituality' that fascinates the 'intellectual' North Londoners is lost on Haroon's white working-class in-laws, who are embarrassed by his Indian heritage. They would rather erase his Indian-ness than celebrate it. Haroon's brother-in-law hopes to deny Haroon's Indian origins by calling him 'Harry'. To him, being an Indian 'was bad enough ... without having an awkward name too'.[27] Neither the North Londoners nor Haroon's in-laws are willing to accept him as just 'Haroon'; each refashions him and clothes him in an identity of their own choosing.

Haroon's son Karim, a 'born and bred' Englishman, is an aspiring actor, but the first part he gets is to play Mowgli in *Jungle Book*. To fit the audience's expectations of Mowgli, Karim, who has never been to India, is directed to speak with Peter Sellers' version of an Indian accent, and his 'creamy-skinned' body is darkened with brown paint. According to Pike, a highly respected white avant-garde director, an 'authentic' Indian is one who fits white society's stereotype of Indians constructed by Sabu's and Peter Sellers' performances.[28] White power 'reconstructs' and 'authenticates' a Haroon or a Karim; or as Gayatri Chakravorty Spivak puts it,

> an earlier experience is being staged in this new, displaced imperialist scene: the horror of an absolute act of intercultural performance ... 'I perform my life this way because my origin stages me so.' National origin, ethnic origin. And, more pernicious: 'You cannot help acting this way because your origin stages you so.'[29]

STRATEGIES OF SURVIVAL/EMPOWERMENT IN THE METROPOLIS: INVISIBILITY, COMPLICITY, RESISTANCE

Both Haroon as Buddha and Karim as Mowgli stage their performances to suit their white audiences. Haroon shows great dexterity as he moves back and forth between 'Harry', 'Haroon' and 'The Buddha of Suburbia'. Similarly, complicity with white racist constructions of migrants is evident in Gibreel Farishta's strategic rejection of his companion Saladin (in *The Satanic Verses*) when the police arrest the latter for being an 'illegal alien'.

The complicity of Srinivas' son Laxman in *The Nowhere Man* is of a different kind. Having grown up among whites in an Imperial Britain of pre-war and war years, Laxman has accepted the notion of 'inferiority' of the South Asian. Like his English wife he feels uncomfortable in the presence of his parents; but unlike her he also feels ashamed:

> Accustomed to the flip repartee of his father-in-law's factory, at which he was adept, [he] could not help looking askance at his prim, not to say provincial parents. Sitting side by side on the divan with the ridiculous roly-poly bolsters, and his mother with her bun. And her clothes, like the robes Jesus Christ wore, only worse with the cardigan. Her face scrubbed clean and yellow like a plain deal table, and her English, which was not the country's or his own. As if, complained Laxman silently, after all these years she could not; although when his wife asked him he could not say exactly what he wanted of his mother. But something: anything that she could do that would sink her indistinguishably into England, instead of sticking out like a sore thumb.[30]

Laxman wants to blend into white society but his parents, by clinging to the marks of their origin, make it difficult for him to do so. He hopes for invisibility through complicity.

Their Indian ways and their Brahmin-ness give Laxman's parents, Srinivas and Vasantha, a sense of authenticity and provides them with an alternative discourse within which to resist their construction by white racism. This is especially evident when Srinivas becomes a victim of racist attacks; he withdraws into himself rather than retaliate. Generations of Brahmin upbringing have instilled in him an aversion to the use of violence.

Unlike Srinivas, Saladin Chamcha (*The Satanic Verses*) is not inscribed by Brahmin marks of self-control and non-violence. It is easy for him to express his anger and hate at the South Asian British 'Club Hot Wax' where his young Black British friends take him. Indeed, violence has a cathartic effect on him. While Srinivas' Brahmin-inscribed 'self', incapable of expressing anger, hatred or pain, is eventually killed by his internalised suffering, Saladin's violent outbursts help him lose his 'horns and hellsbreath' and restore him 'to his old shape'. He is 'humanised – is there any option but to conclude? – by the fearsome concentration of his hate'.[31]

Club Hot Wax performs an important role among the fictional South Asians living in Rushdie's London. It provides angry young Black British with opportunities to vent their anger and hatred safely against the white British by watching wax forms of 'hate-figures' melt down:

> Attendants move towards the tableau of hate-figures, pounce upon the night's sacrificial offering, the one most selected, if truth be told, at least three times a week. Her permawaved coiffure, her pearls, her suit of blue. *Maggie-maggie-maggie*, bays the crowd. *Burn-burn-burn*. The doll – the *guy* – is strapped into the Hot Seat. Pinkwalla throws the switch. And O how prettily she melts, from the inside out, crumpling into formlessness. Then she is a puddle, and the crowd sighs its ecstasy: *done*.[32]

But unlike the young Black British at the club, Saladin's hatred, as mentioned earlier, is not directed against white racism that is responsible for his transmogrification, but towards his former companion, the post-colonial Indian Gibreel Farishta who had refused to intervene when the police arrested Saladin. Saladin's hatred of Gibreel is reminiscent of Frantz Fanon's description of colonised people in *The Wretched of the Earth*: 'The colonised man will first manifest this aggressiveness which has been deposited in his bones against his own people. This is the period when the niggers beat each other up.'[33]

At the Club Hot Wax, Saladin sees Gibreel's face in each of the wax forms and it is against this face that he vents his rage and fury:

> Mr Gibreel Farishta . . . Who should the Devil blame but the Archangel, Gibreel? . . . The face on every one of the waxwork dummies was the same now, Gibreel's face. . . . The creature [Saladin] bared its teeth, and let out a long, foul breath, and the waxworks dissolved into puddles and empty clothes, all of them, every one. The creature lay back satisfied.
>
> Whereupon it felt within itself the most inexplicable sensations of compression, suction, withdrawal; it was racked by squeezing pains. . . .
>
> When Mishal, Hanif and Pinkwalla ventured into the clubroom several hours later they observed a scene of frightful devastation . . . and at the centre of the carnage, sleeping like a baby, no mythological creature at all . . . but Saladin Chamcha himself . . . mother-naked but of entirely human aspect and proportions . . .[34]

Deconstruction of the 'Archangel' is necessary if the 'Devil' is to regain his human form. Destruction of binary divisions of 'black' and 'white', 'God' and 'Satan', 'superior' and 'inferior', 'self' and 'other', which depend on and feed off each other, is necessary for 'human-ness' to emerge.

In contrast to the middle-aged post-colonial migrant Saladin's confusion about the source of his troubles, and Srinivas' strategy of avoidance through withdrawal, young Mishal and her political activist friend Hanif in *The Satanic Verses* are actively involved in resistance against white racism. They see themselves engaged in a battle of epic proportions. Drawing on the vocabulary of the Indian epics, Mishal talks about:

> the Street as if it were a mythological battleground and she, ... the recording angel and the exterminator too. From her Chamcha learned the fables of the new Kurus and Pandavas, the white racists and black 'self-help' or vigilante posses starring in this modern *Mahabharata* [Great India], or, more accurately, *Mahavilayet* [Great Britain] ... These days the posses roamed the nocturnal street, ready for aggrava- tion. 'It's our turf,' said Mishal Sufyan of that street without a blade of grass in sight. 'Let 'em come and get it if they can.'[35]

Assuming the role of Sanjaya who describes the great war in the *Mahabharata* to the blind Dhritrashtra, Mishal explains the war between white racists and blacks to Saladin who, though sighted, is 'blind' to white racism.

But unlike Sanjaya, she is not just an observer; indeed, in preparation for the struggle Mishal and her sister practise 'karate kicks and Wing Chun forearm smashes'. To them the entire 'system', constructed *by* the dominant *for* the dominant, is suspect. The only course open for the disempowered is to reject the meanings the language of the powerful inscribes by reclaiming the words it uses to describe the marginalised. Mishal and her sister admire the 'moonlighters, shoplifters, filchers: scam artists in general ....' Though neither sister 'would ever steal a pin' herself, nevertheless to them such persons represented the 'gestalt, of how it was ... wickedness unpunished made them laugh'.[36] Mishal, her sister and their friends consider those resisting or cheating the system as heroes comparable to 'Dick Turpin, Ned Kelly, Phoolan Devi'. By admiring 'anti-establishment' activities they legitim- ise them and put 'otherness' on its head; admiration allows for a re- construction of the 'other' from the perspective of the marginalised. By problematising 'criminal' and 'legal', Black British like Mishal and their friends aim to renegotiate the political boundaries of inclusion and exclusion.

Mishal and her lover Hanif speak the language of Black Power rather than Gandhian or Brahmin non-violence; they see the hand of white racism everywhere. To them the goat/devil Saladin is one of racism's many 'casu- alties'. They and other young Black British claim him. Signs and T-shirts expressing 'Sympathy for the Devil' appear in enclaves in London territorial- ised by South Asian Britons. Mishal tells Saladin,

> you're a hero. I mean people can really identify with you. It's an image [the devil] white society has rejected for so long that we can really take it, you know, occupy it, inhabit it, reclaim it and make it our own.[37]

Characteristically, Saladin is not interested in the role Mishal wants him to play. Two centuries of colonisation have blunted his political consciousness. He does not want to fight white Britons; indeed, he admires them and has spent thirty years trying to be like them. His hatred and anger are directed towards Gibreel whom he thinks he is destroying at the Club Hot Wax.

While Saladin's hatred of Gibreel blinds him to the real authors of his monstrous construction, Mishal's resistance to British institutions heightens her awareness of other sites of oppression. Her battles are not confined to white racism, they include resistance to her family's strict sexual code and notions of gender. Her political activism against white racism is paralleled by her manoeuvres for sexual freedom. Both oppressions are maintained through ideologies that devalue the 'other'. Each rests on essentialist notions of superiority – racism claims 'natural' superiority of the white race, and sexism bases male dominance on biology. In the case of migrants, the realities of their lives in Britain work to strengthen patriarchy in at least two ways.

First, the sense of powerlessness that male migrants feel in white racist society often leads them to seek compensation at home by asserting their power over their women. In the safety of their family they seek to redeem the 'masculinity' which they feel is denied them in white society. And second, because they are excluded from the national community, migrants like Hind, Mishal's mother, try to replicate the community they have left behind by transplanting to Britain the cultural practices of the land of their origins. Hind treats her husband 'for the most part, like a lord, like a monarch, for in her lost world her glory had lain in his ...'. She expects her daughters to help her with the chores, 'think of marriage, attend to studies', and be obedient, chaste and soft-spoken.[38] As a good Muslim daughter Mishal is expected to stay a virgin until her wedding night. But for Mishal, her parents' expectations of her bear little resemblance to the lives of young people growing up in London in the 1980s. She resists her parents by engaging in sexual liaisons with her friend Hanif, cutting her hair, wearing provocative clothes, and speaking out.[39]

A Mishal figure reappears as Jamila in Kureishi's *The Buddha of Suburbia*. We are told that:

> Under the influence of Angela Davis, Jamila had started exercising every day, learning karate and judo, getting up early to stretch and run and do press-ups. ... She was preparing for the guerrilla war she knew would be necessary when the whites finally turned on the blacks and Asians and tried to force us into gas chambers or push us into leaky boats.[40]

She is assertive and courageous. When told by a young white man to 'Eat shit Paki' she runs after the 'greaser', 'throwing the bastard off his bike and tugging out some of his hair, like someone weeding an overgrown garden'. Her heroes are Fidel Castro, Che Guevara and Angela Davis. She reads Kate Millett and is interested in 'anarchists and situationists and Weathermen'.[41]

It is anti-imperialist, anti-racist and anti-patriarchal discourses that appeal to Jamila; it is within them that she negotiates her identity and shapes her resistance. However, her radical politics seem to lose out in the face of her father Anwar's hunger strike (à la Gandhi) to coerce her to marry Changez, a groom Anwar has 'imported' for her from India. Love for her mother and what her cousin Karim considers 'perversity' finally make Jamila relent. Or perhaps, as Karim thinks, she sees marrying Changez as 'a rebellion against rebellion, creative novelty itself'.[42] In her compliance is also her resistance; the marriage is not consummated. Neither does her marriage affect her sexual behaviour; Jamila continues to engage in 'casual sex' with her cousin and best friend Karim, and over a hundred pages later she joins a commune where she lives as a radical feminist bi-sexual political activist. However, her 'husband' Changez lives in the commune too, though they do not live as a couple. Changez is a reminder of her 'origins', the trace that persists as she reconstructs herself.

While in *The Buddha of Suburbia* Jamila devises a creative resistance to her arranged marriage, in Dhondy's plays *Romance, Romance* and *The Bride* this resistance acquires both a sad and a funny edge. In *The Bride*, Jaswinder, a young Sikh woman and her Muslim lover, Junaid, kill themselves on the night of Jaswinder's arranged marriage to her neighbour's cousin.[43] In *Romance, Romance* the protagonist, Satinder, is successful in illustrating to her parents how the cultural practice of arranged marriages commodifies and demeans women. Upon learning that she is being 'viewed' by a prospective suitor chosen by her parents she adopts the role of a 'show girl'. She appears in a scanty dress and does a risqué routine. As she explains to her stunned father later, 'I was on show. How do you expect a show girl to behave?' [44] In Dhondy's plays and Kureishi's *The Buddha of Suburbia* it is the fathers and not the mothers who try to preserve the institution of arranged marriages. Attention is thus drawn to the link between the authority of fathers and the continuation of the institution.

Though incidents of overt and covert racism and tensions with parents, are a common occurrence for South Asian families in the fictional worlds of Kureishi, Rushdie and Dhondy, all three writers portray Black British[45] as finding much love, warmth and friendship within a growing 'immigrant' community. While in Markandaya's *The Nowhere Man* Laxman has no young friends of South Asian origin, in *The Buddha of Suburbia* young Karim and his cousin Jamila have each other as confidantes, friends, lovers and counsellors. It is with each other that they share their grief and their triumph. Similarly, in *The Satanic Verses*, Mishal and Hanif are comrades-in-arms. Dhondy's short stories 'Iqbal Cafe' and 'Come to Mecca' describe the tenderness and camaraderie that exist among the young working-class South Asian Britons who frequent the café in Brick Lane.[46] Alienated from white society, and often at odds with their parents whose demands on the young are shaped by discourses developed in different social and political formations, young Black British in these stories cling to each other for love, support and understanding.

Mutual understanding and support are also evident among the 'monsters' at the hospital to which Saladin is taken after his transmogrification in *The Satanic Verses*. Though the 'monsters', mutations of dark-skinned people, hail from different parts of the world, their common experience of white racism in Britain creates a bond among them. Their 'great escape' is staged with the complicity of Hyacinth (an 'immigrant' physiotherapist) and through a strategic alliance among the dark-skinned 'mutated' beings.[47] After they make their escape from the hospital, the 'monsters' go in different directions. Clearly, the alliance does not continue after the escape.

The goat/devil Saladin Chamcha finds companionship and shelter after 'the great escape' among his 'people' in a South Asian enclave that resembles Southall in West London. Ironically, the 'Bed and Breakfast' house to which he is taken is owned by Hind (translation: India) and Sufyan (translation: purity). Though Saladin continues to be bothered by what he has become, Hind and Sufyan and their family accept his goat-like appearance as 'normal'. To their older daughter Mishal, Saladin is no different from other casualties of racism they have known. From her window she identifies some of them for Saladin:

> – there, a Sikh ancient shocked by a racial attack into complete silence; he had not spoken, it was said, for nigh on seven years, before which he had been one of the city's few 'black' justices of the peace . . . – over there, a perfectly ordinary-looking 'accountant type' . . . this one was known . . . to rearrange his sitting room furniture for half an hour each evening, placing chairs in rows . . . pretending to be the conductor of a single-decker bus on its way to Bangladesh, an obsessive fantasy in which all his family were obliged to participate.[48]

To Sufyan what matters is not Saladin's physical form but his eternal soul. As he explains to Saladin, 'Your soul, my poor dear sir, is the same. Only in its migration it has adopted this presently varying form.'[49] Father's and daughter's reactions to Saladin highlight the different discourses within which the two negotiate their respective identity. Indian notions of Atman – the eternal soul – and transmigration structure Sufyan's thinking, but Mishal speaks the language of Black Power. Free of 'memories' of the lands of her 'origin', she is open to other discourses than the ones within which her parents move.

Such tension between parents and children on the appropriate response to covert and overt racial attacks appears as one of several themes in South Asian fiction of the last two-and-a-half decades. Unlike their parents, young Black British are assertive and resourceful. Faced with white racism, South Asian parents want to maintain a low profile but their children insist on fighting for inclusion in the British 'nation' on terms set by Black British. In Kureishi's play *Borderline*, Yasmin wants Amina to keep the lights on at night 'So people can know we are here to stay.'[50] The simple message of Kureishi's play is that South Asians like Yasmin and Amina are in Britain and will remain so. They cannot go 'home' to Pakistan as Amina's mother did, for Britain is their home. They are 'Black' British not Pakistani. As 'Black' British they demand the right to their own place; they want to add their 'colour' to the Union Jack, and their story to the nation's (his)story.

To Anwar, a young South Asian Briton in *Borderline*, being British means the right to a voice; the right to 'describe'. He resents British newspapers' representations of South Asians, and the media's reluctance to criticise Britain's racist police, or to give full coverage to the National Front's attacks on dark-skinned Britons. He is cynical of what the white liberal press can accomplish for South Asians living in Britain. Indeed, he accuses Susan, a white liberal reporter, of appropriating the pain of his people: 'you take our voice. Use our voice. Annex our cause ... for a few days you've borrowed our worry. ... You change its nature as it passes through your hands.'[51] To Anwar, the 'immigrants'' pain belongs to the 'immigrants'; and it is they who must articulate it; and fight to eradicate it.

While Anwar and Yasmin (*Borderline*), Mishal and Hanif (*The Satanic Verses*), and Jamila (*The Buddha of Suburbia*) want to reconstruct 'Britishness' so that their faces and their voices are recognisable and integral parts of it, V.S. Naipaul's Ralph Singh (*The Mimic Men*) and the narrator of *The Enigma of Arrival* wish to reconstruct *themselves* to approximate what they believe is 'British'.[52] Twice 'removed' from their 'origins', both Ralph Singh and the narrator come to England as third-generation South Asian Caribbeans. Neither protagonist has any connection with the land of his ancestors, nor has either ever truly adopted the Caribbean. Indeed, in Trinidad, located between powerful whites and marginalised blacks, they want to be like the whites. They epitomise 'colonised intellectuals' for whom becoming 'English' is the logical progression of their development. But what is 'becoming English'? Their definitions are dated, located in textbooks, poems and literature studied in their colonised 'homeland'. Ralph Singh's futile search for his England in London is continued by the narrator in *The Enigma of Arrival*. This time the search for 'Englishness' is undertaken in the English countryside of the Salisbury Plain, the location of 'real' England. He hopes to find 'heritage' and 'roots' by establishing a connection between himself and the trees and plants in his adopted country. He tries to learn the 'language' of England by immersing himself in the English countryside. But he soon realises that he can never be like his neighbour Jack who is 'a man in tune with the seasons and landscape'.[53] Jack's instinctive sense of harmony with the environment is denied the 'migrant' narrator. The latter has to learn and, more importantly, experience the trees, plants and the seasons to approximate that harmony. Though he longs to 'become English', he cannot. As Mrs

Deschampsneufs, a rich, landed French creole points out to Ralph Singh in *The Mimic Men*:

> where you are born is a funny thing . . . you are born in a place and you grow up there. You get to know the trees and the plants. You will never know any other trees and plants like that. You grow up watching a guava tree, say. You know that browny-green bark peeling like the paint. . . . Nobody has to tell you what a guava tree is. You go away. You ask, 'What is that tree?' Somebody will tell you, 'An elm.' You see another tree. Somebody will tell you, 'That is an oak.' Good; you know them. But it isn't the same.[54]

'Origins' tend to persist.

In addition to the special connection to the geography of the place where one grows up, Naipaul's narrator realises that even his excellent command of the English language is inadequate if he is to be 'English'. For the mental pictures evoked in his mind by words, events and places are quite different from those shared by indigenous people. To him (and Naipaul) a gardener conjures images of a bare-foot plantation worker in the Port of Spain, not someone who looks like a country gentleman.[55] Given these differences in imagery it seems that knowing the English language is not enough for the narrator to become 'English', he has to shed the past that structures his images, and learn the 'past' of England. To do so he leases a cottage on the grounds of an old English manor house within sight of Stonehenge, the oldest monument in England.

But for Kureishi's Karim Amir, 'a funny kind of Englishman, a new breed as it were, having emerged from two histories',[56] 'heritage' and 'origins' are contested terrains. His English mother reminds him, 'you are not an Indian. You've never been to India. You'd get diarrhoea the minute you stepped off that plane. . . . You are an Englishman, I'm glad to say.'[57] But when Karim goes for his first audition, Shadwell the director assumes he knows an Indian language and is knowledgeable about India. Clearly to Shadwell, Karim is not an Englishman. Neither is he English in the eyes of 'Hairy back', Helen's working-class father who insults and threatens him when he comes to visit her.[58]

To British-born Karim and his cousin Jamila it is confusing not to be accepted as British. We learn that as young teenagers, they pretended to be someone other than themselves:

> Yeah, sometimes we were French, Jammie and I, and other times we went black American. The thing was, we were supposed to be English, but to the English we were always wogs and nigs and Pakis and the rest of it.[59]

But none of these pretend 'identities' quite fits. Finally, Karim, an actor, 're-creates' and stages a character based on Changez, a newly arrived immigrant and his cousin's imported husband. In doing so, he fills the missing gaps in his own identity; in reinventing Changez' past he invents his own:

> I uncovered notions, connections, initiatives I didn't even know were present in my mind. I became more energetic and alive. . . . I felt more solid myself. . . . This was worth doing, this had meaning, this added up the elements of my life.[60]

By staging an immigrant, which he is not, since he is a 'born and bred' South Londoner, Karim claims his 'origins' and thus locates himself. For male Karim and the male narrator of *The Enigma of Arrival* 'origins' cannot be ignored. But for Jamila, Mishal (*The Satanic Verses*), Satinder (*Romance, Romance*), and Jaswinder (*The Bride*), 'origins' also spell patriarchal control; it is to discourses of radical black feminism, ecology and egalitarianism that they turn for empowerment and liberation.

CONCLUSION

Works of South Asian British fiction present us with a series of representations of the South Asian diaspora in Britain. Markandaya, Rushdie, Kureishi, Dhondy and Naipaul each highlight the power of white discourses of racism and colonialism in structuring British constructions and re-constructions of Britons of South Asian origin; but they also underscore the importance of looking at alternative discourses to which South Asian migrants and their children have access. Only by considering these two sets of discourses can we understand the complex ways in which identity is negotiated by South Asian migrants and their Black British children. Older migrants like Srinivas and Laxman (*The Nowhere Man*) and the narrator in *The Enigma of Arrival* are out of step with the young Black British we meet in the fictional worlds created by Kureishi, Dhondy and Rushdie. Mishal and Hanif, Yasmin and Anwar, Jamila and Karim, and the bevy of youth in Dhondy's short stories and plays, are engaged in different ways of redefining the boundaries of the British nation, and reconstructing the 'system' by turning it on its head. But the wrath and criticism of Black British youth is not confined to the institutions which reproduce racism. Young Black British women, in particular, are equally critical of, and resistant to, institutions of patriarchal oppression brought across the oceans and carefully preserved and reproduced by male South Asian migrants. Unlike their parents, young male and female Black British are averse either to harmonising their voices with those of the Anglos in a chorus written by the latter, or to playing only the music of the lands of their origin. Instead, they see themselves engaged in making a postmodern music of discordant notes and multilingual voices. From such music we can expect a re-articulation and reconfiguration of the discourses of identity.

NOTES

1 I am grateful to Vivek Bald, Virginia Bothun and John Connell for their valuable comments and suggestions on an earlier version of this chapter. Funding for research was provided by an Atkinson Summer Research Grant, Willamette University, Oregon.

2 S. Rushdie, *Shame*, New York, Aventura Vintage, 1984, p. 64.

3 G.C. Spivak, 'Acting bits/identity talk', *Critical Enquiry*, 1992, vol. 18, p. 781.

4 For an extended discussion of this point see S.R. Bald, 'Images of South Asian migrants in literature: differing perspectives', *New Community*, 1990, vol. 17, pp. 413–31.

5 See S.R. Bald, *Novelists and Political Consciousness: Literary Expression of Indian Nationalism, 1919–1947*, Princeton, NJ, Humanities Press, 1982, especially pp. 1–6, for the uses of fiction for gaining insights into the social reality in which the writers are rooted and about which they write.

6 See R. Visram, *Ayahs, Lascars and Princes*, London, Pluto Press, 1990.

7 Pakistan's Eastern wing seceded from West Pakistan to become Bangladesh in 1972. People who came to Britain from East Pakistan now claim Bangladesh as their land of origin.

8 For a discussion of this Act and its impact see M. Anwar, *Race and Politics*, London, Tavistock, 1986; also M. Anwar, *The Myth of Return*, London, Heinemann, 1979.

9 P. Gilroy, *There Ain't No Black in the Union Jack: The Cultural Politics of Race and Nation*, Chicago, University of Chicago Press, 1987.

10 For insights into the power of the definer see E. Said, *Orientalism*, New York, Vintage, 1979; and F. Fanon, *The Wretched of the Earth*, New York, Grove Press, 1968, especially pp. 35–43.

11 K. Markandaya, *The Nowhere Man*, New York, John Day Co., 1972, p. 44.

12 See Mr (Lord) Macaulay's Great Minute, 2 February 1835, in T. B. Macaulay, *Prose and Poetry*, Cambridge, Mass., Harvard University Press, 1954, pp. 721–9.

13 Markandaya, op. cit., p. 44.

14 H. Kureishi, *The Buddha of Suburbia*, London and Boston, Faber and Faber, 1990, pp. 26–7.

15 ibid., p. 26.

16 Markandaya, op. cit., pp. 22, 69, 74–6.

17 R. Doty, 'Sovereignty and national identity: constructing the nation', unpublished paper, 1993, p. 40.

18 ibid., pp. 38–9. The sanitation reference was by Smyth, *Parliamentary Reports, House of Commons*, 17 February 1961, p. 1955.

19 For a discussion of how the dark-skinned colonised native is dehumanised and animalised by the coloniser see Fanon, op. cit., especially p. 42.

20 Markandaya, op. cit., pp. 171–2, 251–5.

21 ibid., p. 6.

22 ibid., p. 312.

23 S. Rushdie, *The Satanic Verses*, New York, Viking Penguin, 1989, pp. 133, 140.

24 ibid., p. 142.

25 ibid., p. 166.

26 ibid., pp. 167–8, my emphasis.

27 Kureishi, op. cit., p. 33.

28 ibid., pp. 140, 146–9, 157.

29 Spivak, op. cit., p. 781.

30 Markandaya, op. cit., pp. 34–5.

31 Rushdie, *Satanic Verses*, op. cit., p. 294.

32 ibid., p. 293.
33 Fanon, op. cit., p. 52.
34 Rushdie, *Satanic Verses*, op. cit., p. 294.
35 ibid., pp. 283–4. *Mahabharata* is the Indian epic which recounts the war between the Kurus (evil) and the Pandavas (good). The battle is immortalised in the Hindu philosophical and religious text *The Bhagavad Gita*.
36 ibid., p. 263.
37 ibid., pp. 286–7.
38 ibid., pp. 250-l, 276.
39 ibid., pp. 243–4.
40 Kureishi, op. cit., p. 56.
41 ibid., pp. 53, 82.
42 ibid., p. 82.
43 F. Dhondy, *Romance, Romance and The Bride*, London, Faber and Faber, 1985, p. 47.
44 ibid., p. 28.
45 The term 'Black British' is used by politically sophisticated young Britons of South Asian descent to describe themselves. To them it accurately expresses their political positioning *vis-à-vis* White British. Though the term is criticised by some South Asian Britons who consider themselves 'brown' rather than 'black', in this chapter 'Black British' is used since it best captures the politics of young British-born 'South Asian Britons' represented in the fiction under study. Black is used here not to describe 'colour' but to describe the history of exclusion and oppression. It is a political, not a descriptive term. As such, 'Black British' also suggests an 'identification' with Afro-Caribbean Britons and other marginalised people of colour.
46 F. Dhondy, *Come to Mecca and Other Stories*, London, Fontana, 1978, pp. 9–26, 45–79.
47 Rushdie, *Satanic Verses*, op. cit., pp. 170–1.
48 ibid., p. 283.
49 ibid., p. 277.
50 H. Kureishi, *Borderline*, London, Methuen/Royal Court Writers, 1981, p. 43.
51 ibid., p. 23.
52 V.S. Naipaul, *The Mimic Men*, London, Penguin, 1987; *The Enigma of Arrival*, London, Viking Press, 1987.
53 Naipaul, *Enigma of Arrival*, op. cit., p. 33.
54 Naipaul, *Mimic Men*, op. cit., p. 171.
55 Naipaul, *Enigma of Arrival*, op. cit., pp. 205–8, 219.
56 Kureishi, *Buddha of Suburbia*, op. cit., p. 3.
57 ibid., p. 232.
58 ibid., pp. 39–41.
59 ibid., p. 53.
60 ibid., p. 217.

6

PERCEPTIONS OF PLACE AMONG WRITERS OF ALGERIAN IMMIGRANT ORIGIN IN FRANCE

Alec Hargreaves

INTRODUCTION

The literary representation of Algerian immigration in France is riddled with contradictions. Those who have been most directly involved in the migratory process have written virtually nothing about it. The most telling accounts of the Algerian experience in France are the work of young authors who in most cases are not immigrants at all. Though Algerian by birth and upbringing, they are – in a literal sense, at any rate – at home in France. It is when they 'return' to Algeria (a country in which few of them ever in fact resided) that they experience the migrant condition in its most evident forms. Yet even within France, the most tiny of spatial shifts may trigger mental turmoil directly comparable to that experienced by international migrants.

These young men and women, popularly known as *Beurs*, are the sons and daughters of primary migrants. Brought up amid the poverty of colonial Algeria, the great majority of the older generation lack any formal education and are consequently unable to read or write. As founding members of the Algerian community in France, they are sometimes referred to as first-generation immigrants, an indirectly misleading term if it is taken to imply that the children they have raised in their country of adoption constitute a second generation of migrants. Reflecting on one of the most accomplished novels so far produced by a writer from an immigrant family, Azouz Begag's *Le Gone du Chaâba*,[1] a young woman from precisely such a background has vividly illustrated some of the many paradoxes which permeate the perception of place among the younger generation of Algerians in France:

> As I refuse to accept any ready-made label, let me just say that I am a student of Maghrebian immigrant origin. 'Beur' and 'second-generation' are words that seem to say everything but which mean nothing. . . . My main concern right now is not with words, but with existing – and that is something conditioned by time and space.

The town where I live, Alès, is a nice, quiet place. I've always lived there. My earliest childhood years were spent amidst enormous tower blocks which, at the time, children living in other apartment blocks called Chicago, as if this was a more suitable name than 'Les Cévennes'. Although the district was less poor than Le Chaâba, as described by Azouz Begag, during the 1970s it, too, was like a little piece of Algeria in the heart of France. . . . For this reason, outsiders have never been able to describe the reality of everyday life in my district.

Even at primary school, most of us were of Algerian origin, and our absence when Muslim festivals took place was a way of asserting our Arab culture. The distinction between Algerian and Arab meant nothing to us. Algeria was basically a piece of countryside to which each family returned during the summer holidays, but it was also a kind of ideal land.

The poverty which my parents had left behind gave me the feeling that I was rich; living in France seemed to be a privilege. Yet if Algeria was equated with material privation, it was also the place where my roots lay, and to which I expected to return. When, still very young, I saw it for the first time in my life, I felt like an exile returning home after a long absence: I recognised the place, but things seemed to have changed. Today, I realise that I had been living with an image of the mountains of Kabylia as my father had left them in the 1960s.

As time passed, I soon realised that I had been nostalgic for a place that didn't really exist. But Algeria, a real, unmythologised place, remains deeply embedded in my mind, where it has the same symbolic importance as in Azouz Begag's novel. The fact is that although we are never directly shown Algeria in the text, the land of the author's origins is intensely present as a vital landmark in the construction of his identity.[2]

Impatient with labels such as 'Beur' and 'second-generation', the author of these lines, Zimba Kerdjou, longs to exist on her own terms, but she recognises that she is constrained in this endeavour by the external realities of time and space. No less importantly, Kerdjou shows that those external realities are overlaid at every step by mental constructions which make it virtually impossible to view any place in a neutral fashion. Almost invariably, places are seen through an interpretative grid structured by the cultural baggage carried in the observer's head. Often, particular places have been, in the most literal sense, preconceived, for before travelling there the observer may already have heard descriptions passed on by others. The (evidently French) children who called Kerdjou's housing estate 'Chicago' are unlikely ever to have visited that American city. Their notion of it was in all probability formed under the influence of television or cinema images. In transferring those images to a housing estate inhabited by heavy concentrations of

Algerians, these children were, no doubt unwittingly, ensnaring their immigrant-born peers in an extraordinarily complex and misguided web of preconceptions.

A similar, but unmalicious, process informs the images through which the young Kerdjou conceived of her parents' native land. The 'nostalgia' she felt for a country in which she had never lived depended on an internalised representation of Algeria which her parents had transmitted to her by word of mouth. Even in the minds of her parents, who were born and brought up in Kabylia, the memories formed by their first-hand experiences of the region were caught in a time warp which, as the place changed during their absence in France, became a screen between them and their native land. The powerful mental hold of places which are hardly known, if at all, is reflected in Kerdjou's internalisation of the parental myth of 'return' to an ideal(ised) land across the Mediterranean. It is embodied no less strikingly in Begag's novel where, as Kerdjou perceptively comments, Algeria helps to shape every page without ever being directly encountered in the text.

Le Chaâba is the Arabic name given to a *bidonville* (shanty-town) built by Begag's father and other Algerian immigrants on the banks of the Rhône not far from the centre of Lyons. It was here that Begag was brought up. Though born on French soil, he was surrounded during his earliest years by people of Algerian origin. Like the housing estate in Alès where Kerdjou grew up, Le Chaâba was to a large extent 'a little piece of Algeria in the heart of France'. It was less the physical geography of Le Chaâba that resembled Algeria than the mental world of its inhabitants. The primary migrants who built its ramshackle huts had spent their formative years on the other side of the Mediterranean, where they had been moulded by the traditions of an Islamic country whose culture they carried with them when settling in France. The Arabic name by which their shanty-town was known was in this sense emblematic of the foreignness of this enclave within the surrounding city. As soon as they stepped outside its confines, the inhabitants of Le Chaâba entered a world structured by radically different norms. At their French primary school, Begag and the other children of the shanty-town were encouraged to internalise those norms. Though school and home were only a few hundred metres apart, in journeying between them each child migrated daily between profoundly different cultural universes. Kerdjou's recollection of the incompatibility between Muslim festivals and the French school calendar is just a small example of the many clashes produced by the juxtaposition of cultures which, prior to the migratory process, were spatially distant from each other.

In considering the literary representation of places encountered during these experiences, the remainder of this chapter is divided into three parts. The first considers the very limited role of outward international migration in the works of immigrant-born writers. The second part focuses on the treatment of spatially contiguous but culturally disparate areas within France.

Finally, we shall examine the difficulties associated with 'return' journeys to Algeria.

OUTWARD MIGRATION

Well over twenty authors of Algerian immigrant descent have now published one or more volumes of narrative prose.[3] Very few of their works describe the act of migration from Algeria to France. Granted that most of these authors were born in France, and that much of what they have written is heavily autobiographical, this absence is hardly surprising. The few works that do describe outward migration are generally by the minority of writers who were born in Algeria and spent part of their childhood there before settling in France with their parents. Yet even these narratives almost never begin with the migratory act itself. The opening chapter of Mehdi Charef's *Le Thé au Harem d'Archi Ahmed* describes the dead-end existence of the protagonist, Madjid, in a working-class Paris housing estate.[4] It is not until the middle of the novel that Madjid's childhood journey from Algeria is recounted in the form of a flashback. Nacer Kettane's novel, *Le Sourire de Brahim* also begins with a scene set in France before recapitulating on the protagonist's earliest years in Algeria.[5] Mohammed Kenzi's autobiographical narrative, *La Menthe Sauvage*, is virtually unique in opening with an evocation of 'Algeria, a distant childhood memory shrouded in mist'.[6] Even here, in the opening words of the text, France implicitly interposes itself between the reader and the land of Kenzi's birth, for these childhood memories of Algeria are filtered through the mind of an older, narrating consciousness situated on the northern shore of the Mediterranean.

This is also true of Brahim Benaïcha's *Vivre au Paradis: D'une Oasis à un Bidonville*, which offers a particularly ironic illustration of the illusions which often inform the migrant perception of place.[7] As a child, Benaïcha shared in the myth that induced many Algerian adults to seek their fortune in France, which, viewed from the poverty of the desert oasis in which he was born, appeared to be a land of golden opportunities. When he arrived in Paris, the shanty-town in which the family settled was anything but the paradise on earth which the boy had anticipated. Amid the mud and misery of his new home, a striking reversal occurred: 'How far away Paradise seemed now, lost in the depths of the Sahara!'[8] Though only 8 years old, Benaïcha was replicating the contradictory sentiments of adult migrants who, as Kerdjou points out, left Algeria because of its poverty but came to look back on it as an 'ideal land'.

Kerdjou also reminds us that even for children born in France, stories of life in Algeria are a staple part of everyday conversation within the family home. Every child in an immigrant family knows that what makes him or her different from native-born peers is the fact that his or her parents came from another land. Although *Le Gone du Chaâba* was conceived from the outset as an autobiographical novel, the first unpublished draft began as follows:

From El-Ouricia to Lyons

It was at the beginning of the 1950s that my father arrived in France. He was already in his thirties. For the first time in his life, he had just left El-Ouricia, a small village situated about a dozen kilometres from Sétif, in eastern Algeria.[9]

Thus while the story he was about to recount was to be that of the author, the first words written by Begag served to acknowledge the fundamental importance of the migratory act accomplished by his father. The remainder of the chapter described the father's life of poverty as a day-labourer on an Algerian colonial farm, his decision to seek better-paid work in France and his eventual arrival in Lyons. In a quite literal sense, everything else in the narrative is contingent upon this founding act. The manuscript thus confirms Kerdjou's intuitive sense of the implicit omnipresence of Algeria, despite its formal absence from the published text.

This preliminary chapter was deleted prior to publication, not because Begag wished to dissociate himself from his Algerian origins, but because the true focal point of the narrative concerned his destiny in France. It is here that, in its definitive form, *Le Gone du Chaâba* begins:

Zidouma is doing her washing this morning. She has got up early so that she can have the only water-point in the shanty-town to herself. This is a manual pump drawing up water from the Rhône: the *bomba* (pump).[10]

When an argument develops between Zidouma and another woman waiting to use the pump, the narrator's mother is quick to join in:

Leaving me to carry on drinking my coffee, she girds up her robust frame, cursing amply.

I make no attempt to stop her. There's no stopping a charging rhinoceros. I quickly down my beverage and jump up to watch the pugilists. I don't know why, but I like sitting on the front doorstep and watching scenes like this at the *bomba* and the *poola* (pool). It's so strange to see women fighting.[11]

Although this opening passage, like much of the novel, is written in the present tense, it is not, as it might first appear, a direct transcription of events as they are seen by the young Begag. When he speaks of his pleasure at watching scenes like this, there is certainly a sense in which he is recalling his childhood reactions. But at another level, the events in this passage, and indeed in the rest of the text, are implicitly presented as a spectacle viewed from a more distant vantage point. Much of the language in which events are narrated ('her robust frame', 'beverage', 'pugilists') is not that of the seven-year-old child witnessing them but is, rather, the construction of an adult self looking back on what are now quite distant memories. The place from which that older self narrates the story is never explicitly defined, but the gentle fun which he pokes at the accent of the Algerian women in the shanty-town – in the original French, the mispronounced words are '*l'bomba (la pompe)*' and '*le baissaine (le bassin)*' – can be enjoyed only because the narrator perceives their accent to be deviant from the standard pronunciation. The adult Begag is clearly at home in the socio-cultural norms which predominate in France. Much of the humour generated by the novel springs from an implicit contrast between the mature narrator, for whom French society holds no secrets, and the naïvety of his younger *alter ego* as he fumbles his way towards those norms. The unspoken place from which the older Begag narrates the story thus forms a crucial pendant to the other invisible location intuited by Kerdjou. In this way, the whole of *Le Gone du Chaâba* is framed by two unspoken but omnipresent places: the Algeria in which the author's father spent his formative years and the France out of which Begag's mature self speaks. The experiential itinerary connecting the two together forms the backbone of the text.

FRENCH ITINERARIES

The title of a recent book by Mehdi Lallaoui, *Du Bidonville aux HLM*,[12] encapsulates the itinerary followed not only by the author, but also by Begag and many immigrant families in France. While a significant improvement in the housing conditions of these families has gradually been achieved – the public housing (HLM) apartments where many of them now live are a far cry from the shanty-towns in which they first settled – their marginalisation within the housing market has remained a more or less constant factor. The most obvious reason for this is economic. Immigrants from North Africa have generally been among the worst-paid workers in France. As their families tend to be larger than those of French nationals, disposable income per capita has been extremely low. They have therefore been obliged to live at the cheapest and least desirable end of the housing market. During the 1950s and 1960s, when there were rapid inflows of immigrant workers to meet the demands of the expanding economy, shanty-towns mushroomed around many French cities. As they were thrown up without authorisation on unoccupied land, their inhabitants did not have to pay rent; no less importantly, they also went without most of the basic facilities (sewers, electricity, domestic water supplies and so on) enjoyed by the vast majority of the population. Government initiatives launched in the 1970s have gradually led to the virtual elimination of shanty-towns. Most of their occupants were eventually offered homes in HLM estates, places which in the 1990s have become synonymous with what the French refer to as *la banlieue* – sprawling areas of poor-quality housing in the outer suburbs of major cities. For many immigrant families, *cités de transit* (transit camps) served as half-way houses. These temporary constructions were of somewhat better quality than the shanty-towns they had left, but they deteriorated rapidly as families waited for ten, fifteen or even twenty years before being allocated to HLM apartments.

If immigrants tended to congregate in particular areas, this was not simply for economic reasons. Important practical and psychological factors also played a role. As primary migrants were venturing into unknown places, their path would be smoothed if they were able to make contact with friends or relatives who were already there. There are numerous tales of illiterate immigrant workers arriving in France clutching a scrap of paper on which the address of a brother or cousin had been written by a friend or relative. Begag's father bore such a paper on arrival in Lyons:

> He had held it throughout the journey, for it constituted his only guide in this foreign country. . . . With the help of this scrawled note, which he showed to everyone in the street as he asked his way, he went up and down the streets of Lyons for hours on end.[13]

It was a huge relief to find the villagers from El-Ouricia whose address he had been given, for they would give him an initial base and, no less importantly, the psychological reassurance of being among people of his own kind. His son would later liken the shanty-town to

> a kind of decompression chamber for families who had decided to take the plunge and enter France. Here they were in a kind of safe area among their own people, and had the chance to adapt to their new life.[14]

Benaïcha, who in 1960 was 8 years old when his father brought him and the rest of the family to live in France, was horrified to discover the shack in which he was to live in Nanterre, which at that time was one of the largest shanty-town areas in the whole of France. Yet ten years later, when the family moved to a *cité de transit* on the other side of Paris, it was not without a certain nostalgia that Benaïcha left Nanterre:

> The shanty-town is now behind us. What has it given us during the last ten years? It has enabled us to preserve our language, our religion, our identity. It has enabled us to remain ourselves amid all the tempestuous clashes of civilisations through which we have been.[15]

As noted earlier, Benaïcha is unusual in having spent the first eight years of his life in Algeria. This gave to him, as to other youngsters brought to France after first spending a significant period in Algeria, a more deep-rooted identification with North African culture than is generally found among the children of immigrant parents. For those who, like Begag, were born in France, the influence of the dominant culture has been more difficult to resist. At school, which may begin as early as the age of 4, thanks to the widespread provision of nursery education, the dominant language and values of France begin to displace the cultural heritage of immigrant children even before it has been fully internalised. Farida Belghoul's *Georgette!*, like *Le Gone du Chaâba*, is primarily about the battle between these rival cultures fought in the minds of children in their daily migration between home and school. The 7-year-old protagonist in *Georgette!* finds it impossible to reconcile the conflicting imperatives issued by her Algerian father and her French school-mistress. As she wanders the streets, unable to decide whether she should go home or to school, the girl fails to observe a set of traffic lights and is killed by a passing car. Though spatially very limited, the loci which structure her experience are like the tips of cultural icebergs: beneath the immediately apparent level of the girl's personal experiences lie two cultural universes.[16]

While Benaïcha claims to have remained faithful to his Algerian roots, he recognises in the passage already quoted that he, too, has been subjected to major cultural contradictions. Some of these are delineated in more detail when he recalls having pretended to live in a block of flats in order to conceal from his French schoolfriends that his home was really in a shanty-town:

> If I said I lived in a block of flats, it was simply so that I could at least give the appearance of sharing in my classmates' lifestyle. That way, they wouldn't try to exclude me from their circle at school. By talking their language and living like them (in theory, at any rate), I could get closer to them and integrate more easily. They were nice to me, and some of them even invited me home. I always found a pretext for refusing the invitation. It wasn't that I didn't want to go home with them, but because sooner or later I would have had to return the hospitality – and there could be no question of my taking them back to the shanty-town. I knew what they thought about places like that.[17]

Here we see the psychological pressures (the desire for peer respect, etc.) pushing towards the acculturation of immigrant-born children, and at the same time the enormous barriers of prejudice standing between children whose homes are separated by no more than a few hundred yards. Small wonder that Benaïcha speaks of his mind having been 'torn in two, practically opposite directions'.[18]

Lil, the protagonist in Tassadit Imache's *Une Fille sans Histoire*, experiences very similar feelings: 'Again and again, she was shaken by the feeling that she was about to break into two warring halves called France and Algeria.'[19] In Lil's case, the tensions between home and school are replicated within the family home itself, for she is the daughter of an Algerian father and a French mother who met while working in the same factory. Their marriage took place at a time when the Algerian war of independence was inflaming national passions and prejudices. Her father wanted to bring Lil up as an Algerian, while her mother, anxious to protect the family from anti-Arab sentiments stirred up by the war, did everything possible to conceal the

North African dimension of the girl's origins. The mother largely succeeded in this, but in the process provoked her husband's slide into alcoholism, leading ultimately to their divorce, together with a profound sense of loss within Lil.

Near the end of *Une Fille sans Histoire*, Lil, now grown up, plans to visit Algeria for the first time, in the hope that she will somehow find there the peace of mind which has so far eluded her. The novel closes with her possible departure shrouded in uncertainty. We cannot know for sure what such a visit might bring, but as we shall see in the next section, the experiences of other 'returnees' do not augur well.

JOURNEYS OF 'RETURN'

Surrounded on all sides by prejudices and constraints, youngsters of Algerian origin in France have every reason to seek a remedy elsewhere for their ills, and there is no more natural place for them to do this than in the homeland of their parents. The hopes invested in such a 'return' are not without their paradoxes, however. As Kerdjou points out, primary migrants left Algeria precisely because they found it an unsatisfactory place in which to live. Almost without exception, though, they planned to return home after saving enough money to live there in the way they wished. Algeria remained in this sense an 'ideal land'. The material privations and racial prejudices frequently encountered in France made migrants all the more ready to filter out unhappy memories of Algeria. Their projected return came to serve as a mythical compensation for everything they suffered during what was seen as their 'exile' north of the Mediterranean. Internalising these images, their children could easily imagine Algeria as an escape hatch from the difficulties besetting them in France.

In reality, those difficulties have often been replicated and in many cases compounded when youngsters from immigrant families have attempted to resettle on the southern shore of the Mediterranean. Spurned by many Frenchmen as 'Arabs' (many of them are in fact Berbers), the children of emigrants are apt to be seen within Algeria as Frenchified outsiders. Many speak little or no Arabic; virtually none are able to read or write in that language. In France, their ties with Islam have often become loose; in these circumstances, the religious norms prevailing in Algeria may come as an unwelcome shock. Administrative inefficiency and corruption may make official dealings with the state thoroughly distasteful.

The shock of feeling foreign in a place which from afar was regarded as home often triggers a reappraisal of life in France, and in some cases this has been the starting-point for the act of writing itself. This was the experience, for example, of Mehdi Lallaoui, who was born in Argenteuil, a northern suburb of Paris. At the age of twenty-two, he attempted to resettle in Algiers, but in less than a year he realised the enterprise was doomed to failure and returned to France, having begun work on a semi-autobiographical narrative which was eventually published as *Les Beurs de Seine*.[20] An extended holiday in Algeria at the age of twenty-nine provided the starting-point for *Les ANI du 'Tassili'* by Akli Tadjer, who, like Lallaoui, was born in Paris. The novel begins with Tadjer's alter ego, Omar, passing through customs and immigration control in Algiers just before boarding the ferry-boat *Tassili*. The migratory act which he is about to accomplish is redolent with irony. Though Algerian by nationality, Omar is thoroughly homesick for France after an unsuccessful attempt to put down fresh roots on the other side of the Mediterranean. 'Oh, yes,' he remarks, 'I'm glad to be leaving "home" so that I can get back "home"';[21] an explicit contrast is drawn between the eagerness with which he departs and the sense of exile felt by immigrant workers returning to France on the same ship after a brief visit home.

Ferudja Kessas, another Paris-born author, was just nineteen when her parents took her back to Algeria for the purpose of an arranged marriage. Only a few months after this had taken place she became anorexic and persuaded her new husband to take her back to France. In the meantime, she had begun writing a novel about young women brought up in France by immigrant parents which was to be published under the title *Beur's Story*.[22] In it, one of the two female protagonists commits suicide rather than accept a marriage which her parents have arranged for her in Algeria.

Among writers of Algerian immigrant descent, none is more outspoken than Sakinna Boukhedenna in her criticisms of French society. Boukhedenna was born and brought up in north-eastern France. In her autobiographical narrative, *Journal. 'Nationalité: Immigré(e)'*, the French are repeatedly reproached for continuing to exploit Algerians just as they did during the colonial period. In a typical journal entry, she describes her acculturation through the French educational system as an act of cultural confiscation:

> I hate all those people who say we are now culturally part of France, that bitter country which has torn us from our true culture. . . . I am an Algerian. Yes, I have been colonised culturally, but I will not rest until I have found my true roots.[23]

In an attempt to recapture those roots, Boukhedenna begins taking lessons in Arabic and takes to wearing henna as an outward mark of defiance towards the French. At the age of twenty-one, she sails on the *Tassili* from Marseilles to Algiers. 'I was happy,' she recalls, 'for I was at last going to see the most beautiful of my illusions, the fantasy, the dream that was Algeria.'[24] Within minutes of landing, those illusions – built from a mixture of second-hand descriptions and personal inventions – suffer a cruel blow. The customs and immigration officials treat her and other impecunious Algerians as rabble while fawning on French tourists amply loaded with foreign currency. Wherever she goes within Algeria, she is subjected either to unwelcome restrictions on her movements, in conformity with Islamic teachings on the role of women, or to indecent advances from men who regard her as a Westernised sex-object. Everywhere, she feels like an immigrant. Eventually, the only people she regards as friends are Palestinian exiles living in Algiers: 'I've decided they are the only people I am going to spend time with because, like us immigrants, they no longer have a country.'[25] Yet even Aïssa, the most sympathetic of the Palestinians she has befriended, cannot help feeling, like many Algerian men, that as a woman brought up in France Boukhedenna is little more than a sex object:

> I really liked this guy, because he was sincerely committed to the struggle of his people. But now what was I to believe in? What was there left for me to hold on to? I was a wog in France or a tart in Algeria, an underling in France or an immigrant in Algeria.[26]

Boukhedenna eventually returns to France, a country which now appears as the lesser of two evils. The final lines of her narratives are pervaded by a profound sense of exile:

> As an Arab woman, I've been given a life sentence. Having chosen the path of freedom, I am now cast out, a migrant woman in exile. No one will accept my true identity as a woman. I am condemned to roam the world in search of where I belong.[27]

Boukhedenna's feeling of belonging nowhere springs from a seemingly unbridgeable gulf between the places which fill her imagination and those actually encountered in the world around her. These tensions are exacerbated by the images projected on to her by those who inhabit these places. 'In France, I learnt to be an Arab,' she writes. 'In Algeria, I learnt to be an Immigrant.'[28] In both learning processes, she was internalising hostile images projected on to minorities by the majority population. In each case, a profusion of preconceptions and inventions seems to have condemned both sides to a dialogue of the deaf. While it is perhaps too much to hope that any work of literature can fully transcend such a divide, the writings of immigrant-born authors nevertheless offer compelling insights into the many paradoxes permeating the condition of young Algerians in France.

NOTES

1 A. Begag, *Le Gone du Chaâba*, Paris, Seuil, 1986.
2 Z. Kerdjou, 'Alès, ma ville, est un lieu paisible et agréable', *Impressions du Sud*, nos 27–8 (Spring 1991), p. 13.
3 For a detailed bibliography see A.G. Hargreaves, *La Littérature Beur: Un Guide Bibliographique*, New Orleans, Celfan Edition Monographs, 1992.
4 M. Charef, *Le Thé au Harem d'Archi Ahmed*, Paris, Mercure de France, 1983.
5 N. Kettane, *Le Sourire de Brahim*, Paris, Denoël, 1985.
6 M. Kenzi, *La Menthe Sauvage*, Lutry, Bouchain, 1984, p. 11.
7 B. Benaïcha, *Vivre au Paradis: D'une Oasis à un Bidonville*, Paris, Desclée de Brouwer, 1992.

8 ibid., p. 47.
9 Azouz Begag, unpublished manuscript of *Le Gone du Chaâba*. I am grateful to the author for allowing me to consult this.
10 Begag, op. cit., p. 7.
11 ibid., pp. 8–9.
12 M. Lallaoui, *Du Bidonville aux HLM*, Paris, Syros, 1993.
13 Begag, unpublished manuscript of *Le Gone du Chaâba*.
14 A. Begag, *North African Immigrants in France: The Socio-Spatial Representation of 'Here' and 'There'*, University of Loughborough European Research Centre, 1989, p. 5.
15 Benaïcha, op. cit., p. 301.
16 F. Belghoul, *Georgette!*, Paris, Barrault, 1986. Like many similar narratives, *Georgette!* is in this respect characterised by a metonymic mode of representation. See A.G. Hargreaves, *Voices from the North African Immigrant Community in France: Immigration and Identity in Beur Fiction*, Oxford/New York, Berg, 1991, pp. 73–8.
17 Benaïcha, op. cit., p. 259.
18 ibid., p. 131.
19 T. Imache, *Une Fille sans Histoire*, Paris, Calmann-Lévy, 1989, p. 123.
20 M. Lallaoui, *Les Beurs de Seine*, Paris, Arcantère, 1986.
21 A. Tadjer, *Les ANI du 'Tassili'*, Paris, Seuil, 1984, p. 20.
22 F. Kessas, *Beur's Story*, Paris, L'Harmattan, 1990.
23 S. Boukhedanna, *Journal. 'Nationalité: Immigré(e)'*, Paris, L'Harmattan, 1987, pp. 70–1.
24 ibid., p. 75.
25 ibid., p. 86.
26 ibid., p. 91.
27 ibid., p. 126.
28 ibid., p. 5.

7

FROM *FRANCITÉ* TO *CRÉOLITÉ*
French West Indian literature comes home
Robert Aldrich

La Caraïbe est terre d'enracinement et d'errance.
(Edouard Glissant, *Poétique de la Relation*)[1]

Edouard Glissant's statement that the Caribbean is a land of *enracinement* (rooting) and *errance* (wandering, drifting) is appropriate to both the history and the literature of the West Indies. It also forms the central image in his analysis of aesthetics, extended by the metaphor of the deep underground 'root', the plant which gets its sustenance from the soil in which it is planted and which wilts or dies when pulled out, versus the 'rhizome', a horizontal stem which sends out shoots both above and below the surface to enlarge its field of nourishment. For Glissant, novelist, poet, playwright and philosopher, West Indian culture is a hybrid, enriched by the grafting of various civilisations onto the islands of the Caribbean basin.

MIGRATION AND THE HISTORY OF THE WEST INDIES

More than many areas of the world, the societies of the Antilles have been created by migrations, both voluntary and forced. Speaking of his native island, Glissant points out that 'Martinican society did not exist before colonialism.'[2] Amerindians – Caribs, and the Arawaks who preceded them – had moved into the islands from the Amazonian basin, but most succumbed to the European conquistadors. Europeans installed themselves as the new masters – the Spanish in Cuba, Puerto Rico and the eastern side of Hispaniola, the French in western Hispaniola, Martinique, Guadeloupe and elsewhere, the English in an arc of islands stretching from Bermuda in the western Atlantic to Trinidad and Tobago off the coast of South America, the Dutch in two separate island groups, the Danish in three islands of the northern Caribbean, the Swedes on Saint-Barthélémy.

The Europeans turned the Antilles into giant sugar plantations. In the seventeenth and eighteenth centuries planters earned great prosperity from the cane-fields thanks to the work of African slaves. From around 1680 until the early nineteenth century, slavers transported black Africans by the millions to the Caribbean and to other colonies in North and South America. The way in which they were bought or captured in Africa, exiled to the New World and pitilessly worked for European gain is a well-known and horrifying story. Africans soon formed the majority of the population in almost all of the Caribbean islands. Recruited from a multiplicity of ethnic groups, speaking different languages and practising different customs, they were moulded towards a homogeneous mass by the cruel exigencies of the plantation system. Slaves preserved some African traditions and adapted others to the new environment; a new language emerged, called *Créole* in the French islands, as the medium of communication between slaves from different regions of Africa and between slaves and plantation-owners.

Slavery in the French empire was abolished during the Jacobin phase of the Revolution of 1789, and the prized French colony of Saint-Domingue won its independence in 1804 as the world's first sovereign black nation under the name of Haiti. But pressure from planters, including the family of Napoleon's first wife, the white Martinican Joséphine de Beauharnais, led to the re-establishment of slavery in the remaining French possessions under Bonaparte's regime. Britain abolished slavery in the 1830s, but not until 1848, after another revolution in Paris and outright insurrection by slaves in the Antilles, was slavery definitively abolished in France.[3] The fortunes of the plantations were now in decline, but cane-growers still needed a workforce, which they sought in free African migrants, Chinese and, more importantly, in contract labourers recruited in India. So great was the migration of Indians that they became the majority in the British colony of Trinidad and Tobago; in present-day Guadeloupe one in six residents counts an Indian ancestor. Migrants later arrived from other shores as well; in Martinique and Guadeloupe, the most significant were merchants from Syria and Lebanon. Europeans continued to arrive; priests sent by the Catholic church, administrators, soldiers and sailors posted by the French government, entrepreneurs, adventurers and ne'er-do-wells eager to make their fortune in the sunny tropics.

The result of generations of migration to the French West Indies was a cosmopolitan and multi-ethnic, but highly stratified population, including a large number of people with mixed ancestry. At the apex of the pyramid in the French West Indies, although more numerous in Martinique than in Guadeloupe, was the white caste, the *Békés*, who controlled the economy; in the middle, the *mulâtres* or *métis* of mixed European and African parentage; and, at the bottom, the *nègres*, the descendants of the slaves, as well as the *coulis*, or Indians.[4]

Migration was not only in one direction. Some colonists and *Békés* returned to France. Thousands of West Indians fought for France and died on the battlefields of Europe in the First World War. The *mulâtre* middle class gave birth to hundreds of officials sent to rule France's African and Asian domains.[5] Students, both *Békés* and *mulâtres*, went to Paris to study and work as doctors, teachers and lawyers. Those with the financial means, no matter what their skin colour, moved easily back and forth between Europe and the Antilles. After the Second World War, larger numbers of less privileged French West Indians moved to Europe, many under the auspices of a special migration bureau established to find workers for the French postal services, hospitals and other lower-level white-collar sectors.[6] At present, over 500,000 men and women of West Indian ancestry live in France, more than that of either of the two French islands in the Caribbean.[7] The constitutional status of Martinique and Guadeloupe favoured such movement. Since emancipation and the institution of universal male suffrage in 1848, residents of France's *vieilles colonies* of Martinique, Guadeloupe, Guyane (French Guiana) and Réunion have been fully fledged French citizens with the right of migration and abode in the metropole. Since 1946, all these islands have been *départements d'outre-mer*, constitutionally as much a part of the French Republic as Paris or Provence.[8]

A further type of *errance* exists in the French West Indies: movement within each island. In the era of slavery, many slaves escaped the harsh conditions of the plantations (though facing severe punishment if tracked down and captured). The *marrons* (maroons), or runaway slaves, took refuge in the hills and up-country districts of Martinique and Guadeloupe or in the dense jungles of Guyane; *marronage* became the symbol of resistance to colonialism. With the decline of the plantation system, many West Indians moved to the small towns which dot the landscape of Martinique and Guadeloupe or flocked to the burgeoning cities of Fort-de-France and Pointe-à-Pitre, the main cities of the respective two islands. Economic necessity, distaste for rural life associated with the hardships of slavery and the plantation, and personal ambitions combined in this migration, the final step of which was sometimes a trans-Atlantic move. As the Martinican novelist Raphaël Confiant explains, 'Each of us rose by steps from the countryside to the town, from the town to the city and from the city to France.'[9]

The history of Martinique and Guadeloupe, therefore, is one of movement – the Amerindian migration from the Amazon, the initial waves of men and women who came from Africa and Europe, from India, the Middle East and China, the move from the cane-fields to the city, and from the West Indies to France (and sometimes to other domains of the French empire). This coming and going continues as a prime trait in islands which are political and, to a great extent, economic dependencies of France. For the cultures of the *Créole* islands, the movement of populations among several continents, and the mixture which it has produced, is a lynchpin of island identity. It is no coincidence that the St Lucian Nobel-Prize-winning poet Derek Walcott took Homer and his stories of classical voyages as the central theme in his brilliant *Omeros*, in which West Indians are pitched between Africa and the Western Isles in fantasy as in reality, while their descendants are cast between the Americas and Europe in a modern odyssey. Walcott's black Homer, blind Seven Seas, says simply that 'a drifter is the hero of my book.'[10]

MIGRATION AND FRENCH WEST INDIAN LITERATURE

Errance has been a feature of French West Indian literature, either because authors were peripatetic or because the works involve the wandering of men and women from countryside to city and from one continent to another.[11] Several of the most famous works on the Antilles, from Father Labat's chronicles at the beginning of the eighteenth century to Lafcadio Hearn's reports at the beginning of the twentieth, are the accounts of expatriates. The oral literature (or *oraliture*), the cultural voice of the slaves, survived well into the twentieth century. It has now been resuscitated by *Créole* novelists, who draw on the traditions and narrative style of Africa, inherited from what Patrick Chamoiseau and Raphaël Confiant term the *cri du cale* (the 'cry from the hold' of slaving ships which transported Africans into exile).[12] The only West Indian to be awarded the Nobel Prize for literature before Walcott was Alexis Saint-Léger, a *Béké* who left Guadeloupe at the age of twelve and never returned. Written under the name of Saint-John Perse, his first collection, *Eloges* (1911), reflects on his youth in the Antilles. A later collection, *Anabase* (1922), was inspired by his diplomatic service in China, while *Exil* (1942) is an anguished rendition of yet another exile, this time from defeated France. In counterpoint, Aimé Césaire's *Cahier d'un Retour au Pays Natal* (1939), the seminal work of the French West Indies' most prominent writer, represents the return of the author to his native land, still manacled by colonialism, which he had left to pursue his studies in Paris. René Maran's *Batouala* (1921), the first novel by a black French man-of-letters to be widely heralded, drew on the experiences of Maran, a Guyanese official in the French colony of Oubangui-Chari in central Africa, to paint a devastating portrait of the abuses of imperialism. Thirty years later, the work of Frantz Fanon, a Martinican psychiatrist who worked in French Algeria, bespoke a searing indictment of the effects of colonialism on 'native' peoples. More recently, the work of the Guadeloupean novelist Maryse Condé is a literary and historical peregrination across the black diaspora: *Hérémakhonon* (1976) tells of a black West Indian woman who goes to Africa to find her 'roots' but is soon disabused of her idealism; *Ségou* (1984) is a chronicle of Africa itself; *Moi, Tituba, Sorcière* (1986) is the story of a black woman accused of witchcraft in colonial America's Salem; and *La Vie Scélérate* (1987) is a fresco of West Indian migration over several generations in a family whose members go to Panama to build the canal, to New York to see the bright lights of the big city and to other destinations as widely separated as Haiti, Jamaica and San Francisco. Edouard Glissant's *Mahagony* (1987) is a triptych of different types of *marrons* pictured in 1831, 1936 and 1978. The narrator of Raphaël Confiant's *Eau de Café* (1991) is a man who returns to his native village in northern Martinique after years of absence in France partly to solve his curiosity about an event from his childhood, the mysterious arrival and enigmatic disappearance of a young woman with the pregnant name Antilia.

The examples could be multiplied, and the theme of *errance* could be analysed in any of these works. But it is possible to go one step further and suggest that all French Antillean literature, indeed the whole literary tradition of Martinique, Guadeloupe and Guyane, is an example of cultural *errance*.[13] Those whose literary expression found voice in books, rather than *oraliture*, first journeyed spritually to France to find the appropriate themes and styles, and in so doing imitated the culture of their white masters: the fate of writers in the French West Indies from the beginnings of the colonial era until the third decade of the twentieth century. This was part of the cultural 'assimilation' dear to imperial doctrine of the time and embodied in the *francisation*, or Frenchification, of natives. However, beginning in the 1930s with the movement of *Négritude*, associated with Aimé Césaire, authors looked to Africa for inspiration. Finally, from the 1970s, Edouard Glissant demarcated the concept of *Antillanité*, and novelists and linguists whose works embody a manifesto of *Créolité* anchored their literary enterprises firmly in the *Créole* islands of the West Indies and championed cultural and linguistic syncretism as their identity.[14] This sequence of genres – *Francité*, *Négritude*, *Antillanité*, *Créolité* – form the framework of the rest of this chapter.

THE CULTURAL VOYAGE TO FRANCE: ASSIMILATION AND *FRANCITÉ*

'Assimilation' was the official policy of Paris towards the colonies in the late nineteenth and early twentieth century. This held that 'natives', if properly instructed, could progress into *évolués*, whose social and political evolution was proven by their becoming increasingly French in language, beliefs and behaviour. This policy represented an extension of the centralising tendencies of metropolitan France and the ideals of the Third Republic. Moreover, governments made efforts to extinguish particularistic cultures (including regional languages such as Breton) and forge a new militant nationalism. The aim was to turn natives into 'black Frenchmen'. In the West Indies, this meant the maintenance of the same school curriculum and educational standards as in France; French culture was vaunted as superior culture. Not surprisingly, *Créole* was dismissed as a barbaric pidgin, and students were punished for speaking it. Oral literature, African traditions and any cultural deviance from the French norm were devalued if not forbidden except when relegated to colourful and entertaining folklore.[15] *Mulâtres*, anxious to redeem themselves from a servile past, to distance themselves from the illiterate blacks surrounding them and to gain political and cultural strength (even when economic power largely escaped them), were receptive to the argument. French education conferred upward social mobility. Literate *mulâtres* spoke French rather than *Créole*, quoted French verses rather than African stories and looked to France as the promised land.

The literature of the West Indies was a copy of French high culture of an academic variety. A number of local authors left the West Indies to live in France. Local writers imitated French culture with classical sonnets or Alexandrine verses. West Indian themes and motifs appeared in soft focus. They offered a note of exoticism or 'local colour' in literature that was at best regionalist and certainly not the type of writing that could offend *Békés* or colonial authories. *Doudouisme* (from *doudou*, dear, sweetheart or child) was the label later critics pinned on this school.[16]

Créole expressions, when used at all, were an ornament to literature. *Créole* might, however, be used to instruct illiterate blacks. For instance, in 1869 appeared *Les Bambous, Fables de La Fontaine travesties en patois créole par un vieux commandeur* (Bamboo: La Fontaine's fables travestied in patois by an old plantation foreman). The very title reveals the general attitude towards *Créole*, reduced to a dialect, and the propagandistic and edifying value of the work is attested by the version of La Fontaine's fable of the wolf and the dog. In the French version, a famished wolf meets a well-fed dog, who offers to take the wolf to his master. On the way, the wolf notices a scar on the dog's neck; the dog reveals that he is kept on a leash all day. The wolf turns on his heels, saying that he prefers hunger to servitude. In the *Créole* version, the ending is reversed, and the wolf gratefully accepts the leash in return for food, for there is 'no better master than the *Béké*'![17] When *Créole* threatened to be subversive, it was simply forbidden; as happened in 1842 when a catechism in *Créole* was outlawed. The *Créole* language, and the narratives the *Créole* story-tellers recited, remained a subterranean, even clandestine *oraliture*. The works of the Antillean literary elite of the 1800s and early 1900s are not without literary merit, and some authors, such as Daniel Thaly and Gilbert Gratiant, achieved renown. However, their work was neither culturally innovative nor politically confrontational.

The sole author of international stature to emerge from the French West Indies was Saint-John Perse. The opaque and incantatory nature of Saint-John Perse's works made them decidedly *avant-garde*, but they seemed to fit none of the usual 'schools' of literature: symbolism, surrealism or other movements. This white Antillean had incorporated the style and devices of black West Indian folktales and *oraliture* into his poetry. His verses were a rather lonely foreshadowing of later, more experimental *Créole* literature.[18]

With the exception of Saint-John Perse, French West Indian writers were looking towards a metropolitan 'model' of culture ill-suited to many of the ideas and ideals of their Caribbean countrymen. Universalistic, Eurocentric, colonial culture was hardly capable of permitting an indigenous voice. In particular, it could not admit the authenticity of the *Créole* language or African heritage. France was civilised, Africa was savage, *Créole* a bastard of the two. Acceptance of cultural mimesis was part of what Fanon saw as the phenomenon of 'black skin, white masks', marked by the efforts of blacks and *mulâtres* to become white (what he termed 'lactification').[19] Such obediencè was satirised by Léon-Gontran Damas in 'Hoquet' ('Hiccough'):

My mother wanted a memorandum son
If you haven't learned your history lesson
you won't be going to mass
on Sunday
in your Sunday clothes
This child will bring shame on our name
this child will be our name of God
Be quiet
Didn't I tell you that you must speak French
the French of France
the French of the Frenchman
French French.

The effect for Damas was, according to 'Solde' ('Balance' or 'Pay'):

I felt ridiculous
in their shoes
in their dinner-jacket
in their shirt-front
in their detachable collar
in their monocle
in their bowler hat.[20]

Undoubtedly, many Antilleans did not feel ridiculous in French cultural garb; some assimilated well to French norms and gained success in French society. The price was the denial of the African branch of their family tree, pushing aside their fellow islanders who were poor and illiterate, working hand-in-hand with a colonial power, which, though it gave them the vote, still treated West Indian islands as colonies to be exploited and West Indians as would-be Frenchmen. The next 'school' of French Antillean writers challenged both the political bases of *Francité* and the cultural forms in which it expressed itself.

THE CULTURAL VOYAGE TO AFRICA: *NÉGRITUDE*

For most colonialists, and the general public as well, all that was black and African was deprecated – black was the colour of night and sin, African religions were heathen superstitions, African art consisted of fetishes, and the 'dark continent' was the site of European fears and ambitions but not of any civilisation worthy of the name. By the beginning of the twentieth century, such views were changing, at least in the intelligentsia. Cubist painters, for example, found inspiration in 'primitive' art. Works on art and history published in the 1920s suggested greatly revised interpretations of African culture. Meanwhile, a so-called 'indigenist' literary movement was taking place in Haiti, and, to the accompaniment of jazz, the Harlem Renaissance moved into full swing in the United States with the works of Langston Hughes, Claude MacKay, Countee Cullen, Marcus Garvey and others. Some of these writers were West Indians and many had spent time in Paris. Such authors described the plight of 'black folk' in America and celebrated the culture of Africans and Negroes.[21]

The 'New Negro' movement appeared in France through the works of black immigrant writers associated with the Paris literary *salon* of the Martinican Nardal sisters and the six issues of a journal they published in 1931–2, *La Revue du Monde Noir*. It was not an avowedly political organ but a forum of African and black culture and social issues. For the first time, coloured men and women in France, most of whom were West Indians, proudly proclaimed their black ancestry, attempted to disabuse the white world of its racist misconceptions about Africa and analysed questions of import to the black diaspora.[22]

A more radical manifesto written by black Frenchmen who read the Nardals' journal was the one and only issue of a journal called *Légitime Défense*, published by Martinican students in Paris in 1932. The editorial acknowledged a triple intellectual influence: André Breton and surrealism, Freud and Marx. The lead article was a bitter attack on the *mulâtre* class of the West Indies, its social pretensions, political corruption and cultural imitation of the French:

> The children of the coloured bourgeoisie are reared in the culture of fraudulence. There are those who, after their secondary studies, go to France to try, generally with success, to 'merit' the title of 'doctor' or 'lawyer'.... Their desire 'not to be noticed', to 'assimilate', given that they carry around with them the indelible marks of their race, gives a tragic character to each of their efforts.

Two years later a group of students published another short-lived journal, *L'Etudiant Noir*: Aimé Césaire, a Martinican; Léon-Gontran Damas, a Guyanese; and two Africans, Léopold Sedar Senghor and Birago Diop. This cosmopolitan group found common ground in their blackness: 'We ceased being essentially Martinican, Guadeloupean, Guyanese, African or Malagasy students, in order to be just one and the same Black Student.'[23] This crystallised the new movement of *Négritude*, an exaltation of the shared traits of those whose ancestral roots lay in Africa, even when their progenitors had been transported to the new worlds of European conquest. It was to Africa, they argued, rather than to Europe, that black men and women should look for their culture: Africa was a culturally unified, potent and refined civilisation constituted on different bases from the European civilisation foisted upon black people by colonialism.

Négritude was an exceptionally important movement in both cultural and political terms: it reconceptualised the history, society and culture of Africa, refocused the view of Francophone blacks away from the model of *Francité* towards Africa, represented an often explicit denunciation of colonialism, and mandated literary experimentation to find an appropriate narrative form for black writers. Simply the act of bringing together black writers, as the poet – and later president of Senegal – Senghor did in his seminal 1948 anthology of black poetry, was an important cultural statement.[24] The works of some *Négritude* authors were filled with anger – for instance, the jazz-like poems of Damas, *Pigments*, published in 1937, and outlawed two years later as a danger to state security. Damas proudly proclaimed his blackness and caricatured the imitation of Frenchness by Antilleans; he predicted an uprising against the colonialists. The most strident poem in the collection is 'Si Souvent' ('So Often'):

> So often my feeling of race frightens me
> as much as a dog howling at night of
> some second-rate
> coming death
> I feel myself ready, always full of rage, to boil
> against that which surrounds me
> against that which keeps me
> forever from being
> a man
> And nothing
> nothing would be able to calm my hate
> other than a fine pool
> of blood
> made
> by the slashing knives
> which strip
> the rum-hillsides.[25]

Aimé Césaire became the most eloquent spokesman for West Indian *Négritude*. Born in 1913 in Martinique, the son of a respectable but poor customs inspector and a dressmaker, Césaire finished his secondary studies in the Fort-de-France *lycée* where Damas was a fellow student. He then went to the elite tertiary Ecole Normale Supérieure in Paris, where Senghor was a classmate. In France, Césaire read the *Revue du Monde Noire* and *Légitime Défence*, though he did not frequent the Nardal *salon*. In 1936, he began writing a long free-verse poem, the 'Cahier d'un retour au pays natal' ('Notebook of a return to the native land') which was published in a French journal, *Volonté*, in 1939.

The word *Négritude* was first used in Césaire's 'Notebook', the author (a very dark-skinned West Indian rather than a lighter-skinned *mulâtre*) intentionally choosing a word with a denigratory connotation – in French, *nègre* is closer to 'nigger' than 'Negro' – to signify his belonging to an international race of downtrodden people. Returning to Martinique after a stay in France, the narrator enters a teeming port to discover 'this squalling throng so astonishingly detoured from its cry', this 'inert town and its beyond of lepers, of consumption, of famines, of fears squatting in the ravines, fears perched in the trees, fears dug in the ground' while a statue of the white empress Joséphine towers over the town-square. He returns to his family to find his mother peddling away on her Singer sewing-machine by day and by night, 'the shack chapped with blisters, like a peach tree afflicted with curl, and the thin roof patched with pieces of gasoline cans ... and the bed of boards from which my race arose'. He remembers his old urge to go away, to flee the poor island. Yet he recounts how in a French tram, he hypo-critically joined other by-standers in smirking condescendingly at an old, hobbled black man, dressed in rags, his gnarled hands clutching a walking stick, a figure of ridicule in chic Paris. He sees his own homeland diminished, 'this land without a stele, these paths without memory, these winds without a tablet', and he takes a decision to embrace the blackness which is the emblem and the cause of this impoverishment, just as he accepts the heritage of deportation and slavery which is his birthright:

> I accept ... I accept ... totally, without reservation ...
> my race that no ablution of hyssop mixed with lilies could purify
> my race pitted with blemishes
> my race a ripe grape for drunken feet ...
> and the flogged nigger saying: 'Forgive me master'
> and the twenty-nine legal blows of the whip
> and the four-feet-high cell
> and the spiked iron-collar
> and the hamstringing of my runaway audacity.

The poet asks, 'Who and what are we? A most worthy question!' He finds that the real patrimony of his people lies in Africa. He acknowledges, without shame, 'heathen idols and obscene songs, parrot plumes and cat-skin garm-ents', and with this burden on his shoulders, he stands tall. Here is 'the seated nigger scum [*négraille*], unexpectedly standing', and he beckons his black countrymen:

> Rally to my side my dances
> you bad nigger dances
> the carcan-cracker dance
> the prison-break dance
> the it-is-beautiful-good-and-legitimate-to-be-a-nigger-dance.[26]

The 'Notebook' is a powerful reclamation of a cultural heritage demeaned by colonisers and viewed with shame by assimilated *mulâtres*. In his later poetry, as well as in his essays and plays, Césaire never abandons a primal reference to Africa, although in the half-century he has been writing, Césaire's style and ideas, not unexpectedly, have changed. Since the late 1970s, he has seldom used the word *Négritude* in any but an historical sense, but that historical and cultural movement was indeed momentous.[27]

The ideas of *Négritude* as expressed at the time of the 'Notebook' were, perhaps intentionally, a simplistic view of Africa. Writers such as Senghor lauded the rapport between Africans and nature, heralded Africans' musical sense, and celebrated their supposed emotiveness. What exactly constituted *Négritude* was unclear – Senghor seemed to favour a biological interpretation, Césaire a geographical and historical one. Later commentators legitimately took *Négritude* and its theorists to task on a number of points.

THE VOYAGE TO THE REVOLUTIONARY COUNTRY: FRANTZ FANON

Some of the ideas of *Négritude* provided a point of departure for later writers, including those who disagreed with Césaire. Most famous among them was Frantz Fanon. Born in Martinique in 1925, Fanon trained as a medical doctor and psychiatrist. In 1952, he published *Peau Noire, Masques Blancs*, an analysis of the cultural alienation Césaire had evoked in the 'Notebook'. Fanon described the inferiority complex which the black man had learned (and internalised – or, as he said, 'epidermalised' because of his skin colour) from colonialist whites; this extended, he added, to the 'deification' of the Antillean who had achieved success in Paris. The middle classes, yearning for respectability, deny their blackness, repudiate *Créole*, and pay ritual obedience to a white culture which would never fully accept them. Blacks turned into strangers both at home and abroad. As a result, Antillean society was 'neurotic'.

Fanon investigated the psychological aspects of racism, including the strong component of sexual fantasising about different races. He denounced racism and the self-hatred which it induced. Yet Fanon did not see *Négritude* as a solution, and here he parted company with Césaire:

> In no way should I derive my basic purpose from the past of the peoples of colour. In no way should I dedicate myself to the revival of an unjustly unrecognized Negro civilization. I will not make myself the man of any past. I do not want to exalt the past at the expense of my present and my future.[28]

Fanon's solution to the dilemma facing Antilleans was not to take refuge in the concept of blackness but to work in solidarity with other oppressed people for liberation and the construction of a new society.

In 1953, Fanon became head of the psychiatric department in a hospital in Algeria; the following year the Algerian war of independence broke out. Over the next few years, Fanon worked tirelessly for the rebels. He became one of the chief theorists of Third World revolution.[29] Fanon himself did not live to see Algerian independence, for he died of leukemia in 1961.

Fanon's ideas do not project a geographical *errance* into West Indian thought, for he fervently rejected both cultural assimilationism and *Négritude*: his intellectual journey had not been a voyage from the West Indies to France, or from the Caribbean to black Africa, but from the Antilles to the global plane of world revolution. His thought represents faith in the gospel of revolution and liberation that was as strong, indeed more vehemently expressed, than previous writers' faith in France or Africa. Yet Fanon's brand of revolutionary ideology was no less reductionist and universalistic than *Francité* and *Négritude*; there was little in the later doctrine of *Les Damnés de la Terre* which was specific to the Antilles, although many general statements in *Peau Noire, Masques Blancs* were based on the West Indian situation. It remained for a later generation to articulate a theory and a literature which attempted more precisely to fit the case of the West Indies.

THE VOYAGE BACK TO THE CARIBBEAN: *ANTILLANITÉ*

By the 1970s, the French West Indies enjoyed a lively and varied literary life; the influences of Césaire and Fanon were still strongly felt, although now somewhat out of date. New novelists emerged, foremost among them two Guadeloupean women. Simone Schwartz-Bart's *Pluie et Vent sur Télumée Miracle* (1972) and *Ti-Jean L'Horizon* (1979) were literary successes which described the lives of ordinary people in Guadeloupe, while Maryse Condé's *Hérémakhonon* (1976) brought to readers' attention an author whose novels chronicled the history of the Caribbean, Africa and the black diaspora.[30] In Martinique, a group of intellectuals in 1971 founded a journal of the social sciences called *Acoma*. Their leader was Edouard Glissant, novelist and founder of the Institut Martiniquais d'Etudes.[31] Although it published only five issues over the next two years, *Acoma* represented an important stock-taking in French West Indian culture and pioneered new lines of analysis. In the pages of the journal, Glissant looked at the socio-historical bases of cultural alienation and what he called 'mental disequilibrium' in contemporary Martinique, and he investigated social groups and tensions in the island.

Acoma examined the West Indies at a time when the plantation economy (and agriculture in general) had plunged into crisis; Martinique and Guadeloupe no longer lived from sugar-cane but increasingly from French government subsidies and the inflated salaries paid to bevies of public servants. According to some observers, they had become 'consumer colonies' with an 'artificial economy' which guaranteed a relatively high standard of living and social-security benefits while maintaining economic dependence. Césaire's *Négritude* and Fanon's calls for violent revolution seemed anachronistic and not entirely applicable to the French West Indies, despite significant minority support for independence, particularly in Guadeloupe. The old theory of cultural assimilation had been abandoned (although constitutional union had been effected through *départementalisation*, a status which continued to be contested). Instead, Glissant warned of a new danger of political integration, economic dependency and Europeanisation:

> One must suppose that French colonisation in Martinique will soon threaten to attain the 'final stage' of all colonisation, which is completely to depersonalise a community and to absorb it into an *exterior* body; in this sense, the colonisation of Martinique would then show itself to be one of the rare 'successful' colonisations in modern history.[32]

Such a scenario inspired Glissant to search for a new basis of identity.

This, for Glissant, was *Antillanité*, 'more a vision than a theory'. Glissant had first used the word in the late 1950s and his ideas had been fermenting since that time. In *Le Discours Antillais*, Glissant surveyed West Indian history and culture, and rejected a linear, conjunctive view of Antillean history:

> The Antilles are the site of a history formed by ruptures, the beginning of which was a brutal tearing-out (*arrachement*), the slave-trade. Our historical conscience could not 'stratify', if one can call it that, in a progressive and continuous manner, as with those people who created a sometimes totalising philosophy of history, the Europeans, but aggregated under the pressure of shock, contraction, painful negation and explosion.[33]

Wounded historical conscience prohibited the formation of a unitary, clear identity and created uncertainty about the direction in which to search for it. The first two generations of slaves in the seventeenth century might have hoped for a return to the mother-country, Africa, but that was manifestly impossible. They were also unable to preserve the bulk of their own culture. The African transported to the New World was a 'naked migrant': 'He could not carry with him his utensils, the images of his gods, his usual instruments, nor could he send news of himself back to his neighbours, hope to be reunited with his kinfolk, or reconstitute his old family in the place of his deportation.'[34] The forced migrant to the West Indies was unlike other migrant groups who transplanted their cultures or hoped someday to return to the motherland. The African had to create a culture in the new world, a process which replaced the mythical 'return' and which Glissant called a 'detour' (*détour* rather than *retour*).[35] The 'detour' included the *Créole* language, religious syncretism, the folktale, migration to France and even the ideas of Césaire and Fanon. Their works were, in fact, extremist positions, one in literature, the other in politics. Césaire's *Négritude* and Fanon's revolutionary message were generalisations, but they retained pertinence in that they 'permit us, via a detour, to return to the only site where our position is really an issue'.[36] That is, to the Antilles.

Glissant called for West Indian culture to repatriate itself to the Caribbean rather than search for salvation in colonialist France, mythical Africa or utopian revolution. This meant a discovery (or rediscovery) of indigenously created Antillean culture. True Antillean culture, that which already existed and that which was as yet only potential, posited a synthesis of the diverse civilisations which had come together in the islands:

> Today the Antillean no longer rejects the African part of his being; but he no longer needs, in reaction, to laud it exclusively. It is necessary that he *recognise* it. He understands that all our history ... has resulted in *another reality*. He is no longer constrained tactically to reject Western elements of it, even though they are still alienating, because he knows that he may choose among them. He sees that alienation lies first and foremost in the impossibility of making a choice, in an arbitrary imposition of values, and perhaps in the very notion of 'values'. He believes that synthesis is not the act of bastardisation that he was led to believe but rather a fertile practice through which the constituent parts are enriched. He has *become* Antillean.[37]

For Glissant, *Antillanité* was a political charter for collaboration and perhaps eventual unity among the islands of the Caribbean basin, and a cross-fertilisation of cultures from Cuba and Haiti to Trinidad and Curaçao, a multinational and multicultural synthesis of the West Indies. Glissant, however, did not call for the collapse of various Antillean cultures into one another but for their cohabitation within the commonality of traits and roots which form a distinctive Antillean experience. This was not a *mulâtre* culture of mimesis (of the French or others), but one which drew on the intermingling of European, African and other traditions which marked the Antilles. It could incorporate the 'verbal delirium' characteristic of Antillean story-telling, and explore indigenous themes such as *marronage*; Glissant said that the *marron* was the true folk hero of the West Indies.[38]

Glissant's 1975 novel *Malemort* provides a crucial link between his own view of Antillean history and the literary movement of *Créolité* then developing, partly under his influence, through a panorama of the common people of the West Indies. One chapter examines electoral politics and corruption in Martinique, another the aspirations (and pretentions) of the local bourgeoisie; yet another pays tribute to Antillean women. The style is experimental – a chapter written in free verse is followed by one of nineteen pages composed of a single sentence – and the book contains a glossary of *Créole* words and phrases. Haunting the book is the search for identity, 'the lost rediscovered lost trace' of ancestries and personal itineraries. There is the memory of the first voyage, from the interior of Africa to the slave-trading port of Gorée, and there is the voyage to France where 'you find the land which provides bread and sausage, and ham and cheese and lots of sodas and really good champagne.' The novel, rambling backwards and forwards from the eighteenth century to the present, is a complex journey through West Indian history and culture and an exploration of the links of Antilleans with Africa, France and the Caribbean.[39]

THE RETURN HOME: *CRÉOLITÉ*

Cutting across Glissant's theoretical work on *Antillanité* in the 1970s and 1980s was a current of renewed interest in the *Créole* language. *Créole* had remained a largely oral language. Various attempts were made to write *Créole* works, but authors faced the problem of how to transliterate the spoken word into standardised written style; works in *Créole* were inaccessible to those who did not speak the language and, ironically, to many of those who did, as they were often illiterate. In 1975, the first modern novel was written in *Créole*. This was *Dézafi*, the story of a *zombi*: in West Indian belief, the wandering 'living-dead'. The author was a Haitian, Franketienne. The work attracted literary attention and its influence spread to Martinique and Guadeloupe, where others began writing in *Créole*.[40] Notable among them was Raphaël Confiant, who appeared on the literary scene with five novels published in *Créole* between 1977 and 1987. Others, such as Patrick Chamoiseau, employed *Créole* expressions in their writings, though without publishing entire books in the language. The year after the publication of *Dézafi*, a Guadeloupean writer, Dany Bebel-Gisler, wrote a treatise on *Le Langage Créole, Force Jugulée* (1976), which sought to revalidate the language as a literary medium. In the mid-1970s, scholars at the University of the Antilles-Guyane in Martinique organised a research group on *Créole* and Jean Bernabé wrote an authoritative grammar of Antillean *Créole*.[41]

In 1988 Bernabé, Chamoiseau and Confiant issued a manifesto of *Créolité*, dedicated to Césaire, Glissant and Franketienne (or Franketyèn, as they spelled his name in *Créole*). It opened with the words: 'Neither Europeans, nor Africans nor Asians, we proclaim ourselves *Créoles*.' They reviewed the cultural history of the West Indies: political, economic and cultural dependence and the need to see the world 'through the filter of Western values'. The authors paid homage to Césaire, whom they said gave to the world a new image of 'Africa the mother, the matrix, black civilisation' and who used his pen to denounce colonialism. Yet, gently, they criticised Césaire for replacing the European myth with an 'African illusion'; both directed West Indian culture away from the West Indies. Glissant paved the way in situating Antillean civilisation in its Caribbean environment; but his political and cultural project minimised the differences between the French West Indies and neighbouring islands, and the links between the French islands and other *Créole* regions.[42]

Bernabé, Chamoiseau and Confiant charted a geography of *Créolité* that was not restricted to territories where French served as the linguistic foundation but included *Créole* regions throughout the world. French *Créole* was part of a larger phenomenon. This was its advantage: *Créolité* is a 'synthesis, not simply a *métissage*, or any other unity. . . . *Créolité* is an annihilation of false universality, of monolingualism and of purity.' In the case of the West Indies, it was 'the adaptation of Europeans, Africans and Asians to the New World [and] the cultural encounter between these people on the same site which leads to the creation of a syncretic culture labelled *Créole*'.[43]

The authors of the *Créole* manifesto, inspired by Glissant, set out five demands for a new *Créole* literature. It must be firmly rooted in the oral traditions of the islands. It must draw on the collective memory of the *Créole* people rather than just the official history of colonisation: hence, it should stress the resistance of the *marrons*, and the heroism of slaves. It should take its themes from the daily life of ordinary people, for there is 'nothing in our world which is too small, too poor, too useless, too vulgar or too inappropriate to enrich a literary project'. Fourth, *Créole* literature must commit itself to political change, though the manifesto authors refused to subscribe to a specific programme. Finally, *Créole* writers must accept the basic bilingualism of the *Créole* islands and draw on this wealth. Some authors may choose to write in *Créole*, others in French. Those who do the former 'will have for their first task that of constructing this written language', while '*Créole* literature in French will have as its urgent task to invest in and rehabilitate the aesthetics of our language,' to achieve a renewal of French itself.[44]

What this creed means in practice is best seen in the works of Confiant and Chamoiseau: contemporary Martinican authors who write in a variety of genres. Their writings, along with the works of Glissant and books by others including Daniel Maximin, Catherine Lépront and Ernest Pépin, establish a new literary style for the French West Indies. Their works are firmly anchored in the West Indies themselves; they are set neither in Africa nor in France, and relatively few place extended episodes overseas. The typical setting is a real or fictitious village in the Martinican countryside, or the poorer areas of Fort-de-France, or the Texaco neighbourhood of Chamoiseau's novel of the same name.[45] Moreover, they concentrate largely on the *petit peuple* – the working class, whether urban (the *djobeurs*) or rural – even if the occasional *Béké*, administrator, schoolteacher or *blanc-France* (a Frenchman from the metropole) makes an appearance. The main characters are a roster of traditional, if disappearing, Antilleans. *Marrons*, whether in slave times or in a modern incarnation, populate the novels of Glissant.

Much attention is given to indigenous Caribbean social structures. One example is sexual relations and romantic attachments, reflecting on the idiosyncratic structure of the family in the Antilles compared to bourgeois France: episodic liaisons, a relatively low rate of legal marriage and a high rate of illegitimacy. This was often explained as a survival of the slave heritage, when plantation-owners encouraged slave women to bear as many children as possible (and fathered half-caste progeny themselves) yet broke apart families. In the twentieth century, irregular unions continue to be numerous, perhaps enforced by Caribbean *machismo* and relative lack of concern with pre-marital virginity. Rather than being hidden away, 'illicit' sexual intrigues become omnipresent in recent novels, which describe (often indiscretely) sexual couplings, irregular partnerships and infidelities.[46]

Another theme is the religious syncretism of the Antilles: ardent if somewhat formalistic Catholicism combined with superstitious practices, many inherited from African ancestors. These are more important in Haiti, where they have developed into a full-blown religion (*vaudou*, voodoo), than in the French West Indies, but even there, the power of the *quimboiseurs* (wisemen or wisewomen, shamans, soothsayers, healers) remains strong, and novels play on such practices. They may also draw on the traditional folktales of the Antilles, in which a character representing a common man or woman (*Compère Lapin*, Brer Rabbit) uses his or her wiles when pitted against a dominant individual, often a European, a *Béké* or a black plantation foreman (*Compère Tigre*, Brer Tiger).[47] Novels also feature the pre-Lenten Carnival (*Vaval* in *Créole*), with its street festivities, costumes and licence for merry-making and social transgression – an event with origins in both Western Christian beliefs and African celebrations.

The time-periods in the novels vary, but generally turn on points of great change: historical episodes which form the foundation of the plots, the effects of past events or phenomena (particularly slavery and *marronage*) on the lives of contemporaries, or the memories of the past in the stories and reminiscences of story-tellers and other characters. Fact and fiction intersect in many novels. Novelistic description of Martinique's past concurs with that of recent historians, particularly the social history which emphasises political and cultural challenges mounted by the *nègres* against the overwhelming power of colonialism. Novelists such as Glissant, Chamoiseau and Confiant are far too astute and critical simply to produce fiction which is 'politically correct' history, but the capacity for survival of the underclasses against the dominant elite in the islands is a constant theme in their works.

The style of the novels is both innovative and an exemplification of the call by Glissant for the introduction of the oral word into the written word. Glissant himself does not write in *Créole*, but the gradual unveiling of plot, the circularity of the narrative and the changes of voice accord with his view of the Antilleans' disjunctive experience of their own history. His language, 'a synthesis of written syntax and that of oral rhythm, tries to situate itself at the junction of writing and speaking'.[48] Confiant and Chamoiseau also employ a narrative style which departs from standard European usage – including run-on sentences, extremely long paragraphs, frequent interjections, accounts which digress and meander and which, indeed, are closer to convoluted oral recitations than to old-fashioned written literature. Confiant, even in his French-language novels, uses a wealth of *Créole* words and expressions, as does Chamoiseau. These words are not incorporated merely to add local colour but to represent phenomena which have no parallel in France. They introduce the reader to the universe of the Antilleans: for example, the differences between the locally born Frenchmen (*blancs-goyave* or 'guava whites') and those from France (the *blancs-France*), and the various gradations in skin-colour, and often of status as well, of non-whites (*mulâtres*, *chabins*, *nègres*).

Chamoiseau's most recent novel, *Texaco*, winner in 1992 of France's most prestigious literary award, the Prix Goncourt, is a fine example of the new *Créole* literature. The book is presented as the reconstructed memoirs of Marie-Sophie Laborieux, a now-elderly Martinican who was the first settler of a poor neighbourhood next to a Texaco oil refinery on the outskirts of Fort-de-France. *Texaco* portrays Fort-de-France's shanty-towns and the ways in which its people live and survive:

> What was this In-Town [*En-Ville*]? . . . It's the bottle into which our histories flow. Time, too. The 'tation [*habitation*, plantation] pulled us apart. The hillsides [of *marronage*] forced us into a life of drifting. In-Town puts into action ties together moors kneads and rekneads us at top speed. . . . It's not a place of happiness. It's not a place of un-happiness. It's the calabash of our destiny.[49]

There are also references to France. For the residents of Texaco, France is the country of hope and prosperity. Esternome, Marie-Sophie's father, sees France as the country which brought emancipation and suffrage to the slaves in 1848. For his daughter, France is the country of culture and learning – a neighbour reads Saint-John Perse and Césaire to her. Marie-Sophie dreams ecstatically of a trip to France which she has been promised by an employer, but it is cancelled by the war. France is also the dreamland of one of her lovers, who sends her a postcard of a wheatfield and a mill in the French countryside when he has finally made the journey. Her own closest brush with France is de Gaulle's visit to Martinique. She decides to invite the French president to dinner, polishes her modest furniture and lays out her best plates and cutlery, prepares a feast of Antillean delicacies, dolls herself up in her best dress and takes off to the Savane (Fort-de-France's central park) to invite de Gaulle. Yet she barely catches a glimpse of the great man, falls in the rush and tears her dress before returning home disappointed. The episode is perhaps intended as a parable of the mythification of France by West Indians and the deceptions induced by trust in the 'mother-country'.

IDEOLOGY, CULTURE AND SOCIETY

The ideas associated with *Créolité* constitute a newly syncretic conceptualisation of West Indian identity and represent a significant departure from earlier notions of *Francité*, *Négritude*, the revolutionary doctrine of Fanon and, to some extent, *Antillanité*. However, the break is not so great as first appears. Even at their most militant, ideologists like Césaire and Fanon did not totally reject French culture or language nor preach a return migration to Africa. Neither wrote in *Créole* but in eloquent French. Moreover, the heritage of French and European thought is omnipresent in the works of West Indian writers, even the most contestatory. The contributors to *Légitime Défense* explicitly acknowledged their debts, and Césaire is considered the most important black surrealist poet. The ideas of psychology and psycho-analysis permeate the works of Césaire and Fanon in their discussions of identity, and Glissant took his idea of the 'rhizome' and 'root' from the contemporary French psychiatrists Félix Guattari and Gilles Deleuze. Perm-utations of Marxist thought are evident in all French West Indian literature from the 1930s through to the 1980s. The form of the 'manifesto' – from Césaire's *Discours* in 1955 to the proclamation of *Créolité* in 1988 – is a very French literary device; the theorising of Glissant's works, to an Anglo-Saxon reader, is certainly an extremely French approach.

The idea of cultural *métissage* is not entirely new; somewhat ironically, the *mulâtres* of the Antilles who were subject to such harsh criticism by writers in the 1930s (and later) were pioneers of cultural syncretism, even though they minimised (or rejected) their African patrimony. In his 1955 pamphlet on colonialism, Césaire spoke about the importance of bringing civilisations into contact with each other and blending cultures – but the ideology of domination, racism and the violence of colonialism, he thought, were precisely what kept such a synthesis from happening.[50] His promotion of cultural intermixing foreshadowed *Antillanité* and *Créolité*; Césaire himself used the word 'detour', later employed as a central metaphor of Antillean culture by Glissant, in the 'Notebook'. Even in language and style, the links are greater between the different literary 'schools' than at first apparent. Saint-John Perse's and Césaire's literary experimentation led the way for the non-traditional narrative styles of Glissant and the supporters of *Créolité*.

The historical context in which the various theories of West Indian culture were formulated moulded their different approaches. 'Assimilation' and *Francité* were products of the colonial era and its views on the hierarchy of races and the superiority of European values. *Négritude* was part of a rediscovery of African civilisation that owed much to the anthropology, ethnography and history of the 1930s; in later decades it reflected and to some extent merged with the idea of Pan-Africanism. *Négritude* owed much, too, to the broad challenge to European culture mounted by the *avant-garde* in the early years of the twentieth century: artistic movements such as Cubism and literary movements such as Dadaism and surrealism. By the 1930s, the Depression, with its economic repercussions in the empire, was putting the question of colonialism into higher relief, while colonialists celebrated the virtue of empire with renewed ardour. Publication of *Légitime Défense* followed only one year after the gigantic celebration of French overseas expansion at the Colonial Exhibition of 1931, an event attacked by many intellectuals, socialists and communists who issued an anti-colonial manifesto and (albeit largely unsuccessfully) called on the public to boycott the fair.

Césaire's *Discours sur le Colonialisme* was written only eight years after the French put down – at the cost of tens of thousands of lives – a revolt in Madagascar in 1947, an event to which it bitterly refers. Césaire's work, like Fanon's *Peau Noire, Masques Blancs*, was published at the time of the Indochinese war, which saw the defeat of France in 1954 and the independence of Vietnam, Cambodia and Laos. The same year saw the start of the Algerian war of independence, which continued with untold savagery until 1962 and which formed the background for Fanon's writings and activism.

Glissant's theoretical writings of the 1970s show the mark of the New Left but also the failure of the more radical Trotskyist and Maoist ideas which came to the forefront. The era of decolonisation had seemingly come to a close, and the political and economic records of many of the countries which had gained independence in Africa were often disappointing. By the end of the 1980s, with the emergence of *Créolité*, ideas of world revolution and the building of new societies based on socialist or communist principles had been discredited; economists and political scientists emphasised political and economic interdependence. Notions of racially based societies, let alone ethnic purity, were widely unacceptable, although they would horrifically return in some parts of the globe in only a few years. In his most recent writing, Glissant has called for *Créolisation* – contact between and intermixture of cultural systems – as an alternative to the dangers of racial purity, xenophobia and religious fundamentalism.[51]

The ideas which have been prominent in French West Indian literature not surprisingly mirror broader currents in the world, as well as charting the socio-economic history of the Caribbean territories. The new literature of *Créolité* has shown great promise, but it has also set a great task for itself.[52] Can *Créole* culture avoid the pitfalls of exoticism and folklorisation? To what extent can the *Créole* language serve as a vehicle for modern culture? For how long will *Créole* be the dominant vernacular in the French West Indies, a region increasingly marked by urbanisation, migration, higher levels of education, the disappearance of old ways of life and continued Europeanisation and Americanisation? Will the sort of literature pioneered by Glissant, Chamoiseau and Confiant treat the realities of the urban Caribbean, divorced as it now is from the world of the story-tellers, *quimboiseurs* and *djobeurs*? Will it be able to render into prose and poetry the world of the Antillean diaspora, that of several hundred thousand Martinicans and Guadeloupeans who live in France, as well as growing numbers who are born there and whose contacts with the West Indian islands are but tenuous? Will it, in the long term, be more than another 'regionalist' literature, a new 'indigenist' school of writing? Is there a danger of simplistic populism? Can the differences and disparities between various countries of the Caribbean basin be surmounted in the name of *Antillanité*? Does the future of Martinique, Guadeloupe and Guyane lie in continued attachment to France, independence, a multi-island federation or some other political arrangement – or is this question simply too difficult to be solved once and for all? Is *Créolité* a strong enough identity to provide cultural unity to islands as diverse as poverty-stricken and crisis-beset Haiti, the tiny former British possessions of St Lucia and Dominica and the French *départements d'outre-mer* of the West Indies, let alone *Créole* islands as distant from the Caribbean as Mauritius, Réunion and the Seychelles in the Indian Ocean?

THE PERIPATETIC ANTILLEAN

Errance – drifting, wandering, the pilgrimage, the search for roots, migration – is a characteristic of West Indian intellectuals. For some, such as Césaire, the first trip to Paris was a desired liberation from the insularity of the Caribbean. Yet there – as Césaire, Fanon and Glissant all testify – Antilleans became aware of their blackness when they discovered other black men or when they encountered the racism of the French. 'Furthermore', as Mireille Rosello remarks, 'exile is not necessarily geographical, and [even] the Martinicans or Guadeloupeans who remain on their "native" land can indeed suffer a syndrome of exile.'[53] Antillean writers were propelled back on a return voyage, to the 'native land', but they often ricocheted back to France or to Africa, discovering other, historical exiles, recovering lost homelands, perhaps condemned, like *zombis*, to perpetual itineracy. As Césaire put it in the 'Notebook':

> To go away.
> As there are hyena-men and panther-men, I would be a jew-man
> a Kaffir-man
> a Hindu-man-from-Calcutta
> a Harlem-man-who-doesn't-vote.[54]

Personal *errance* became a theme for Antillean literature. For Rosello:

> Exile was long the raw material of the works of Aimé Césaire, Edouard Glissant and Maryse Condé and would continue to be a source of inspiration. . . . The texts which 'came from' the islands and which 'returned' were ill not from Exile but from a series of exiles, they suffered from departure and an impossible return; they were marked by the ambiguity of an eternal strategy of 'detour'. It was not necessary to pursue the quest of the writers of *Négritude* in their search for an 'authentic' 'black' voice because the Antilles were not the inheritors of just the deportation of the slaves. For Edouard Glissant, the 'return' and the 'detour' thus became inseparable, and the quasi-homonymy of the two words [*retour, détour*] suggests that we look toward *language* to find a possible definition of the 'native land' for Antillean writers.[55]

Thus, *errance* is the very history *of* Antillean literature, and the homecoming occurred with the acceptance of an Antillean literature rooted in the West Indies rather than Africa or Europe, but open to the influences of other continents, the traditions of all of the populations of the islands, the language of *Créole*.

NOTES

1 E. Glissant, *Poétique de la Relation*, Paris, Gallimard, 1990.
2 E. Glissant, *Le Discours Antillais*, Paris, Editions du Seuil, 1981, p. 209.
3 For an account which emphasises the role of the slaves' revolt, see E. de Lépine, *Questions sur l'Histoire Antillaise*, Fort-de-France, Editions Désormeaux, 1978.
4 There is a large literature on the history of the West Indies, including a growing body of works by West Indian specialists. On the French West Indies, see in particular the works of the Martinican historian A-P. Blérald, *Négritude et Politique aux Antilles*, Paris, Editions Caribéennes, 1981; *Histoire Economique de la Guadeloupe et de la Martinique du XVIIe Siècle à Nos Jours*, Paris, Karthala, 1986; *La Question Nationale en Guadeloupe et en Martinique*, Paris, L'Harmattan, 1988. On contemporary issues, see F. Constant, *La Retraite aux Flambeaux: Société et Politique en Martinique*, Paris, Editions Caribbéennes, 1988; Henri Bangou, *Voies de la Souveraineté: Peuplement et Institutions à la Guadeloupe*, Paris, Editions Caribbéennes, 1988. There is little recent material in English except W. Miles, *Elections and Ethnicity in French Martinique: A Paradox in Paradise*, New York, Praeger, 1986.
5 The best-known was the Guyanese Félix Eboué who, as governor of Chad, was the first colonial governor to rally to General de Gaulle after the German defeat of France at the beginning of the Second World War. See B. Weinstein, *Eboué*, Oxford, Oxford University Press, 1972.
6 See A. Anselin, *L'Emigration Antillaise en France: La Troisième Ile*, Paris, Karthala, 1990.
7 The population of Martinique in 1990 was 359,597; that of Guadeloupe 387,000. Approximately 528,000 Martinicans and Guadeloupeans live in France.
8 See R. Aldrich and J. Connell, *France's Overseas Frontier: Départements et Territoires d'Outre-Mer*, Cambridge, Cambridge University Press, 1992.
9 R. Confiant, *Eau de Café*, Paris, Grasset, 1991, p. 163.
10 D. Walcott, *Omeros*, London, Faber, 1990, p. 283.
11 M. Rosello, *Littérature et Identité Créole aux Antilles*, Paris, Karthala, 1992, pp. 91–3.
12 P. Chamoiseau and R. Confiant, *Lettres Créoles: Tracées Antillaises et Continentales de la Littérature, 1635–1975*, Paris, Hatier, 1991.
13 This chapter will only discuss the literature of Martinique and Guadeloupe (and, occasionally, Guyane). There is also a particularly rich Haitian literature.
14 An excellent recent overview of French West Indian writing, in theory and in practice, is R. Burton, 'Ki Moun Nou Yè? The idea of difference in contemporary French West Indian thought', *New West Indian Guide/Nieuwe West-Indische Gids*, 1993, vol. 67, pp. 5–32. There was a special issue of *Callaloo: A Journal of African-American and African Arts and Letters*, 1992, vol. 15, no. 1: 'The Literature of Guadeloupe and Martinique', guest-edited by M. Condé, includes theoretical articles, a lengthy bibliography of works in English and French, and essays on Glissant, Confiant, Maximim, Chamoiseau, Marie-Magdeleine Carbet and Condé herself. See also B. Ormerod, *An Introduction to the French Caribbean Novel*,

London, Heinemann, 1985, which has chapters on Césaire, Glissant, Zobel and Schwarz-Bart, as well as several Haitian writers. In French, the best works are Chamoiseau and Confiant, op. cit., Rosello, op. cit., and Régis Antoine, *La Littérature franco-antillaise: Haiti, Guadeloupe et Martinique*, Paris, Karthala, 1992.

15 This can be seen in music as well. The African-influenced *zouk* of Martinique was domesticated into the polite and acceptable *biguine*.

16 See various examples of writing from Martinique in a two-volume anthology, covering the period from the 1600s through the 1800s, edited by A. Joyau, *Panorama de la Littérature à la Martinique*, Fort-de-France, Éditions des Horizons Caraïbes, 1974.

17 The book is discussed by Chamoiseau and Confiant, op. cit., pp. 76–9.

18 E. Yoyo, *Saint-John Perse et le Conteur*, Paris, Bordas, 1971. In English, see the biography by E. Ostrovsky, *Under the Sign of Ambiguity: Saint-John Perse/Alexis Leger*, New York, New York University Press, 1985.

19 F. Fanon, *Black Skin, White Masks*, London, Pluto, 1986 (originally published in French in 1952).

20 L.-G. Damas, *Pigments – Névralgies*, Paris, Présence Africaine, 1972, pp. 36–7, 41 (first published in 1937).

21 M. Fabre, *From Harlem to Paris: Black American Writers in France, 1840–1980*, Urbana, University of Illinois Press, 1991.

22 The six issues of the rare *Revue du Monde Noir* were republished by J.-M. Place in Paris in 1992 with a new introduction by L.-T. Achille, one of the surviving contributors to the journal.

23 Quoted in Chamoiseau and Confiant, op. cit., p. 120

24 L.S. Senghor (ed.), *Anthologie de la Nouvelle Poésie Nègre et Malgache de Langue Française*, Paris, Présence Africaine, 1948.

25 Damas, op. cit., p. 49.

26 These quotations are all taken from A. Césaire, 'Notebook of a return to the native land', in *The Collected Poetry*, Berkeley, University of California Press, 1983 (trans. C. Eshleman and A. Smith).

27 Césaire continued to be a critic of colonialism, especially in his *Discours sur le Colonialisme* of 1955. He has also had a noteworthy political career as *deputé* from Martinique and mayor of Fort-de-France since 1946.

28 Fanon, op. cit. p. 226.

29 See D. Caute, *Fanon*, London, Fontana, 1970. Fanon's most famous book was *The Wretched of the Earth*, London, MacGibbon and Kee, 1965.

30 The remainder of this chapter concentrates on Martinican literature.

31 The works of E. Glissant include *Les Indes, Le Sel Noir, Poèmes* and *Pays Rêves, Pays Réel* (volumes of poems); *Soleil de la Conscience, L'Intention Poétique, Le Discours Antillais* and *Poétique de la Relation* (essays); *Monsieur Toussaint* (theatre); and *La Lézarde, Le Quatrième Siècle, Malemort, La Case du Commandeur* and *Mahagony* (novels).

32 E. Glissant, 'Introduction à l'étude des fondements socio-historiques du déséquilibre mental', *Acoma*, 1971, no. 1, p. 83.

33 Glissant, *Le Discours Antillais*, op. cit., pp. 130–1.

34 ibid., p. 66.

35 ibid., p. 32.

36 ibid., p. 36.

37 ibid., pp. 17–18.

38 ibid., pp. 104, 242.

39 E. Glissant, *Malemort*, Paris, Editions du Seuil, 1975.

40 At this time, approximately 10 million persons spoke some variety of French

Créole: 6.9 million in the French West Indies and Haiti, 0.5 million in the Commonwealth Caribbean (Dominica, St Lucia and islands with a migrant population from these islands, notably Trinidad and Jamaica), 0.6 million in France and Britain, 0.5 million in the Unites States, Canada and Africa, and 1.5 million in the Indian Ocean (Mauritius, the Seychelles, Réunion). Data from Glissant, *Le Discours Antillais*, op. cit., p. 343.

41 J. Bernabé, *Foindal-Natal: Grammaire Basilectale Approchée des Créoles Guadeloupéenne et Martiniquaise – Approche Sociolittéraire, Sociolinguistique et Syntaxique*, Paris, L'Harmattan, 1983, 3 vols.

42 J. Bernabé, P.Chamoiseau and R. Confiant, *L'Eloge de la Créolité*, Paris, Gallimard, 1989.

43 ibid., pp. 28, 30.

44 ibid., pp. 34–48.

45 P. Chamoiseau, *Texaco*, Paris, Gallimard, 1992.

46 On the Antillean family, see Glissant, *Le Discours Antillais*, op. cit., pp. 293–302.

47 ibid., p. 243.

48 ibid., p. 256.

49 Chamoiseau, op. cit., p. 322

50 A. Césaire, *Discours sur le Colonialisme*, Paris, Présence Africaine, 1973, 6th edn, p. 9 (originally published in 1955).

51 E. Glissant, 'Le Cri du monde', *Le Monde*, 5 November 1993, pp. 27–8.

52 The literature is joined by various other cultural productions which show the influence of *Antillanité* and *Créolite*. For instance, there is the music of various Martinican and Guadeloupean groups, the best known of which is Kassav, who sing in *Créole* and use rhythms and instruments which draw on traditional Antillean music (such as the Martinican *zouk* and the Guadeloupean *gwo-ka*) yet attempt to avoid folklorisation. There is also the beginnings of a new French West Indian cinema, best seen in the works of the Martinican director Euzhan Palcy. Her first film, *Rue Cases-Nègres*, was a dramatisation of Joseph Zobel's novel about a heroic woman who sacrifices and struggles so that her grandson can gain a French education. She followed this with an English-language film, *A Long Dry Summer*, about the struggle for racial justice and democracy in South Africa. Her most recent work is *Siméon*, the story of an old Antillean musician. Intentionally or not, Palcy's films are a nice triad about West Indian attachment to French culture, the hopes of Africa, and the Creole culture of the islands.

53 Rosello, op. cit., p. 89.

54 Césaire, 'Notebook', op. cit., p. 43.

55 Rosello, op. cit., p. 90.

8

PIED-NOIR LITERATURE
The writing of a migratory elite
Rosemarie Jones[1]

COLONIAL BEGINNINGS

Conquest and settlement: upon these twin policies the French presence in Algeria was founded. Guarantors of French political, economic and social dominance, they underlie the discourse which will become the literary tradition of the *pieds-noirs*, generating both a praxis and a *politique* in that noun's dual sense of politics and policy. The particularly intense nature of French involvement in Algeria renders representations of that experience especially interesting for the study of colonial literatures.

From the very first days of the invasion in 1830, Algeria provoked a host of writings of various kinds, from the official reports sent back to France by superior officers, to more intimate letters and accounts.[2] With territorial expansion and pacification, the appeal of the newly annexed and exotic country drew, as visitors, some of the most illustrious names in contemporary letters: Chateaubriand, Daudet, Fromentin, Gautier, Gide, Maupassant. They came, they saw, they wrote ... and they departed, and this first wave of writings about Algeria has come to be referred to as a 'tourist' or a 'postcard' literature.

Gabriel Audisio[3] distinguished between a literature inspired by Algeria and one produced by Algeria. It took nearly seventy years for a settler literature to emerge, and when it did, significantly enough, one of its founding fathers, Louis Bertrand, was neither born in the colony nor died there. He was profoundly marked by the years 1891 to 1897 he spent in Algeria, which in turn left their trace upon the entirety of his subsequent writing. Bertrand is now best remembered for his concept of Latin Africa, as set out principally in his essays and prefaces. 'I have introduced a new conception of North Africa, which is in fact nothing more than the former Roman province of Africa.'[4] The horizontal movement of expansion across Algeria and the in-depth probings of archaeology combine, for Bertrand, to prove that the true face of the country is revealed in its Roman settlements – now ruinous golden towns – which nevertheless continued to shape and dominate the whole civilisation of the country. Civil and domestic architecture, forms of art and

125.

artefact, dress: all are traceable to Roman sources. The original inhabitants, the Berbers, are themselves Latinised, and in the first centuries of our era founded the glorious tradition of the African church. Regrettable incursions of invaders – Vandals, Turks, Arabs, barbarians all – are mere surface phenomena, not troubling the deep cultural continuity. From this vantage-point, the French invasion is seen as a rediscovery, at most a reconquest, more aptly as a homecoming. 'In coming back to Africa, we simply recuperated a lost province of Latin civilisation.'[5] One conquest is invoked to justify another. Its object, however, through Bertrand's Latinising eyes, acquires higher status: not just any dusty old colony, Algeria has monuments and sites superior to those of Rome and equal to those of Egypt. The past value of the colony is employed to justify its present re-evaluation.

The cast-list for Bertrand's essays are those indispensable *personae* of the colonial enterprise, the soldier and the priest, albeit represented here in a new guise: the soldier as discoverer and preserver of the monuments of the past, and the priest as prophet of the future. His novels, however, foreground a new type of hero, fit inhabitant of a new world. The drivers of the wagons carrying materials from the capital to the south have the mentality of frontiersmen: thriving in the harsh conditions of their employment. Proud, violent and truculent, they find in the conquest of plateau and desert both an outlet for their bubbling energies and fulfilment of their passion for freedom and independence. Each journey is an undertaking of epic proportions, pitting the individual against the elements and the temptations of comfort in a process of self-discovery and self-expression.

The forms in which these first texts are cast will prove to be at the same time enduring models and means of appropriation. It is without doubt fiction that dominates the *pied-noir* tradition: the most spacious of genres and one which is characteristically European, foreign to the Arabophone tradition. But fiction has tended, in Algeria, and very frequently in the work of the same authors, to be accompanied by the essay, as though the work of imagination acquired thereby some justification, in the form of a genre which permits both the general statement and – since *pied-noir* essayists often adopt a quite intrusive first person – a forum for the assertion of personal belief and identity.

FRENCH, ALGERIAN OR . . . ?

The new hero incarnates in a different form in the character Cagayous, protagonist of immensely popular novels and broadsheets which appeared fairly regularly from 1895 to 1920, written by 'Musette' (Auguste Robinet).[6] Cagayous is leader of a band of layabouts: urban, streetwise, fiercely anti-Jewish, they enter fully and vigorously into the political life of the capital as it filters into the crowd scenes: they are the first to chant the slogan of the day, they find a good position at the port from which to observe new arrivals from France. In the absence of such events, they patrol their territory of Algiers constantly in search of a new *coup* or farce. The city is an arena for fight – Cagayous has a curious arm, a large ball attached to a cord, with which he managed to 'get four Arabs in one go' – or for flight from police; (Jewish) shops exist to be broken into; walls and windows are spaces to be covered with slogans or graffiti. The youths speak a language which Musette is the first to record – *pataouète* – the French of the working-class quarters of Algiers, based on popular French but studded with expressions drawn from Spanish, Italian, Maltese, Arabic. Paul Siblot has analysed how use of this language defined the *pied-noir* for himself and in contradistinction to the 'outsider':

> These expressions, few on the whole, take on a decisive function. Unfamiliar to the reader outside the colonial community, they deliver no meaning and constitute a mark of otherness; to those familiar with them they offer an opportunity for complicity and give rise to a feeling of belonging to a particular group: they become signs of identity.[7]

'Are you French? We're Algerian!' This most famous line of Musette's[8] highlights one of the tenets of *pied-noir* belief. Not French, but certainly not Arab either. 'Algerian' denoted, at the time, the community of European origin. This is the first sense in which it was used by the 'Algerianist' movement which took up the literary baton from Bertrand. Still lingering on, the movement was most active and influential during the years 1920 to 1935, and was represented principally by Robert Randau (Robert Arnaud), Jean Pomier and Louis Lecoq. Unlike Bertrand, these writers were born in Algeria, and made a virtue of *pied-noir* necessity. In 1920 the *Association des Ecrivains Algériens* (Association of Algerian Writers) was established, leading to the creation of a *Grand Prix Littéraire de l'Algérie* in 1921, and the launch of the review *l'Afrique* in 1924. The ambition underlying this three-pointed foundation was clearly to legitimise the title of Algerian letters, and indeed Randau announced the aim of aesthetic autonomy for the colony.[9] Again, the essay was an indispensable adjunct to fiction: Randau's preface to *De Treize Poètes Algériens (1920)*[10] constituted a form of Algerianist manifesto, and the non-fictional writing amounted to fifty or so novels that could be called Algerianist, including Randau's own *romans de la patrie algérienne* (novels of the Algerian homeland): *Les Colons* (1907), *Les Algérianistes* (1911) and *Cassard le Berbère* (1921).

However, if, as Gabriel Audisio was later to claim, the Algerianists did not succeed in their objective of creating an autonomous Algerian literature, remaining simply a literary school, this was in part because failure was pre-inscribed in the attempt itself. The Algerianists exhibit a highly ambiguous attitude towards France, composed on the one hand of sulky resistance which asserts 'Algerianeity' over Frenchness, and on the other of an ostrich-like pretence that France is not really 'there'. But in fact they depend upon the metropolis for their very existence as Algerianists: take France away, and the 'Algerians' are nameless multitudes of Muslims. Towards these latter the attitude is again ambiguous in the extreme. More 'liberal' than Bertrand, the Algerianists recognise that the Algerian identity must perforce include a certain Arabo/Berber component; these people, however, are represented as unequal to the Europeans and as needing to be initiated into the superior civilisation of France. Obscurely aware of their contradictions and of their vulnerability, the group is ceaselessly preoccupied with self-definition, titles and protagonists proclaiming their status and categorising the other, on a scale running from the accolade of 'thoroughbred Algerianist' down to the epithet 'vermin'.

.Unresolved, the question of identity continued to ferment. Whilst it is true that a large number of *pieds-noirs* were not actually French in origin, it was France that guaranteed their presence, and the gap between, in André Rosfelder's terms,[11] the *patrie française* (French homeland) and the *terre algérienne* (Algerian soil) remained uncomfortably wide. One possibility of negotiating the terms, initiated by Randau, was to bracket off the political domain and to replace and resolve the problem in literary limits. *Patrie* is replaced by *pères* (father-figures): a gallery of spiritual fathers from whom the *pied-noir* writer can claim descent with pride and without troubling thoughts of nationality. Randau cites Apuleius, Tertullian and St Augustine; and Gabriel Audisio, who began writing in the last days of the Algerianists, includes the same names in his genealogy of literary 'African' ancestors, among whom he also numbers Ibn Khaldoun and Senoussi. This appeal to past authority can be seen as a variant of the tendency, apparent already in Bertrand and continually resurfacing in the *pied-noir* tradition, to represent the past in epic terms, as a story of origins and of heroes.

But the new movement over whose birth Audisio presided, and to which he gave the name *Ecole nord-africaine des lettres* (North-African school of literature) – now more generally and familiarly known as the *Ecole d'Alger* (the Algiers School) – was characterised by a turning-away from the past to concentrate upon the present, and by concern with space rather than with time. Beginning in the mid- to late 1930s, and gathering momentum in the 1940s, this was the golden age of *pied-noir* literature, which produced Albert Camus, René-Jean Clot, Claude de Fréminville, Max-Pol Fouchet, Emmanuel Roblès and Jules Roy. This first wave was followed by a second, of near-contemporaries: Raoul Celly, Jean Daniel, Jean-Pierre Millecam, Marcel Moussy, Jean Pélégri, André Rosfelder and the poet Jean Sénac.[12]

The new movement impinged upon public consciousness first of all through the work of Audisio and Camus. Audisio resisted what seemed to him exclusive adoration, on the part of Bertrand and his followers, of what was actually Roman, not properly Latin. His own response to the question of identity was to base both a 'homeland' and a 'people' on the Mediterranean and its coastal dwellers: 'My homeland is the sea, the Mediterranean.'[13] In this he is closely followed by Camus, who frequently hymned the Mediterranean, who found Greece more congenial than Rome, and who spoke of the *pieds-noirs* as 'this race ... indifferent to things of the mind'.[14] But Audisio's criticisms of Bertrand take account of only part of the latter's work, and the angle of critical vision reflects a tension in *pied-noir* writing between a desire for purity, related to the honour of the *mère-patrie* (the French motherland), and the necessity of coming to terms with bastardy, the fact that the *terre algérienne* had brought forth sons of doubtful legitimacy. Admirer of Rome, Bertrand is also the cantor of the races which populate Algeria; Randau would create a 'pure-blooded' Algerian unity superseding racial and religious divisions.[15] Audisio too is not without contradictions: his genealogy referred to above, normally a proof of legitimacy, here contains heterogeneous elements; his exaltation of a Mediterranean race has a long life in his work, but he is forced later to admit that 'the amalgam of these different elements remained superficial, more apparent than real.'[16]

In fact, the elements which will not amalgamate in the European crucible are the Semitic ones. The valorisation of *Barbarie* over against France – and it is significant that the Latinised and 'original', therefore acceptable, Berbers are more easily assimilated by the *pieds-noirs* than the Arabs – hides the impossibility of fitting the Muslim population and civilisation into any of the Eurocentric visions imposed upon Algeria. One of the reactions to the problem, counterpoint to that of pretending not to see France, is to ignore the Arabs. In Camus' texts persons, unless otherwise specified, are of European origin. It has often been noted that, apart from a brief mention of the living conditions of the Arabs, Oran in *La Peste* is presented as though it were a European town, an 'Anywhere'.[17] Despite Audisio's fraternal feelings for the Muslims, in his 'map' of the Mediterranean the countries of the eastern seaboard have disappeared in a species of black hole.

The thrust of the Algiers School is towards universalism, towards a generalised humanism which is itself Eurocentric in nature.[18] As the origins of France's Algerian adventure become more remote, or more preferably forgotten, so myth replaces epic, and the mythic hero enacts the fantasms of the *pied-noir* imagination. For both Audisio and Camus, Ulysses is a favoured figure, but he appears as the traveller, the scourer of the Mediterranean, not as Ulysses the cunning. For Camus, Sisyphus the resistant symbolises man's daily struggle in and with life. The creative energy not connected with existing figures tends to create myths for today: *L'Etranger*, *La Peste*, *La Chute*, *L'Exil et le Royaume*, even *Le Premier Homme* that Camus was working on at the time of his death. The novels rewrite, in secular terms, the primary stories of the human condition. A traveller in space, the *pied-noir* hero is liberated in time, or more exactly through timelessness. The obligation to rediscover and reiterate the past has been lifted; all empires end in ruins, and the history composed of battles and colonisation dissolves into eternity. 'Eternity of the sea and eternal youth of the Mediterranean. Nothing has changed since ancient times, long before Rome.'[19] For Camus too, the sea and sky are the same ones that sparkled in the eyes of Sisyphus; like Audisio's peasants who perform the unchanging seasonal and ritual gestures, Camus' young people on the beaches of Algiers form a living frieze like the athletes of Delos reproduced on a Greek vase. Since antiquity is alive, caught up in the eternal present, its monuments become irrelevant, discardable, and the towns of Algeria are 'without a past'.[20]

The clearing of the ground effected by Audisio and Camus freed the Algiers School from the narrow particularism of Randau's circle. No longer obliged to recite an Algerian *credo*, confident henceforth that to be a *pied-noir* writer was not necessarily to be a provincial curiosity, their contemporaries were free to engage with the present and the supra-Algerian. The 1940s and the early 1950s, in effect, were the period when *pied-noir* literature appeared on the face of it to be least 'Algerian', when it seemed as though the colonial tributary might blend unobtrusively with the main current of metropolitan writing. Emmanuel Roblès, rather like Audisio, was a great traveller, principally at this time in the Mediterranean, siting his novels in and off the coast of Spain, Sardinia, Italy.[21] His 'flawed' heroes, ordinary men passionate in the defence of freedom, justice and sometimes women, combat authority, exploitation and the elements in whatever situation or country they find themselves. Jules Roy's first works engage with the Second World War during which he himself was a pilot with the British RAF; he too celebrates courage in danger and adversity, the fraternity of those battling to survive in the face of enemy attack, their own fear and the continual threat posed by the elements and by technical failure.[22]

But if *pied-noir* production at this time was able to show such a universal face, this was largely due to the operation of certain processes of displacement. Whether the *pied-noir* hero becomes involved in a European-inspired war or absents himself from Algeria, he still does not depart from what has become the tradition of his race. Commentators on the Algiers School are almost unanimous in referring to certain values endorsed within these texts. To take a random but typical example, Pierre Grenaud identifies the characteristics of 'rigour, courage, virility, renunciation'.[23] Whether the individual critic ascribes these characteristics to a racial blending, to the Mediterranean or to the sun, it remains the fact that they are primordially the virtues of the soldier and the frontiersman. It will have been obvious from the above comments how central to Roblès' and Roy's work of that time is the notion of conflict; indeed, looking back, Roy describes landing in Algeria in vivid and significant terms: 'it was with the joy of a conqueror that I would land my plane on the runways of El Aouia.'[24] Conquest, albeit displaced, is rarely far from the surface of these texts; take the examples of 'absurd man' in Camus' *Le Mythe de Sisyphe*:[25] Don Juan, *le conquérant* and the artist are conquerors all, whether of women, terrain or form. Sisyphus himself, grimly pushing his rock towards the summit of a mountain, only to have it roll back again, is eternally denied the conqueror's vista he seeks.

For this is a world of men, based on masculine, even macho, priorities, in which the greatest insult consists in questioning another's masculinity. Camus describes the distinctive 'morals' of the *petit blanc*: [26]

> You honour your mother. You see that your wife is respected in the street. You show consideration for pregnant women. Two of you don't attack a single opponent, because 'that's mean.' If anyone doesn't keep these simple commandments, 'he's not a man,' and that's that.[27]

If woman appears in these texts, it is most frequently in a minimalist function as the *repos du guerrier* (warrior's rest); desire tends to follow more circuitous paths, and may surface as the desire for conquest itself, sometimes as a (suppressed) desire for the foreign woman, or indeed as desire for Algeria – Martine Astier Loutfi has written of the 'curiously feminized vision of conquered countries'.[28] There are and have been women *pied-noir* writers: a list of the better-known would include the names of Magali Boisnard, Marie Cardinal, Isabelle Eberhardt, Lucienne Favre, Maximilienne Heller, Lucienne Jean-Darrouy and Elissa Rhais. But it is the masculine vision which has dominated, to the extent that writing the story of *pied-noir* literature imposes a choice of male heroes.

More commonly than with the sentimental adventure story, the combative energy of the male protagonist engages with the natural Algerian world, and one finds in texts depicting the landscape and climate of the colony a Manichaean division. The beaches offer the sensual delights of exposure of the body and of swimming: Camus' narrator speaks of 'possessing the waves'.[29] The littoral in particular is often represented as a *pied-noir* paradise of pre-lapsarian innocence and physical pleasure. But an equally constant and widespread tendency portrays Algeria as a land of malignant natural disasters. Characters in Bertrand and Randau die in flash floods, Camus sets his plague in Oran, Roy describes the hordes of locusts settling on vegetation in the Sahel, previously a mosquito-infested swamp. Truly Algeria is the equal of Egypt. In the absence of calamity, the European inhabitant still has to struggle with the stone and desert that cover much of the country, with its sheer magnitude, and with the inhospitable seas off the Barbary coast.

Discovering the motif of conquest, in its displaced forms, underlying the journeyings of the *pied-noir* hero, who even within the colony is constantly 'on the move', serves to substantiate for the Algerian context the point made by Ashcroft, Griffiths and Tiffin that:

> During the imperial period writing in the language of the imperial centre is inevitably, of course, produced by a literary elite whose primary identification is with the colonizing power... their claim to objectivity simply serves to hide the imperial discourse within which they are created.[30]

But while these authors are able to affirm that in the English literature which is the subject of their study, the texts of the literary elite 'inevitably privilege the centre, emphasizing the "home" over the "native", the "metropolitan" over the "provincial" or "colonial", and so forth,'[31] these latter distinctions are, as we have seen, less clear-cut in *pied-noir* writing where there have been attempts to promote 'Algerian' cultural independence and a disinclination to render unto France all her due. Much of the *pied-noir*'s insecurity stems from his or her status 'betwixt and between', and in fact the diverse characters of the literary tradition – the soldier, frontiersman and their successors the driver, traveller, wanderer, all statements of fictional identity – could be seen as surface manifestations of an underlying type: the mobile hero, caught between two countries yet 'at home', at ease, in neither. These are distinctive features of *pied-noir* writing, reflecting the distinctive circumstances of Algeria as a settler-empire in which full hegemony was never totally established.

It would appear, however, that there is one unfailing means of consolidating even so fragile an identity as that of the *pied-noir*.

> A number of recent analyses of the reciprocity between colonizer and colonized have concluded that colonial discourse establishes the colonized as the repressed and rejected 'other' against which the colonizer defines an ordered self and on to which all potentially disruptive psycho-sexual impulses are projected. . . . The process of self-fashioning required the continued presence of an 'other' so that the maintenance of subtle points of differentiation from the colonizer would continue to reproduce, not only the subordination of the colonized, but the superordination of the colonizer.[32]

The settler is able to colonise only by displacing the indigenous.

A striking feature of *pied-noir* literature is the constancy of the treatment meted out to the 'Arabs' over much of the 132-year period of colonial rule. They appear in subordinate, function-dominated roles: as goatherds, players of the flute, shoe-shiners, sellers of lemonade – mostly silent, frequently nameless, darkly loitering, their women 'dancers' or prostitutes. Curiously, these stereotypes are not properly *pied-noir*: they first become established in the pages of 'tourist literature'. This foreign model is itself clearly out of place in the colony; if the traveller can enjoy the picturesque, the settler ignores at his peril the demographic, political and social realities. But the perpetuation of the stereotypes is testimony to the utility of the purpose they serve: preserving distance between coloniser and colonised, and presenting the indigenous as fixed, trouble-free units.

However, if the figures remain the same, the attitude towards them evolves over time. In Bertrand, prejudice is openly, shamelessly displayed: the indigenous, called *bicots*[33] to their faces, are said to poison wells, kill and dismember Europeans; they look and smell dirty; they are stupid. Rafael, protagonist of *Le Sang des Races*, expresses a common sentiment: 'seeing all these people in their rags gave him the sensation of vermin crawling on his body.'[34] As already discussed, the Algerianists' need to recuperate Arabs into an Algerian unity is already a source of unease, if not sometimes of vermin-carried disease: the language used about the indigenous population exhibits a controlled restraint, in comparison to that of Bertrand and Musette, and the more violent expressions of racism are directed into anti-Jewish feeling.[35]

By the 1940s, however, the indigenous population has become an overt fictional problem, a source of embarrassment. Meursault's solution in Camus' *L'Etranger* is to gun down the Arab who is in his way. René-Jean Clot's mayor of Tipasa is content to verbalise a like desire: 'clear out!' he commands an Arab who may have seen too much.[36] But already with this novel, as in the narratives of Marcel Moussy, fictional expression of *pied-noir* attitudes is becoming critical portrayal of *pied-noir* prejudice. An even more radical departure from the norm is effected with Emmanuel Roblès' *Les Hauteurs de la Ville*,[37] whose indigenous narrator-protagonist, Smaïl, in an intertextual reply to *L'Etranger*, shoots a European Nazi agent who is displacing Arabs to work in Germany. Smaïl's 'revolt' is not entirely convincing as a general statement, being based largely on a rather adolescent injured pride; moreover he is hostile and alienated, constantly observing himself in the mirror as if for reassurance. Nevertheless, he is a highly significant figure in initiating a movement which will continue to gain ground: the indigenous, no longer expendable, able to be conjured away, will henceforward occupy more literary space, in a fictional reversal of the colonising process itself. Richard Zrehen has noted how the Algiers School reached the limits of the land: 'literary attention, previously occupied explicitly with the land, finds a new objective. Gradually turning away from both Black and White Africa, literature is seduced by the call of the sea.'[38] If the Europeans are not literarily pushed off the edge, it would yet seem that the only other place for their 'story' to continue would be on the other side of the sea.

Meanwhile, it appeared that Paris had turned to Algeria. After the collapse of 1940, French literary life found a refuge on the other side of the Mediterranean. The review *Fontaine* had been founded in 1939; it was the first to protest against capitulation in 1940, and during the Occupation it published some of the most celebrated of metropolitan writers: Aragon, Eluard, Mauriac and the white French West Indian St-John Perse. *Fontaine* was only one among a cluster of journals founded at this period: *Rivages* in 1938, *L'Arche* in 1944, *Forge* in 1946, *Soleil* in 1952, the regrettably short-lived *Terrasses* in 1953 – the titles generally carry a symbolic significance. Typically Janus-faced, the reviews looked two ways: towards France, obviously, even after the focus of literary activity had shifted back to the capital, but also towards a generation of indigenous writers, and it was here that the names of, for example, Mohammed Dib, Mouloud Feraoun, Mouloud Mammeri and Kateb Yacine first came to the notice of the *pied-noir* public. Under European licence, admittedly, but in what was to prove the final phase of French presence in Algeria, these reviews, the final element in generic *pied-noir* literary production, manifested a vitality and a richness which seemed to have actualised the liberal dream of Algeria as a cultural crucible. The 'manifesto' of the single issue of *Terrasses* epitomised, in 1953, such hopes for the future.[39]

> At a time when the surest values of Western civilisation are being questioned, it is the aim of the review *Terrasses* to represent this country in its unique situation as a cultural crossroads, and to contribute by all means within its power to free man from his present confusion.
>
> Bringing together the thought of the Mediterranean and that of the desert, the Oriental message and the Roman, European and Islamic structures, Algeria is proving to be one of the most ample crucibles of contemporary literature.[40]

DEATH OR EXILE?

One year later came the beginning of the end, with the outbreak of the war of independence. From that point onwards the war within, whose name would not be spoken, came to dominate *pied-noir* writing into the 1960s and 1970s. The initial fictional reaction took various forms: a stunned silence, as it were, as though the text could not comprehend or make sense of what was happening – often the writing-out is postponed until after the war. Some novels of the war years foreground the disquiet and discomposure of the *pied-noir* character or narrator, exploring why the relationship between European and indigene might have failed, attempting to heal the rift, to re-establish, at least in fictional terms, understanding and justice. Cases in point are the story 'L'Hôte' in Camus', *L'Exil et le Royaume* (1957), Jean Pélégri's *Les Oliviers de la Justice* (1959) and Emmanuel Roblès' rewritten, republished version of *Les Hauteurs de la Ville* (1960), in which Smaïl appears as more sympathetic a character, more clearly representative also of the subjugated, humiliating situation of his people.

But the selfsame authors may turn their texts away from Algeria. Roblès, as the conflict draws on, ranges further afield: to Mexico with *Les Couteaux* (1956), to Japan with the title story of *L'Homme d'Avril* (1959), back to Italy and the Second World War with *Le Vésuve* (1961). More significant a textual departure is made by Camus, the greater part of whose previous work had been Algerian-based. Fiction is now set for the first time in Paris (with the story 'Jonas' in *L'Exil et le Royaume*), the Netherlands (*La Chute*, 1956) and Brazil ('La Pierre qui Pousse' in *L'Exil et le Royaume*). Interestingly, these are all ex-colonial contexts. The tone in Paris and Holland is sombre, the only colour in the greyness of the Netherlands being provided by the golden dream of the Dutch, those nostalgic colonisers. 'They are dreaming. . . . They have gone thousands of miles away, to Java, the distant island.'[41] A narrative colonial dream in counterpoint depicts Brazilian Iguape as a place untainted by colonial conflict, where the French spirit in the person of the engineer D'Arrast has a distinct mission to fulfil and is welcomed both by the notables and by the humblest and poorest of the people.

Both as eschewal of painful colonial reality and as displacement of the colonial dream, the search for a new world may also be interpreted as a positive gloss put upon the necessity for eventual exile, a fictional acknowledgement that the saga of the *pieds-noirs* will end in diaspora. Like the Algiers School on the edge of Algeria, these late texts hover at the point of departure, looking out towards France but without naming her, without admitting that the 'French leave' may be definitive. Camus' Yvars, in the conclusion of the story 'Les Muets', gazes out across the Mediterranean: 'he wished Fernande and he were young again, and they would have gone away, to the other side of the sea.'[42] The narrator of Pélégri's *L'Embarquement du Lundi* does not in fact embark. 'It would be too easy to leave! To go and seek salvation elsewhere . . . First of all I had to keep writing until I had filled this exercise book. Find salvation in the town where I live.'[43] However, in Pélégri's later *Les Oliviers de la Justice*, the child, symbol of the *pied-noir* future, absent in France during the distressing narration-time, does not return to Algeria.

The fictional alternative to exile appears to be death in Algeria. Already in 1949 Clot's *Fantômes au Soleil* had clearly expressed the options – the mayor declares: 'I'm not leaving. Other people can take the suitcase. I've chosen the coffin.'[44] Throughout *pied-noir* literature runs the tendency to write two stories from the same point of departure, which take opposite directions, and in which the one questions the other. Bertrand looks backwards to the ruined grandeur of Rome at the same time as he maps out a future for the races of the colonial empire in Algeria. Doubt is cast on Randau's 'Algerianism' by his own characters. Sophie Peterhof (Si Yahia), modelled on the traveller and writer Isabelle Eberhardt, realises how different from her own is the attitude of her Algerianising friend Cassard:

> He had woven a conspiracy around her: a fanatical Algerianist, he would prevent her from uttering the truth about the lowest strata of the indigenous proletariat, from taking pity on the weak, from denouncing the dramas of the indigenous village, from joining, spontaneously, in the lament of the despairing.[45]

Camus continually presents *l'envers et l'endroit*[46] of experience, both in Algeria and in general terms. Overall, parallel to the story of a new race, of expanding frontiers, of self-confidence and faith in the future, runs a story of death and decomposition which questions identity, legitimacy of presence and the very possibility of a future.

As the tension of the political situation mounts in the mid-twentieth century, this sub-story comes to be written out all the more clearly. To take the example of Roblès who, as we have seen, emphasises the humanist values of dignity, courage, fraternity, in all his very considerable literary production

Roblès based only five texts in Algeria: *Les Hauteurs de la Ville*; the story 'L'Attentat de la Banque Levasseur' in *La Mort en Face*, *Federica* (1954); the story 'Le Rossignol de Kabylie' in *L'Homme d'Avril*; and *Saison Violente* (1974). If one excepts the autobiographical *Saison Violente*, the imaginative work portrays Algeria in the darkest of colours. It is a place to be escaped from if possible, as the characters in *Federica* attempt to, and as Jauffé of the story 'Le Grain de Sable' in *L'Homme d'Avril* has managed to do. To escape, Jauffé has to survive a plane crash, since one escapes only at the risk of death or arrest: Smaïl in *Les Hauteurs de la Ville* is stopped as he waits for a train to Morocco. But remaining in Algeria also brings death (*Federica* and *Le Rossignol de Kabylie*). The *pied-noir* men, in these novels, even the 'heroes', are robbers, thieves, murderers, dealers in contraband and in false papers. The secondary characters, men and women, are frequently grotesque, malformed: in *Federica*, outstandingly, they are stunted, hunchbacked, have lost here an ear, there an eye. But if *pied-noir* status is criminal and abnormality the norm, the Algerian 'story', whether the plot of the novel itself or the story associated with Algeria written into the novel, is often macabre or apocalyptic, stemming from and portraying an imagination 'living in the last days'. Georges, in *La Croisière* (1968)[47] tells the story of a man from Oran who, drunkenly looting, discovered the mummified bodies of some nuns in a crypt and, for a joke, carried one away and placed it where his fellow soldiers would find it in the morning.

The scenario of horror, destruction and apocalypse recurs constantly in Clot's *Fantômes au Soleil*, set, ironically, in Tipasa, Camus' Eden of before the Second World War. The creative energies of Rafael Perez are channelled into the construction of a coffin; Perez is shot by the (*pied-noir*) postman while taking a nap therein, and his undiscovered, putrefying body is eventually incinerated when his workshop is set alight. Fire rages again in Pélégri's *Les Oliviers de la Justice*: it destroys the coloniser's farm, and funeral wreaths made of celluloid ignite spontaneously on a *pied-noir* coffin. And perhaps the most vivid of *pied-noir* nightmares is given form in Camus' story 'Le Renégat' in *L'Exil et le Royaume*, the story of a European enslaved, mutilated, tortured and finally killed by his indigenous masters. If death is omega, it is perhaps because it was also alpha, as the work of fiction frequently acknowledges. The narrator of *Fantômes au Soleil* takes up position in the shadow of the cenotaph; the narrator of Pélégri's *L'Embarquement du Lundi*, lately released from an asylum, begins writing in a cemetery. And curiously, in a reversed retrospective revaluation, burial in Algeria is advanced as a justification for presence there. 'Now, my father was part of this land – there was, and there always would be, this root planted in its native soil,' writes Pélégri, referring to the father's body.[48] Roy, referring to his 'origins', echoes: 'the only ones which are important to me are the farm where I lived as a child and the cemetery where the dead of my family rest.'[49]

AND AFTERWARDS

So the fictional visions materialise, and the *pieds-noirs* in their vast majority leave Algeria with the coming of independence. Yet in the decades following, the exiles return time and again in imagination, as though the literature could not tear itself away from the country. Clearly there are some, like Roblès, who do not look back, and whose references to a past heavy with memories surface only occasionally in a pointed detail or in a constant, haunting 'nostalgia for the sea'.[50] The tendency of the texts which return is to engage with a double-bind situation: the 'looking both ways' of earlier literature has become an *impasse* of guilt and trauma – 'damned if they do, damned if they don't'. On the one hand the destinies of the European and the indigenous are seen as helplessly, hopelessly intertwined, singularly unimaginable; on the other, the end is death. In between, the protagonists, never indifferent, love, challenge, kill one another; even in death their bodies unite.

Georges, the son of the European in Pélégri's *Le Maboul* (1963), kills the nephew of the *maboul* or madman and is killed by the latter; the bodies of the two young men are juxtaposed in the same well. The Arab and the European in Jean-Pierre Millecam's *Sous Dix Couches de Ténèbres* (1968), figurative brothers, love the same woman. Millecam's *Et Je Vis un Cheval Pâle* (1978) develops a similar fraternal relationship between Geoffroy and Salah Eddine, 'best enemies', divided in the war of independence. Salah Eddine entrusts to Geoffroy's care his cousin Nejm'el Arab. Obliged to absent himself from the camp, Geoffroy returns to the mess where he eats, unknowingly, the heart of the boy who has been tortured, served up to him by his fellow soldiers. With this 'naming' of betrayal and cannibalism – however innocent – at the heart of relationships based on colonisation one reaches the last circle of the infernal imaginings of the *pied-noir*. Indeed, both Millecam's and Pélégri's novels return obsessively to the same events, the same scenes, their visions patently apocalyptic, with Millecam's pale horse of Revelation and Pélégri's *Les Monuments du Déluge* (1967).

Jules Roy's originally six-volume *Les Chevaux du Soleil* published between 1968 and 1972[51] is also a novel of return, but more distanced, more serene, a rich and splendid frieze in which the background figures – the soldier, the *colon*, the teacher – act out their parts again in a re-creation of the *pied-noir* epic from 1830 to 1962. This time they are seen, not alone, but in constant relation to the indigenous population, as generation after generation seeks itself and the other in bonds of friendship, admiration, rivalry in love, sexual attraction. It is a series of novels which seeks to write an impossible story, and which is both a form of nostalgia and regret and a tale of repeated failure to meet in the time appointed. Significantly, this fiction is uncertain as to its own end. The two protagonists of the last generation, *Doppelgänger* in relation to each other and to Jules Roy, are Hector König and the Captain de Roailles: the former is shot at the gates of a cemetery (by which 'side' one does not know); the latter, who has fallen impossibly in love with the half-French, half-Arab Raissa and deserted his regiment, observes the euphoria of Independence Day and in his interior monologue anticipates his court-martial. The conclusion is an attempt not to write the end, but to step aside from 'real' ineluctable time and to postpone judgement, since in order fully to understand de Roailles's position one would need to 'go back to the beginning', and in this way too it represents a passage from Revelation to Genesis, a justification in the perspective of future judgement.

> There is not, and never has been, such a thing as Algerian literature. By this I mean that there does not exist, or at least not yet, an autonomous, specific literature defined by the existence of an Algerian language, race and nation.... And yet there are and always have been men born, nourished and formed in this land, who exercise the profession of writer.[52]

So Gabriel Audisio wrote in 1953. With the benefit of hindsight one can distinguish a *pied-noir* literature, understanding by that a literature written by men and women who were, like Audisio himself, of often mixed European race, who wrote in French and who lived in Algeria. Born in the last years of the nineteenth century, it is dying with the twentieth. For now that literature has become a tradition without a hearth, a landless literature deprived of its title and of the need for justification which was its initial and sustaining inspiration. Conquest and settlement had been written out and transformed, in the texts of the *pieds-noirs*, into a fierce desire for cultural distinction and independence; this was the legacy bequeathed to the Algerians. For nearly a century Algeria produced the writing of a minority elite which, like the two-faced Janus, re-enacted its past and imaged its future. As a source for the study of the cultures of colonialism and imperialism, the literature of Algeria has few equals.

NOTES

1 The original manuscript of this chapter was completed shortly before the tragic death of Rosemarie Jones. Her executors, Mr and Mrs D.H. Compton, have given permission for its publication. Minimal editorial amendments have been made by Paul White with the help of Michael Heffernan, University of Loughborough. All of Dr Jones' original material has been retained: notes 17 and 18 and parts of notes 4 and 9 have been added by White and Heffernan.

2 A number of these are quoted in Assia Djebar, *L'Amour, La Fantasia*, Paris, J.-C. Lattès, 1985.

3 G. Audisio, *Jeunesse de la Méditerranée*, Paris, Gallimard, 1935.

4 L. Bertrand, *Les Villes d'Or*, Paris, Fayard, 1921, p. 5. Translations of quotations are by the present author. A monthly journal, *l'Afrique Latine*, was established in Algiers in 1921 'for the service of French art and literature in North Africa'. Although it is not clear who edited the journal, quotations from Bertrand's writings and speeches were given prominence. Only twelve numbers appeared, and publication ceased in November 1922.

5 ibid., pp. 8–9.

6 For recent editions see, for example, 'Musette', *Les Amours de Cagayous*, Algiers, Editions Méditerranée Vivante, 1953; *Cagayous à la Caserne*, Algiers, Editions Méditerranée Vivante, 1952; *Le Mariage de Cagayous*, Algiers, Editions Méditerranée Vivante, 1952.

7 P. Siblot, 'Pères spirituels et mythes fondateurs de l'Algérianisme', in *Itinéraires et Contacts de Cultures, Vol. 7, Le Roman Colonial*, Paris, l'Harmattan, 1987, p. 51.

8 'Musette', *Cagayous à la Caserne*, op. cit.

9 R. Randau, *Les Algérianistes*, Paris, Albin Michel, 1911. See also R. Randau, 'La vie intellectuelle dans le Nord de l'Afrique', *l'Afrique*, 1940, no. 162, pp. 1167–9.

10 R. Randau, *De Treize Poètes Algériens*, Paris, Albin Michel, 1920.

11 A. Rosfelder, *Les Hommes-Frontières*, Paris, Domat-Montchrestien, 1952, p. 3.

12 These are only a selection of the numerous authors published at the time: I have concentrated, probably with injustice, on those I considered to be both representative and quite well-known.

13 Audisio, op. cit., p. 24.

14 A. Camus, *Noces*, 1938; reprinted in A. Camus, *Essais*, Paris, Gallimard, p. 74.

15 Randau, 'La vie intellectuelle dans le Nord de l'Afrique', op. cit.

16 G. Audisio, *Feux Vivants*, Paris, Rougerie, 1957, p. 26.

17 This view has been taken by Conor Cruise O'Brien, who also points out that 'Camus's position does not permit him to look coolly and analytically at the reasons why the Arabs have to be got out of the picture. He therefore leaves them out without admitting that he is leaving them out.' C.C. O'Brien, *Camus*, London, Fontana, 1970, p. 48.

18 However, Edward Said has recently challenged this universalist view, particularly of Camus. Said argues that 'Camus's writing is informed by an extraordinarily belated, in some ways incapacitated, colonial sensitivity' which reflects the French Imperial venture in Algeria: 'To resituate *l'Etranger* in the geographical nexus from which its narrative trajectory emerges is to interpret it as a heightened form of historical experience.' To accept Camus' 'universalism' is to deny the specifically colonial political context of his work. E. Said, *Culture and Imperialism*, London, Vintage, 1994 (the quotations above are drawn from pp. 213 and 224); also E. Said, 'Narrative, geography and interpretation', *New Left Review*, 1990, no. 180, pp. 81–97.

19 G. Audisio, *Sel de la Mer*, Paris, Gallimard, 1936, p. 147.

20 A. Camus, 'Petit guide pour des villes sans passé', in *L'Eté, Essais*, op. cit., pp. 845–50.

21 For example, E. Roblès, *Travail d'Homme*, Algiers and Paris, Charlot, 1945; *La Mort en Face*, Paris, Seuil, 1951; *Cela s'Appelle l'Aurore*, Paris, Seuil, 1952.

22 For example, J. Roy, *La Vallée Heureuse*, Paris, Charlot, 1946; *Comme un Mauvais Ange*, Paris, Charlot, 1947; *Retour de l'Enfer*, Paris, Gallimard, 1951.

23 P. Grenaud, 'La littérature du soleil', *La Table Ronde*, 1955, no. 95, p. 135.

24 J. Roy, *Autour du Drame*, Paris, Julliard, 1961, p. 119.

25 Camus, *Essais*, op. cit., pp. 89–211.

26 A term used to denote the European lower middle class and proletariat, as distinct from the opposite pole of *pied-noir* society, *le gros colon*.

27 Camus, *Essais*, op. cit., p. 72.

28 M.A. Loutfi, *Littérature et Colonialisme*, Paris and The Hague, Mouton, 1971, p. 61.

29 Camus, *Noces*, in *Essais*, op. cit., p. 57.

30 B. Ashcroft, G. Griffiths and H. Tiffin, *The Empire Writes Back*, London, Routledge, 1989, p. 5.

31 ibid., p. 5.
32 D. Cairns and S. Richards, *Writing Ireland: Colonialism, Nationalism and Culture*, Manchester, Manchester University Press, 1988, pp. 8, 10.
33 There is no direct translation: a pejorative term, one of many, used to designate the indigenous. The profusion of epithets used of the self and the other is a worthwhile field of study. Various etymologies have been proposed for the term *pied-noir* itself. Two among the more plausible are: that it referred to the black boots of the French soldiers; and that it described the feet of the settlers, stained from treading grapes.
34 L. Bertrand, *Le Sang des Races*, Paris, Ollendorff, 1899; republished, Paris, Robert Laffont, 1978, p. 202.
35 This hierarchy of scapegoats is found not only in Randau; mention has been made of Cagayous' anti-semitism, and Bertrand's characters can also be more aggressive towards Jews than towards Arabs.
36 R.-J. Clot, *Fantômes au Soleil*, Paris, Gallimard, 1949, p. 357.
37 E. Roblès, *Les Hauteurs de la Ville*, Algiers, Charlot, 1948.
38 R. Zrehen, 'Ecrit au soleil: la littérature de 1830 à 1962', in E. Roblès (ed.), *Les Pieds-Noirs*, Paris, Philippe Lebaud, 1982, p. 109.
39 *Terrasses* was founded by Jean Sénac; its editorial committee included Mohammed Dib, Jean de Maisonseul, Mouloud Mammeri and Jean-Pierre Millecam.
40 *Terrasses*, vol. 1, no. 1 (June 1953), p. 1.
41 Reprinted in A. Camus, *Théâtre, Récits, Nouvelles*, edited by R. Quilliot, Paris, Gallimard, 1962, p. 1482.
42 A. Camus, *L'Exil et le Royaume*, in *Théâtre, Récits, Nouvelles*, op. cit., p. 1608.
43 J. Pélégri, *L'Embarquement du Lundi*, Paris, Gallimard, 1952, p. 117.
44 Clot, op. cit., p. 274.
45 R. Randau, *Cassard le Berbère*, Paris, Albin Michel, 1921. For a study of the actual Isabelle Eberhardt see A. Kobak, *Isabelle: The Life of Isabelle Eberhardt*, London, Chatto and Windus, 1988.
46 'The right side and the wrong side', title of his first published collection of essays, reprinted in *Essais*, op. cit., pp. 1–50. The English translation of 'betwixt and between' occludes the antithesis at the core of the French expression.
47 E. Roblès, *La Croisière*, Paris, Seuil, 1968.
48 J. Pélégri, *Les Oliviers de la Justice*, Paris, Gallimard, 1959, p. 206.
49 Roy, *Autour du Drame*, op. cit., p. 10.
50 E. Roblès, *L'Ombre et la Rive*, Paris, Seuil, 1972, p. 8.
51 J. Roy, *Les Chevaux du Soleil*, revised edition – six volumes in one, Paris, Grasset, 1980.
52 G. Audisio, *Visages de l'Algérie*, Paris, Editions des Horizons de France, 1953, p. 99.

9

FRIULANI NEL MONDO
The literature of an Italian emigrant region
Federica Scarpa

INTRODUCTION

The north-east Italian region of Friuli-Venezia Giulia is a classic example of an emigrant region. Its history of emigration – both to other countries and to other regions of Italy – extends back four centuries. Moreover, its migratory experiences encompass all major forms of emigration, including permanent emigration with no return, temporary emigration with eventual return and seasonal migration.[1]

The two geographical components that in 1948 were united to form the region are very different from the historical, cultural, linguistic, socio-economic and migratory points of view. On the one hand the sub-region of Friuli (see Figure 1), consisting of the two provinces of Udine and Pordenone, joined the new-born Italian nation in 1866 and participated in Italian migrations in their peak years, experiencing the typical phenomena of rural and mountain depopulation. On the other hand the much smaller sub-region of Venezia Giulia, consisting of the two provinces of Trieste and Gorizia, only joined the Italian nation in 1918 after being a very thriving part of the Austro-Hungarian Empire and an area of immigration; Venezia Giulia experienced out-migration, though of a very different nature from Friuli's, only after the Second World War.[2]

With this distinction in mind, this chapter will focus on the sub-region of Friuli, given the uniqueness and length of its migratory experience. Migration being so embedded in the cultural tradition of this corner of Italy, Friulian literature is consequently rich in references to emigration, confirming Gaetano Rando's suggestion that before the Second World War the migration theme did not emerge in mainstream Italian literature but was only found in the work of minor or regional writers.[3] The literature on migration which will be analysed here ranges from novels and poetry where migration features as a main or subsidiary theme, to personal documents such as auto-biographical accounts and letters recording the subjective impressions of migrants themselves. The aim of this chapter is to analyse the extent to which such sources, both in spite of and because of their subjectivity, provide

141

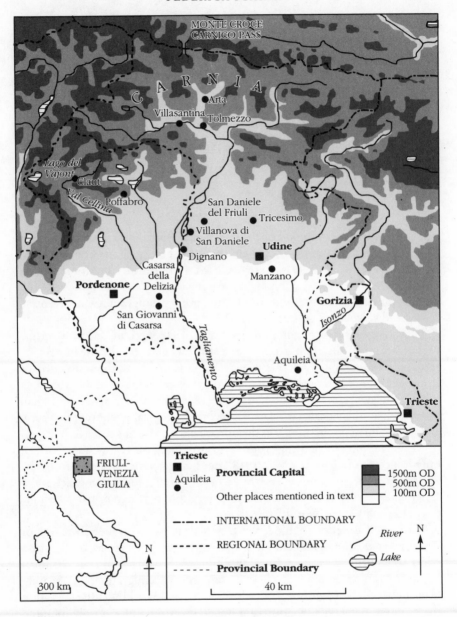

Figure 1 Friuli-Venezia Giulia and the location of places mentioned in the text

insights into the private and individual elements of the social phenomenon of mass migration and, at the same time, into the distinctive features of Friulian migration. This involves not merely using literary evidence to produce an annotated account of Friulian migration, but also examining the more subtle ways in which Friulian identity has been asserted by the creation of a 'heroic' vision of Friulian migrants and of community responses to migration from and back to the region. Migration literature has to some extent played a role in forging Friulian self-identity as a 'submerged' region of Italy from the eighteenth to the mid-twentieth centuries, and in romanticising the 'myth' of *friulanità* (Friulian-ness) through the heroic characterisation of Friulian emigrants as hardy souls ejected from, but forever linked to, their native land.

A BRIEF HISTORY OF FRIULIAN MIGRATION

From the second half of the sixteenth century and for the next 150 years, the Friulian peasants from the northern mountain area of Carnia (Figure 1) migrated seasonally to save enough money to ensure a living in their old age.[4] Their destinations were the countries of the Austrian Empire and Germany where they were active as wool and linen weavers and as itinerant salesmen. A second, although less consistent type of migration, was the flow of largely unskilled labour from the Friulian lowlands to the neighbouring Venetian regions to the west, to the County of Gorizia to the east, and to the coastal lands which had belonged to the Patriarchate of Aquileia until 1420, when its territory was annexed by the Republic of Venice.

In the second half of the seventeenth century these migratory flows increased in scale and became permanent, spreading towards Austrian East Friuli, Istria and Trieste. The reasons for this were two-fold: economic crisis in the Republic of Venice, and demographic growth in the Friulian mountains which strained even further the area's meagre economic resources. When in 1797 Venetian Friuli became Austrian, the migratory flow towards the other countries of the Empire increased. The migrants came from nearly all districts of Friuli and were mostly labourers and itinerant salesmen. Slightly later, mid-nineteenth-century industrial development in the cities of Lombardy and Veneto caused the disappearance of the traditional craftsmanship of Friuli; typical migrants of this period were masons, stone-cutters and kilnmen who worked on building sites and in road and railway construction in various European countries.

Following Friuli's annexation to the Kingdom of Italy in 1866 there was no major change in the destinations of migratory flows, but they became more long-term or permanent. After 1876, however, Friulian migration started to have as its major destination South America, especially Argentina, where farmers and masons were in great demand. The first migrants to the United States and Canada also left during this period. Between 1881 and 1914 the migratory flow from Friuli reached its peak as a result of the deteriorating conditions for agriculture, especially in the uplands. Destinations were many: the traditional countries of Central Europe, the Americas, and, after 1890, Switzerland.

The outbreak of the First World War briefly staunched the Friulian haemorrhage, but it regained momentum after 1918. This new phase of emigration was characterised by fewer but more skilled emigrants, who left on their own or in small groups, thus avoiding the mass recruitments of previous years which had often resulted in exploitation and difficulties over integration in the host communities. Many of these new migrations were permanent. The migration flow out of Italy stopped almost completely after 1925, when Mussolini promoted internal migration of labour to the massive public-works programmes which were initiated to reduce unemployment. Out-migration was only encouraged later through recruitment campaigns to populate Libya and other parts of the 'Italian Empire'.

The economic marginality of Friuli and heavy losses of territory to Yugoslavia along the Italian north-eastern frontier after the Second World War accounted for the resumption of Friulian emigration which was particularly intense during the decade 1950–60. In this new phase, however, the destination countries were predominantly European and the migration was a more temporary phenomenon. Some emigration to Australia took place during the 1950s, but this was heavier from Venezia Giulia than from Friuli.

Returns started to become a constant feature of the Friulian migration scene during the 1960s, and in 1968 returnees from abroad outnumbered outgoing migrants for the first time. The majority of returnees were economically active and found employment in small and medium-sized enterprises in the construction and manufacturing sectors. These enterprises are dispersed over rural areas which in the past provided migrant labour. Such changes in local labour demand have acted both as a pull factor for migrants abroad, who have found jobs near their former places of residence, and as an alternative for those who were about to leave, therefore reducing the necessity to migrate. Diffused industrialisation in Friuli, therefore, contradicts the normal view of the unlikelihood of economic development in areas of migratory exodus and return.[5] The 1976 earthquake has had the effect of accelerating this process of modernisation and return: small building companies have proliferated in response to the demand created by reconstruction work.

SEASONAL MIGRATION

As already mentioned, seasonal migration has characterised Friuli at least since the sixteenth century. Interestingly, one of the earliest documents of Friulian literature contains a suggestive sing-song, which can be dated back to the thirteenth century in its oral form, in which reference is made to a lonely wife whose husband has gone to France and who is only too willing to betray him.[6]

More recently, a detailed account of the pattern of Friulian seasonal migration during the nineteenth century is provided in Lodovico Zanini's book *Friuli Migrante*. Though published in 1937, the book has a section on Friulian migration in the first half of the nineteenth century[7] where the writer, himself a migrant to Bavaria, relates stories heard from his father, a kilnman working in Austria, and from his grandmother, whose children and grandchildren had all been migrants. The protagonists of Zanini's stories are workers (kilnmen, road and railway builders, masons, stoneworkers, woodmen), craftsmen (potters, *terrazzo* makers, knife-grinders) and itinerant salesmen; they come from the upland areas north of Udine, where Zanini's native village, Villanova di San Daniele, is located (Figure 1). The destinations of the migrants were the Austro-Hungarian Empire and other German-speaking countries: Austria, Slavonia, Croatia, Bohemia, Hungary, Prussia and Bavaria. They would leave after Easter and return in early autumn, bringing back their savings to their parents, wives, children and younger brothers. Among Austrian employers, hard-working Friulian masons and labourers were more in demand than Austrian ones who, 'with their imperturbability and the inevitable pipe in their mouths, drove their Austrian employers mad'. During the winter the 'bosses from Germany' would go all the way to Friuli to the 'market-places of men' to recruit workers for the following season. Some of these market-places specialised in specific occupations: Tricesimo for kilnmen, Villasantina and Tolmezzo for builders, Carnia for woodmen. The work contract was verbal, sealed by a handshake and a deposit to the worker: the day usually ended in the local *trattorie* where employers and employees would treat themselves to stockfish, tripe and *polenta*.[8]

The migrants chronicled by Zanini went on foot to their work destinations. Bad weather could put their very lives at risk as they had to cross many mountain passes during the avalanche season. Some villages in Carnia where the through-traffic of migrants was most intense organised a regular cart service to the foot of the Monte Croce Carnico pass (1360 m) on the border with Austria: sturdy local women would then help the migrants carry their sacks to the top of the mountain. The migrants carried the tools of their trade (picks, axes, planes, set-squares) or pushed a wheeled vehicle holding goods to be sold such as pots and pans, ironware and items made of wood. The *terrazzo* makers would even carry a load of carefully selected pebbles for their terraces! The migrants going to Graz or Slavonia had to navigate stretches of the rivers Drava and Sava on rafts and risk their lives on the rapids. When returning home with their hard-earned savings, brigands were a threat. Although in these references migrants are somewhat romantically characterised as heroic adventurers, who can deny the real migrants their untold experiences of hardship and danger?

The theme of seasonal migration is at the centre of a short story by Caterina Percoto,[9] set in Carnia and published for the first time in 1844. The story hinges upon the simultaneous return to the village of Arta of two brothers from different types of migration. Giacomo, a migrant to an unnamed place south of Carnia, returns after three years spent working for a timber dealer. His brother Giovanni, a herdsman, returns from a three-month stay in the mountains north of the village where he had pastured the cattle of a better-off fellow villager during the summer. Giacomo has brought back some money to buy timber in Carnia on behalf of his employer; besides this money, he has a small sum he has managed to save in the hope that, with some help from Giovanni, he might be able to return permanently to Arta, set up a small timber business of his own and marry the girl who has been waiting for him. Giovanni, however, has set his sights on his brother's savings to pay a debt without having to sell the only two cows the family has been left with. As soon as he arrives, Giacomo realises the great poverty of his family and consequently spends both his money and his employer's to relieve it, thereby renouncing his dream of settling in Arta.

Carlo Sgorlon's novel *Gli Dèi Torneranno* (The gods will return), which is set soon after the Second World War, features the summer trade of the *sedonere*, women itinerants from the mountain areas of Friuli who sold wooden utensils – spoons, ladles, egg-cups, whisks – that the men from their village had made during the winter. They would go in pairs, dressed in black, pushing a cart laden with their goods. They were paid both in cash and in kind: 'Fantastic savers, they would never eat in a tavern and practically lived out of people's charity.'[10] The *sedonera* of the novel – whose nickname is *la Clautana*, from Claut, her village of origin – travels on her own as the woman who used to be her companion died after accidentally slipping into freezing river water. She is also a widow whose husband, a woodman, died when struck by a falling tree. Afterwards, she had to leave her young daughter with her sisters-in-law and start her itinerant trade which took her far from her village, to the cities of Verona, Trieste and Vicenza, and even to Vienna. For Simone, the returnee who is the protagonist of the novel and who had been her lover before migrating, *la Clautana* is a symbolic figure representing 'all the *sedonere* of the valley, all the lonely women separated from their men through emigration or death, and who had felt an agonising lust on a Saturday night and sometimes opened their door to let it in'.[11]

THE SOCIO-ECONOMIC CONTEXT OF MIGRATION

Friulian literature abounds in references to the conditions of poverty, un-employment and virtual starvation in the rural and mountain areas of the region. Zanini's grandmother used to tell him of the times when 'in the whole village there were only three pairs of shoes to be counted: and the only publican of the village would sell yearly hardly more than 50 litres of wine.'[12]

As a means to avoid starvation, smuggling was a common practice: one of Zanini's uncles was killed by Italian customs officers whilst crossing the border illegally to buy salt in Austria. After the annexation of Friuli to Italy in 1866, salt became expensive because it was a state monopoly. In Sgorlon's *Gli Dèi Torneranno* old Geremia tells Simone that the consequent lack of salt in the peasants' diet caused the outbreak of the deficiency-disease pellagra which raged among them until the end of the century. The Italian government is portrayed in Geremia's stories as only one of the innumerable oppressors of Friuli. To be under the rule of Venice, Austria or Italy did not make any difference to the peasants except for the different ways they had to pay taxes.[13]

Sgorlon makes further references to smuggling in *Il Trono di Legno* (The wooden throne), set at the beginning of the twentieth century. Were Giuliano not to migrate, his only possibilities of earning a living would be as a smuggler or a woodcutter, as he has no land of his own to cultivate. His main motivation to leave, however, is his romantic desire to live an adventurous life outside the village and to find, somewhere in 'the North', his father, the mythical 'Dane' whom he can hardly remember. Giuliano does not want to cut wood or to end up a migrant working as a kilnman or a mason – the destiny of the migrants who fill the north-bound train he catches, continuing in the new Austria their fathers' travels along the routes of the former Austro-Hungarian Empire.[14]

The nineteenth-century migratory flows to that Empire are a thread running through Friulian literature. In Chino Ermacora's diary of his sentimental journey in the 1950s to the Americas and Africa in search of Friulian migrants who had settled permanently abroad, he found families from Poffabro (a mountain village in the province of Pordenone) who had settled in Philadelphia where they had become the most eminent builders of the city and were said to be descendants of the migrant masons who built the palaces of Budapest and the castles of the Rhine.[15] In *Gli Dèi Torneranno*, Simone remembers the migration of two men from his village, one to work on the construction of a bridge in Prague, the other to work on the Trans-siberian railway.[16] The construction of the latter recurs in Friulian literature on migration: Ermacora calls this 'the legendary period of our migration', when whole villages were working abroad constructing bridges and railroads across steppes and forests overrun by wolves.[17]

Whilst in the conventional 'scientific' literature on Friulian migration departure is overwhelmingly interpreted by reference to economic factors, from Friulian literature we also learn of more idiosyncratic reasons to leave. Autobiographical accounts such as letters written by migrants, though not literary in the strict sense of the term, are characterised by an absolute spontaneity which gives powerful insights into the more personal background of the migrants. In the volume *Il Pane degli Altri* (The bread of others), Bongiorno and Barbina collected letters written by Friulian migrants between the late 1940s and the early 1960s.[18] Among the personal motives for migrating are two cases of women getting married to foreign servicemen whom they met in Friuli during the war: one married an American airman and lives in the United States, whilst the other married a New Zealander.[19] Both women write home to reassure their relatives of their happiness and comparative wealth. The woman living in New Zealand provides an interesting insight into the narrowness of Italian provincial mentality by expressing the desire to pay a visit home one day to show off her husband and children to those 'stupid gossips' who said that she had married a streetsweeper only to avoid ending her days as a spinster.

There are also instances of Friulian migration during the nineteenth century for reasons other than dire necessity. Emilio Franzina's *Merica! Merica!* contains letters written between 1876 and 1902 by peasants from Veneto and Friuli who had migrated to Argentina. Three of these letters were written by Friulian small landowners who had been comparatively well-off back home.[20] Written in 1879 and 1880, the letters reveal an anti-migration stance as they contain attacks against the steamship company and travel agents who manipulated emigration for their own gain by deliberately and unscrupulously exaggerating the opportunity and opulence offered by America. In these letters we learn of other small landowners whose standard of living was higher in Italy than in Argentina, and who are unable to return because they sold all their property to pay for the outward journey. In one letter it is specified that migration is a good option only for 'those poor Italians who starved at home and now, working day and night, can at least eat a piece of bread'.[21]

The spirit of adventure and the curiosity which were the main reasons for the departure of the three migrants mentioned above also provide the motive for the migratory experience of the protagonist of *Gli Dèi Torneranno*, which starts with Simone's return to his native village, the imaginary Jalmis, after twenty-seven years. For ten years he had been a sailor all over the world and had then spent the remaining time working as an itinerant actor and ballad singer in unusual (from the point of view of Friulian migratory experience) places such as Peru and Bolivia, managing to save a small fortune. The reason why he left his village was the appeal of the so-called *Lasimpòn*, a fanciful name, probably deriving from the German *Eisenbahn* (railway), which the peasants of Jalmis used to call any country other than their own, a place which 'was everywhere but stayed always far away'.[22] In Simone's case, migration is not 'the ancient thief' which in Friulian history has stolen so many people's best years, but a way out of his static and old-fashioned native village into a life of wanderings and exciting experiences. Simone's unusual migratory experience is counterbalanced in the novel by the tragic and unfortunately more realistic emigration of Remigio, who went to Australia where he died of overwork, failing to realise his ambition of bringing money back to set up his own carpentry shop in Jalmis.[23]

A unifying feature in Friulian literature is the migrants' resignation to the grim necessity of sacrificing the best years of their lives for the prospect of a better future. The idea of migration as the only means of advancement in life is held even by migrants whose experience abroad has been one of failure: in Pier Paolo Pasolini's novel *Il Sogno di una Cosa* (The dream of a thing), set in San Giovanni di Casarsa in the province of Pordenone immediately after the Second World War, an old drunkard – who had built up an extensive although negative migratory experience before the First World War – still tries to persuade a group of adolescents of the advantages of migrating:

> What is there to do here in Italy? Italy! Ah, come off it! Even if you travel all over the world you will never find another country as corrupted as Italy. Italians are thieves: give them the chance and they will cheat you, the poor even worse than the rich. . . . Off with you, young lads, it's abroad you must go! You must travel around the world as I did, the Americas, Belgium, Germany . . .[24]

A fatalistic resignation to migration, therefore, seems to be a feature embedded in Friulian culture.

Percoto interprets migration in terms of a rural–urban conflict where the corruption of the city is juxtaposed with the idyll of an unspoilt countryside; she also argues that migration is to be blamed for the absence of a poet among the people of Carnia:

> They leave their mountains, wasting the loving years of their life in the turmoil of the cities, and then return to debase their simple home villages with the vices of society, and God has punished them by not granting them a poet.[25]

Such conflict is absent in Sgorlon's view of the necessity of Friulian migration which he traces back to 1420, the 'year of the apocalypse', when Venice conquered the Patriarchate of Aquileia, thus destroying Friulian freedom. Since then Friulians had been drawn to migrate in 'a tragic diaspora': 'it was their destiny, there was nothing to do against it. The villages were being depopulated, the young left, the old died and the only inhabitants of the houses were the wind and the birds of the night.'[26]

This resignation in the face of migration, therefore, is closely linked with another unconscious feeling deeply rooted in the Friulians' psyche, that of being exiled in their own land. Friuli being a 'frontier country', lacking a real routeway function,[27] the feeling of exclusion experienced by its inhabitants is very understandable. It features prominently in Pasolini's literary production about Friuli. Born in Casarsa della Delizia, Pasolini spent only part of his childhood in that small town before moving first to Bologna and then to Rome, where he became a famous film-director. His poem, 'La miej zoventút', written in 1953, describes the young men of the village painfully celebrating their departure in a sad feast which has repeated itself for centuries.[28] These young men, 'the best youth', are seen by Pasolini as being excluded from the world of history which is created by politicians and not by the humble:

Come, trains, take away the young
to look in the world for what here is lost.
Take, trains, around the world, driven out of their village,
these merry young men, not to laugh any more.

The rebellion of the excluded through their increasing political self-awareness as a social class provides the main themes for Pasolini's *Il Sogno di una Cosa*, whose title originates in a quotation from Karl Marx hinting at the peasants' growing consciousness of their state of subjugation to the landlords.[29] Written between 1949 and 1952, although published only in 1962, the novel is set in the postwar years of reconstruction at the time of the first struggles of Friulian peasants against the landowners. The protagonists of the novel are idealistic young peasants whose land-hunger and non-existent possibilities of self-improvement force them to migrate, some to Yugoslavia and the rest to Switzerland. As in other East European countries, in Yugoslavia land-reform laws had been quickly passed after the war: this, together with the young men's communist creed and the geographical proximity of Yugoslavia to Friuli, explain their illegal migration to that country, the idiosyncracy of which is counterbalanced in the novel by the more traditional emigration to Switzerland of the other young men in the group.

THE SITUATION OF THE MIGRANTS ABROAD

Once at their final destination, the migrants' success did not always depend on their entrepreneurial spirit: from a letter written in 1878 by two brothers from Dignano (in the province of Udine) we learn that success in Argentina depended on political decisions which were well above the migrants' heads. In other words, the land concessions granted to European migrants from 1865 had been taken on contract by private colonisation agencies: these were in turn linked to big landowners in Italy who wanted to promote the migration of small landowners in order to take possession of their land at a low price.[30]

This corroborates Ciafardo's view that emigration from north-east Italy around the turn of the century was a clear product of a process of capitalistic restructuring.[31] Ciafardo's aim in collecting letters written between 1921 and 1938 by two brothers from San Daniele del Friuli who had migrated to Argentina is to give an insight into the workings of chain migration and especially into the ambiguous relationship of exploitation/co-operation between the two brothers and the migrants – relatives or fellow villagers – who had settled in Argentina in previous years. Once in Argentina, Fernando and Lino go to Marcos Juarez to find their uncle who owned a soda factory and had lured them into migrating by promising them a job. In letters to his relatives back home, Lino – a typographer by profession and a militant socialist – gives vent to his rage concerning his working conditions which

150

have turned out to be much worse than those promised. Lino therefore moves to Rosario where another of his uncles has settled. Still a 'prisoner' of chain migration, he accepts work as a typographer for a *paesano* (fellow villager) but is yet again exploited with low pay and long working hours. So he looks for another job, writing to other fellow villagers in the province of Jujuy. Following their honest replies concerning the poor working conditions they would be able to offer him, Lino eventually finds a better job outside the migration chain. His letters to his relatives back in San Daniele also point out a widespread scarcity of work, which clashes with the migrants' expectations before departure. In a reply to Lino his father expresses surprise at his son's difficulty in finding a job and at the lack of support from *paesani*, drawing a contrast between Lino's plight and his own earlier migratory experience in America, where he always had money and friends to rely on. Disappointment at the conditions found abroad is also portrayed in letters written by Friulian migrants in the 1950s and early 1960s: in letters written from Lyon, Switzerland and Canada, emigrants inform their relatives back home of the very serious financial difficulties they are experiencing abroad.[32]

For the vast majority of Friulians, however, migration has proved to be a positive experience, resulting in a measure of both professional and social upward mobility. Ermacora's travel diary, though occasionally overlaid by a triumphalist tone in describing the success enjoyed by the migrants and their offspring, provides a very interesting account of the cohesion and entre-preneurial spirit characterising Friulians throughout the world. Written in the 1950s, his account spans a period of approximately eighty years of migration. The key words of the narrative are 'hard work' and 'house', for Friulians whose migration is planned as permanent have always given priority to building their own house in their final destination. One quotation from a migrant in Philadelphia summarises very effectively the mentality of the Friulian communities who have settled abroad:

> everyone of us owns a house here, which has often been built with each other's help; Friulians never queue for the unemployment benefits, have never been to prison; divorce is virtually unknown. Everybody helps everybody else, without any envy ... the law of fraternity reigns amongst us.[33]

This sense of fraternity is also exemplified by the many associations or 'families' of Friulians (*Famee Furlane*, or *Fogolar Furlan*) which are scattered all over the world. Within the *Famee Furlane* of Toronto a mutual-aid association was founded in 1932, the members of which received a subsidy of 10 dollars a week in case of illness and their families 200 dollars in case of death.[34] The close-knit social character of the Friulians abroad, their *friulanità*, is certainly one of the most outstanding features of this group – as also within Italy. The most evident agent of cohesion is their distinctive language, Friulian (*friulano*), which is a non-Italic idiom far removed from standard Italian.[35]

In *Il Sogno di una Cosa*, when the young migrants from San Giovanni di Casarsa are asked in Rijeka where they are from, Nini replies proudly 'Friulians' (and not 'Italians'). And at the depth of their desperation, having realised that their migration to Yugoslavia has been a failure, they find comfort in singing Friulian songs.[36] In *Gli Dèi Torneranno*, the moment Simone recalls a Friulian song he thought he had forgotten after so many years spent in South America, he knows that the time has come to return to his native village.[37]

Another recurrent theme is Friulian food which, alongside their language, features prominently as a means of preserving the migrants' cultural identity when abroad. Letters from migrants in Europe in the 1950s and 1960s abound in detailed descriptions of what they eat abroad (which is generally regarded as far inferior to what they used to eat at home), with requests for Friulian *polenta*, cheese, sausages, ham, salami and wine to be sent from home.[38] In Ermacora's account too, the distinctive feature of Friulian families abroad is seen as their maintenance of the culinary habits they, or their parents, had before migrating, so much so that even the few non-Friulian wives skilfully stir the *polenta* and speak some Friulian.[39] Even where some acculturation has taken place,

> it comes as a touching surprise to discover, among the oranges from California and the bananas from Florida, cheese from Fanna and nuts from Frisanco: a modest memory of the small Friulian homeland grafted on to the 7-kilo turkey, the symbol of American prosperity.[40]

On the other hand, it is interesting to note that in letters written from South America at the end of the last century, food occurs only in reference to its plentifulness in Argentina as opposed to its scarcity back home: thus an emigrant writes to his father stating that 'there's no need to be sorry to leave the *polenta*, as over here there's plenty of food, meat, bread and poultry.'[41]

Besides the Friulians' strong solidarity abroad and their nostalgia for home, another recurrent feature is their need to have an active role in their families' lives back home. The sense of responsibility that migrants feel towards their families seems to be strengthened rather than weakened by distance and is mirrored, especially in the case of younger migrants, by their parents' concern that their emigrant offspring keep alive their native customs and mentality. Thus, for example, parents try to convince their children living abroad to stick to *paesani* as their friends and not to get married to a person who is not from their native village.[42] On the other hand, migrants may entreat their relatives and friends back home to hold to a particular kind of political behaviour. In Bongiorno and Barbina's collection of letters from the 1950s and 1960s there are many examples of migrants trying to convince their relatives to vote for the Italian Communist Party which they consider to be more sensitive to migrants' issues; the same migrants express dissatisfaction with the failure of the party in power (the Christian Democrats) to find a solution to the root causes of emigration.[43]

Turning now to the occupations of Friulian migrants abroad, in all the literature examined Friulians are described as hard-working and thrifty, so much so that in Friulian there is a specific term of abuse, *bíntar*, to designate a migrant who has not managed to save any money to bring back home. Thus, according to the letters from late nineteenth-century Argentina collected by Franzina, farmers fared well in their destination as the land was plentiful and fertile and there were many incentives from the Argentinian government.[44] The descendants of these colonisers are described by Ermacora in the 1950s as having set up thriving farm businesses, especially in the viticultural sector; this writer also mentions the recent establishment of a silk mill which produced its own cocoons.[45] From the same source we learn that the descendants of the masons, *terrazzo*-makers, carpenters and stone-cutters who migrated to Philadelphia, Detroit and Toronto between the 1880s and the 1920s, owned thriving tile-factories and marble-carving businesses by the 1950s.[46] Many claims concerning the Friulians' hard-working and entrepreneurial qualities can also be found in letters written from Argentina during the 1920s and 1930s and from other places during the 1950s and 1960s.[47] For example, one migrant writes from Canada boasting of his thriving sausage business which has branches in the United States and is advertised on television; another has a garage and petrol station in France, and so on. Of course, such documentation should be treated with caution. Migrants tend to exaggerate their prosperity and success in their letters home. An uncritical reading of these letters allows the 'heroic' mythology of Friulian migration to perpetuate itself. The qualities of Friulian enterprise and hard work (to the extent that they are real) are likewise imbued in the regional literature of novels and short stories.

THE SITUATION BACK HOME AND PLANS TO RETURN

As in many other out-migration regions, the impacts of migration on Friuli have been two-fold: on the one hand it has caused the breaking-up of family units and the depopulation of villages, especially in the upland areas; on the other it has injected new blood into the stagnant economic situations of the labour-exporting areas, initially through remittances from the migrants and then through their return from abroad. These two impacts will now be examined.

The sad necessity of breaking up their families has always been viewed with resignation by Friulian migrants: migrant couples have been forced to leave behind young offspring in the care of their own parents or other relatives, parents have had to bid farewell to their migrant children, and wives have been left behind by their migrant husbands. In many cases marital separation has been accepted with resignation by both male and female, especially when they had a brighter future together to look forward to. Typical is this husband's letter to his wife:

> we have sworn to each other that we would wait for each other even a lifetime, so I leave in a happier mood because I know that you are going to wait for me and that when I come back we will start a new life and I'm sure it won't be too late, we will have time to love each other and to have all the children we want.[48]

However, there are also examples of women who rebel against being left on their own: such a rebellion, if not necessarily a physical one, like that of Sgorlon's women[49] who produced illegitimate children, may at least be mental, as exemplified by the wife who wrote to her migrant husband: 'I want you to come back now, tonight, and take me dancing, because I'm afraid that time will go too quickly and the two of us will become old.'[50]

Women wasting away while waiting for their husbands can also be seen from the point of view of their children, as in the two poems by Siro Angeli: 'Nedal' or 'Christmas' and 'La ciasa' or 'The house', written respectively in 1959 and 1974. In the first the poet reflects on his mother's premature ageing by saying:

> The young woman she has been
> where is she now
> where is the sky of the day when in the church
> (it was April, then)
> she stepped taking his arm?[51]

The title of the second poem refers to the motivation for the poet's father to migrate, namely to build a new house for the family, either in Friuli or in France where he worked; however, this dream never seemed to materialise and the poet's parents were wasting their lives for nothing.[52]

Unmarried young women who did not migrate would also experience uneasiness because of the departure of the majority of young males. In *Gli Dèi Torneranno*, Margherita refuses to leave her ancestors' house which was in a state of neglect after the migration of the rest of its inhabitants; she keeps waiting for something she cannot define, which turns out to be the returnee Simone.[53] In both Sgorlon's novels considered in this chapter, the imaginary villages of Cretis, Jalmis and Cleulis are described as 'sleeping and half-empty'; they were villages 'the old did not dare to leave only because of their advanced age', considering them 'an ante-chamber to death', whilst 'the children stayed there only because they had not yet begotten the idea of being able to leave.'[54] The departure of the young strata of rural society is also the main theme of Pasolini's poetry in Friulian. In 'Viers Pordenon e il mont' (Towards Pordenone and the World), the young people who migrated from Casarsa

have remained behind the glass of the cupboards
staring with their light eyes
into the light of the kitchens,
not seeing anymore neither the fireplaces
nor the smoke-blackened beams
nor the greasy table . . .[55]

The depopulation of rural areas by migration is closely associated with the inexorable disappearance of rural culture; this theme is central in the novels by Sgorlon. In *Il Trono di Legno* the seat which gives the novel its title is the oak armchair on which old Pietro sits when telling his stories. These are a mixture of real events (he had himself been a widely travelled migrant), legends, fictional accounts and folktales. A close parallel can be drawn between Pietro and Clementina Todesco, the North Italian story-teller who migrated to the United States in 1930 and whose folktales have been collected by Mathias and Raspa.[56] Throughout Northern Italy story-telling is defined as a central activity in rural community life. Like Clementina, Pietro possesses particular talents of memory and verbal skill to evoke the traditions of the local group, mixing reality and imagination and in so doing dilating the closed space of the village. Through Pietro's story-telling, Giuliano's curiosity to know the world is neutralised, so much so that he eventually decides to remain in the village and not migrate; he marries the widow of a migrant to Canada.

The title of Sgorlon's other novel, *Gli Dèi Torneranno*, is explained in the last pages of the book and refers to the imminent reversal of the trend of depopulation in rural areas: according to the author, such a reversal would be brought about by the renewed vitality of the culture of small local communities as opposed to urban ones. At the end of the book, Simone's return to Jalmis, which took place in the first chapter of the novel, finally acquires a meaning in the context of this wider trend of the migrants' return to their traditions:

> The gods will return. They are returning already. . . . Men left their homes, but the gods were returning, and watched over the things men had left behind . . . the local gods like Tarvos, the god of snow, or Beleno, or Akileia. . . . Yes. The gods would return to consecrate those places again. Maybe they had never left.[57]

Simone's task becomes the preservation of peasant culture, doomed otherwise to disappear, by collecting songs, legends and accounts of his people. In writing this history, a special place would be reserved for the migrants' letters. Simone felt that the tragic stories of those who had left to go to every corner of the world, whose names had been lost and who had lived, silently worked and died, were of central importance in the history of his people.[58]

The renewed vitality of small local communities envisaged by Sgorlon has a strong bearing on the current economic development of the Friulian areas of out-migration and will be further discussed in the conclusion to this chapter. We shall deal at this point with the temporary economic relief brought about by remittances. As mentioned earlier, the Friulian derogatory word *bintar* means a migrant who has returned without saving any money. The vast majority of Friulian migrants were not *bintars* and sent money regularly to their families back home, as illustrated in the following instances.

In the correspondence collected by Ciafardo, Lino regularly sends money to his mother, even when he has to pawn a pair of ear-rings to do so; indeed, his letters seem to be a mere accompaniment to the remittances, as we learn from his family's complaints about his not writing at all when unable to send money.[59] The moral duty of sending remittances home is so embedded in Friulian migrant mentality that Lino continues to do so even after he gets married abroad. In one case, even death is not enough to stop the influx of money to Friuli: in a letter written from Australia in 1960, the Australian widower of a migrant Friulian woman tells his dead wife's family that he will continue to send them all the money he can save.[60]

Remittances basically provided the means of survival for the migrants' families; they did little or nothing to develop the economies of the out-migration areas. Return was always prominent in the minds of the migrants for whom working abroad had been planned only as a temporary move. Although migrants return for a variety of reasons, personal motives are of major relevance in the literature examined here: nostalgia for the native village and its familiar stability, as in the case of Simone whose motives for return are 'kneaded into the landscape' of his village;[61] or family motives, such as the woman deserted by her husband while they are abroad, or the widow going back to the house that had been built with the couple's remittances.[62] In the more specific context of Friuli, the collapse of the artificial dam on the River Vajont on 9 October 1963, a tragedy which obliterated several villages and killed hundreds of people, called many migrants back to the stricken area: in Bongiorno and Barbina's collection of emigrant letters a whole section is dedicated to the touching letters written by the survivors to their migrant relatives abroad asking for help with the reconstruction, and by the migrants begging to know the fate of their relatives.[63]

Usually, however, the pattern of return would be for the migrant to save enough money to buy or finance the construction of a house in the native village. However, the economic stagnation of the upland districts of out-migration offered very scanty working opportunities which, in any case, bore no resemblance to the employment followed abroad. Often the longed-for return would not take place, in some cases for family reasons: Margherita's sister, Mercedes, whose wish to return to Cleulis becomes more intense as the years go by, has to acknowledge, reluctantly, that her life is in Canada with her husband and children.[64] Even sadder is the story of a relative of Simone's who migrated to Nebraska: after detailed planning, over many years, for his return to the village, including consideration of what vegetables to sow in the garden, he died of a stroke after his American-born children had refused to leave and his wife had chosen to stay with them.[65] Of such tragedies are real migrations made.

On the other hand, when return did take place, it could involve realising the end of a world which distance had idealised. On returning to Casarsa, the mythical village of his childhood, Pasolini laments the end of 'the greatest illusion, which seemed the very essence of my life and the world: I return, walking on the collapsed bridges, like an Australian.'[66] On a somewhat more practical level, the disappointment of return could be linked to growing nostalgia for a glittering world outside the village, as with Noemi who had been a chambermaid in a luxury hotel in Milan.[67] Or disappointment could be connected to the returnees' realisation that the money saved was not enough to guarantee security to their families, as in the case of Lino's father whose earlier migration to America had not prevented his own children from becoming migrants.[68] The necessity for the returnee himself to re-migrate represented the worst possible outcome of such scarcity of resources at home.

Another important aspect of return is the social reintegration of the migrants who, after a period abroad, come back 'different'. This has many aspects. At a superficial level their external appearance, accent and clothing might reflect their foreign sojourn, as with Sgorlon's Simone who returns wearing American clothes and who is *expected* to display the returnee behaviour and 'sophistication' which are thrust upon returnees by the non-migrant population.[69] Also significant are the new mentality and value systems returnees bring back. Though such changes can be viewed in a negative light, as when the young men returning from Turin and Germany are labelled 'bandits',[70] they can also contribute to the modernisation and development of the areas of earlier exodus. In Pasolini's *Il Sogno di una Cosa*, the young men's migratory experiences in Switzerland and Yugoslavia prove to be of critical importance in the process of their self-discovery as members of a social class ranged against the wealth and power of the latifundian elites. Migration has enabled them to enter the world of history from which they had hitherto been excluded.[71]

CONCLUSION

The returnees' socially innovative behaviour that Pasolini traces to the late 1940s might be considered the first stage of the economic revolution that has characterised Friulian migration history since the late 1960s. As mentioned at the start of this chapter, in Friuli the migrants have contributed positively to the changing economic structure of the region by setting up or finding a job in scattered small industries in the rural areas of earlier exodus. This change from 'migration as subsistence' to 'migration as development' – generating extra resources for the development of the local economy – is documented embryonically in the literature. In letters written by four migrants in the early 1960s, the writers express their intentions to come back home and set up their own businesses.[72] In three of these the nature of the business – a garage, a carpet trade and a more vaguely worded 'commercial business' – bears some relation to the migrants' work experience abroad (two were in Germany and the third, the carpet dealer, in Morocco).

For a more thorough analysis of this contemporary phenomenon of 'productive return' we must move away from the 'literary literature' which has formed the backbone of this chapter to the social-scientific literature on Friulian migration published over the last ten to fifteen years. A policy perspective is also necessary: since 1970 the regional authorities have offered special resettlement grants to returning migrants, and by 1980 11,000 families – a sizeable proportion of the total return flow – had taken advantage of these return grants. Independent of this policy, however, there occurred a process of spontaneous dispersed industrialisation, particularly in the mechanical and building trades, sectors in which many Friulian migrants had accumulated experience abroad. Particularly during the 1970s but also in the 1980s, these buoyant trades attracted many returnees, both as dependent workers and as the founders of new enterprises. In Saraceno's survey of 1,500 returning Friulians interviewed in 1982, 30 per cent had moved into self-employment – a high percentage compared to other surveys of Italian returnees.[73] The extent to which this has happened in Friuli reflects both the variety of foreign work experiences and the economic opportunities finally offered by the local area, particularly in the aftermath of the 1976 earthquake when a lot of relief money was made available for reconstruction and rehabilitation. Moreover, many returnees have shown an adaptive willingness to mix jobs on return, particularly common being the combination of part-time farming on a residual family holding with work in industry, construction or the service sector. In this, too, they exhibit both a flexibility learnt during the migration process and a long tradition of 'worker-peasant' society which has been woven into the fabric of rural Friuli since the eighteenth century.[74] The openness of Friulians to different work possibilities in different places at different times is part of the region's cultural history which contrasts with the suffocated world of that other great reservoir of Italian migration – the South. Thus Friuli's culture of emigration and of open contacts – which has embedded itself deeply in the region's literature for more than a century – has come full circle and ultimately benefited the economic development of the region. Now, if Friulians migrate, they do so by choice. The long and dramatic migration experience of the people of Friuli is over.

NOTES

1 For a good review of Friulian migration literature see F. Micelli, 'L'emigrazione dal Friuli. Saggio bibliografico', in C. Grandi (ed.), *Emigrazione, Memorie e*

Realtà, Trento, Provincia Autonoma di Trento, 1990, pp. 289–316.

2   G. Valussi, 'Friuli-Venezia Giulia. Il quadro generale', in M. L. Gentileschi and R. Simoncelli (eds), *Rientro degli Emigrati e Territorio*, Naples, Istituto Geografico Italiano, 1983, pp. 41–2.

3   G. Rando, 'The migration theme in Italian literature', *Migration Monitor*, 1991, vol. 3, no. 9, p. 13.

4   The historical summary which follows is largely based on material in G. Bellencin Meneghel, 'Friuli', in Gentileschi and Simoncelli, op. cit., pp. 49–55.

5   E. Saraceno, 'The occupational resettlement of returning migrants and regional development: the case of Friuli-Venezia Giulia, Italy', in R. King (ed.), *Return Migration and Regional Economic Problems*, Beckenham, Croom Helm, 1986, pp. 69–78.

6   D. Virgili, *La Flôr. Letteratura Ladina del Friuli*, Udine, Società Filologica Friulana, 1968, vol. l, p. 26.

7   L. Zanini, *Friuli Migrante*, Udine, La Panarie, 1937, pp. 271–307.

8   Corn-meal mush, a typical dish of Friuli and north-eastern Italy.

9   C. Percoto, 'Lis cidulis', reprinted in B. Maier (ed.), *Novelle*, Bologna, Cappelli, 1974, pp. 43–84.

10 C. Sgorlon, *Gli Dèi Torneranno*, Milan, Mondadori, 1977, p. 40.

11 ibid., p. 253.

12 Zanini, op. cit., pp. 309–10.

13 Sgorlon, op. cit., pp. 131, 273–4.

14 C. Sgorlon, *Il Trono di Legno*, Milan, Mondadori, 1972, pp. 87, 100, 104, 110.

15 C. Ermacora, *Nostalgia di Focolare*, Udine, Doretti, 1967, p. 126.

16 Sgorlon, *Gli Dèi Torneranno*, op. cit., p. 173.

17 Ermacora, op. cit., p. 145.

18 A. Bongiorno and A. Barbina (eds), *Il Pane degli Altri. Lettere di Emigranti*, Udine, La Situazione, 1970.

19 ibid., pp. 10–11, 73–4.

20 E. Franzina, *Merica! Merica! Emigrazione e Colonizzazione nelle Lettere dei Contadini Veneti in America Latina 1876 –1902*, Milan, Feltrinelli, 1979, pp. 102–5, 111–12, 115–19.

21 ibid., p. 111. In this quotation 'Italians' stands for 'Southern Italians' and emphasises the socio-cultural division between Friulians and other Italians, particularly those from the South.

22 Sgorlon, *Gli Dèi Torneranno*, op. cit., p. 116.

23 ibid., pp. 167–72.

24 P.P. Pasolini, *Il Sogno di una Cosa*, Milan, Garzanti, 1976, p. 27. (First published 1962.)

25 Percoto, op. cit., p. 69.

26 Sgorlon, *Gli Dèi Torneranno*, op. cit., p. 156.

27 J. Bethemont and J. Pelletier, *Italy: A Geographical Introduction*, London, Longman, 1983, p. 140.

28 P.P. Pasolini, 'La miej zoventút', in *La Nuova Gioventù*, Turin, Einaudi, 1975, pp. 152–3.

29 Pasolini, *Il Sogno di una Cosa*, op. cit.

30 Franzina, op. cit., p. 114.

31 E. Ciafardo, 'Cadenas migratorias e immigración italiana. Reflexiones a partir de la correspondencia de dos immigrantes italianos en Argentina (1921–1938)', *Studi Emigrazione*, no. 102, 1991, pp. 233–55.

32 Bongiorno and Barbina, op. cit., p. 126.

33 Ermacora, op. cit., p. 126.

34 ibid., p. 147.

35 G. Valussi, 'Linguistic minorities', in M. Pinna and D. Ruocco (eds), *Italy: A Geographical Survey*, Pisa, Pacini, 1980, p. 181. See also G. Barbina, *La Geografia delle Lingue*, Rome, La Nuova Scientifica, 1993, pp. 148–9 for the resistance of Friulian amongst different generations of migrants in different destination countries. The Italian government has only recently recognised the importance of linguistic minorities; in 1992 a bill on bilingualism was passed wherein Friulian features among the eleven 'protected' languages which can be taught in primary and secondary schools and used in public offices, official meetings and public documents issued by local authorities in the regions where such languages are spoken.

36 Pasolini, *Il Sogno di una Cosa*, op. cit., pp. 49, 52–3.

37 Sgorlon, *Gli Dèi Torneranno*, op. cit., p. 16.

38 Bongiorno and Barbina, op. cit., pp. 24, 37–8, 51–2, 66.

39 Ermacora, op. cit., p. 126.

40 ibid., p. 133.

41 Franzina, op. cit., p. 97.

42 Ciafardo, op. cit., pp. 251–3; Bongiorno and Barbina, op. cit., pp. 56, 138.

43 Bongiorno and Barbina, op. cit., pp. 85–6, 89, 91–3, 126.

44 Franzina, op. cit., pp. 96, 101, 103, 111, 119.

45 Ermacora, op. cit., pp. 86, 90. Silk production was also the main industry in rural Friuli until the 1960s: see D.R. Holmes, *Cultural Disenchantments: Worker Peasantries in Northeast Italy*, Princeton, Princeton University Press, 1989, pp. 164–203.

46 Ermacora, op. cit., pp. 125–6, 133–4, 136, 144–7.

47 Ciafardo, op. cit., pp. 241–2, 248; Bongiorno and Barbina, op. cit., pp. 65–6, 70.

48 Bongiorno and Barbina, op. cit., p. 5.

49 Sgorlon, *Il Trono di Legno*, op. cit., p. 132.

50 Bongiorno and Barbina, op. cit., p. 42. On this same theme, see the poem 'Parcè a mi Signôr' (Why to me, Lord?), written in 1962 by the Friulian poet Leonardo Zanier, himself a migrant to Switzerland, in which he interprets the despair of a woman whose husband, a seasonal migrant, has just left: 'Why to me, Lord? / Every year to me, Lord? / This cross / and this void at dusk: / being on my own for so long / in this bed that is so big / and creaks / I get undressed in the dark / Lord / without touching my own flesh / Lord / because I am afraid to cry / to shout . . . / I've changed the sheets / Lord / they still smelled of him / of us / I've put the new ones / those from my trousseau / but my body / Lord / my heart / Lord / those I cannot change / Lord / Are you listening Lord? / ARE YOU LISTENING, LORD? / Goodnight, husband / think of me / be well / write / COME BACK!' L. Zanier, *Libers . . . di scugnî là (poesie 1960–1962)*, Milan, Garzanti, 1977, pp. 65–6. On Zanier's 'migration poetry' see R. Mordenti, 'La poesia di Leonardo Zanier: dal Friuli all'Europa alla ricerca di un nuovo volgare', in J.J. Marchand (ed.), *La Letteratura dell'Emigrazione. Gli Scrittori di Lingua Italiana nel Mondo*, Turin, Fondazione Agnelli, 1991, pp. 283–99.

51 This poem is in Virgili, op. cit., p. 181.

52 S. Angeli, *L'Aga del Tajament*, Tolmezzo, Edizioni Aquileia, 1976, pp. 18–19.

53 Sgorlon, *Gli Dèi Torneranno*, op. cit., pp. 77, 79–80, 107.

54 Sgorlon, *Il Trono di Legno*, op. cit., p. 192; *Gli Dèi Torneranno*, op. cit., p. 76.

55 Pasolini, *La Nuova Gioventù*, op. cit., p. 131.

56 E. Mathias and R. Raspa, *Italian Folktales in America. The Verbal Art of an Immigrant Woman*, Detroit, Wayne State University Press, 1988.

57 Sgorlon, *Gli Dèi Torneranno*, op. cit., pp. 301–2.

58 ibid., pp. 174, 261.

59 Ciafardo, op. cit., pp. 247, 252–3.
60 Bongiorno and Barbina, op. cit., pp. 55–6.
61 Sgorlon, *Gli Dèi Torneranno*, op. cit., pp. 54–5.
62 Bongiorno and Barbina, op. cit., pp. 57, 128.
63 ibid., pp. 133–41.
64 Sgorlon, *Gli Dèi Torneranno*, op. cit., p. 116.
65 ibid., p. 174.
66 Pasolini, *La Nuova Gioventù*, op. cit., p. 187.
67 Sgorlon, *Gli Dèi Torneranno*, op. cit., pp. 50–2.
68 Ciafardo, op. cit., p. 254.
69 Sgorlon, *Gli Dèi Torneranno*, op. cit., pp. 49–50.
70 Pasolini, *La Nuova Gioventù*, op. cit., pp. 242–3.
71 Pasolini, *Il Sogno di una Cosa*, op. cit., pp. 124–5.
72 Bongiorno and Barbina, op. cit., pp. 68–9, 73, 95.
73 See Saraceno, op. cit., and, by comparison, the results set out in the various regional chapters in Gentileschi and Simoncelli, op. cit.
74 Holmes, op. cit., especially pp. 56–9.

10

SUNDAY TOO FAR AWAY

Images of emigrant existence in the literature of Slovenes in the United States, Canada and Australia

Jerneja Petrič

INTRODUCTION

Slovenia has been a land of emigration throughout its history. Nowadays its emigrants can be placed, for practical purposes, in one of the following four categories: political, economic, professional and personal.[1] Political emigration is the oldest of all, its roots reaching back into the seventeenth century, the era of merciless Catholic persecution of Protestants. The first big wave of political emigration, however, came after the Second World War, after the Communist regime had established itself in the country. Economic emigration has been greatest of all. Between the years 1880 and 1914 alone, 300,000 Slovene people emigrated – a number which, at that time, represented one third of Slovenia's natural increase of population. Professional emigration, relatively small in scale, was represented mainly by the Catholic missionaries, especially to the territories of the present-day United States and Canada, in the seventeenth, eighteenth, and nineteenth centuries. Finally, so-called personal emigration (people who emigrated for various personal reasons) has not been limited to a particular time and has consisted mainly of emigrants' families.

Of all the English-speaking countries that lured the Slovenes – the United States, Canada, Australia – only Slovene-Americans have already lived to see their fourth generation. The first Slovene-Canadians settled down in Canada at the end of the nineteenth century, while the earliest Slovenes came to Australia in the immediate aftermath of the First World War.

This diversity in timing and typology of emigration is of course reflected in the literature written by Slovene emigrants: it varies according to the category they belong to, the period of their emigration and their country of destination.

Regardless of the country of immigration, autobiographical literature has been one of the strongholds of Slovene immigrant (and later ethnic) society. In the United States alone, between the years 1891 and 1965, there appeared

162

in print 671 autobiographical narratives, 122 of which were long enough to be serialised.[2] They were all published by Slovene-American newspapers, magazines, almanacs and other publications. Eight autobiographies appeared in print in book form and five more books were written but remained unpublished. The literary output in Canada and Australia has been smaller but the mode of publication has been much the same.

Fiction, poetry and drama, when not downright autobiographical, often made use of emigration and immigration as their main themes. These, however, were frequently overlaid by a consciously rationalised response on the part of the protagonist to the drab reality of 'sticking it out', a response meant to bring home a didactic or moral note, especially to Old Country readers, in order to make the New World more fearful than fascinating.

The first and foremost response to the increasing complexity of the New World society was romantic nostalgia, a tendency to dwell on pleasant memories of the Old World. Following on its heels was a down-to-earth realistic/naturalistic presentation of New World circumstances: in its extreme form mere socialist propaganda. However, it required more than a sensitive observer to do full justice to both worlds. With but a few exceptions, Slovene emigrant literature therefore either exhibited partiality in favour of the Old Country or failed to penetrate beneath the exterior of the New World. This, of course, holds true mainly for the generations of writers prior to the Second World War, and undoubtedly had a lot to do with the general low educational level of the early emigrants.

Generally speaking, emigrant experience is reflected in literature in terms of three major categories: mythological, sociological, and psychological. I shall deal with each of these in turn.

MYTHOLOGICAL: EMIGRANT EXPERIENCE AND THE MYTH OF THE NEW WORLD

According to William Boelhower, 'America was an idea before it became a geographical reality.'[3] This statement may just as well be transplanted to other emigrant destinations. As for America, its myth originated in Slovene folk tradition and it was based on the popular assumption that one did not need to work hard in America, the land of unlimited possibilities, in order to become rich. The oral tradition pictured America as the land that welcomed everyone who was willing to take a chance. Spurred on by the myth, many decided to challenge destiny: 'Then I looked toward America, where streets were paved with gold and there was opportunity at every turn. I knew that Slovenia held no future for me. I must go to another country.'[4]

A controversy soon arose among those who claimed knowledge about America. This disagreement centred around the question of whether America was a true Eldorado or an utterly destructive force which threatened not only to keep the newcomers poor forever but to destroy their souls as well. Thus the projection of America as a mythic ideal confronted its antonym. Both themes were taken up by Slovene authors of the late nineteenth century and became part of the nation's literature. There is little doubt as to which of the two – myth or antimyth – the emigrants, regardless of the time of their emigration or their final destination, believed in. It was a widespread notion that America was little short of Eden. In *Laughing in the Jungle* Louis Adamic expressed the sentiment in the following way:

> My notion of the United States then, and for a few years after, was that it was a grand, amazing, somewhat fantastic place – the Golden Country – a sort of Paradise – the Land of Promise in more ways than one – huge beyond conception, thousands of miles across the ocean, untellably exciting, explosive, quite incomparable to the tiny, quiet, lovely Carniola.... In America one could make pots of money in a short time, acquire immense holdings, wear a white collar, and have polish on one's boots like a *gospod* – one of the gentry – and eat white bread, soup, and meat on weekdays as well as on Sundays, even if one were an ordinary workman to begin with.[5]

The fact is, however, that America figured only as an abstract ideal in the oldest literary records written by 'unprofessional' emigrants. Other English-speaking countries were less idolised, mainly because they were 'discovered' later than America. With the development of communications, the later emigrants to Canada, Australia and even the United States had sensed the conflict between the two worlds before they left home – all the more as their general level of education rose. This does not mean, of course, that the 'American dream' was dead. The more the Old Country's harsh realities weighed on the emigrants-to-be, the more they were keen to project their dreams on an ideal New World. Only the myth itself began to lose its original power and appeal as it slowly turned into a vague anticipation of a better life.

In his novel *The Men Who Built the Snowy* the Australian-Slovene author Ivan Kobal explains the displacement process that took place during the Second World War and immediately afterwards in Europe and in Yugoslavia in particular. People unwilling to participate in the revolutionary war found their way to Australia, which was seen as the land of mythic proportions. Yet deep in their hearts, many were able to anticipate a slightly harsher reality: 'This unknown land, its untouched nature, had been Greek for them; according to all reports it was unused, uncertain, and yet better than any other country on this wobbling planet.'[6] Many emigrant writers' autobiographical or purely fictional narratives treat the myth of the New World as one of the main themes. When doing so, they often balance adroitly on the thin line which divides comedy from tragedy. This ability to perceive events as at once poignant and faintly ridiculous introduces a subtly emotional tension into the narratives. Through the ensuing irony the authors manage to achieve a satisfactory distance from their experiences which, at the time, must have been far from funny. In the words of Frank Magajna,

> I came to America, an eighteen-year old, still wet behind my ears, and, consequently, full of glorious dreams – such as only youth can conjure up. I travelled across the 'pond' toward the immeasurably rich, free, miraculous land, and already on my way I breathed in the romanticism promised to me. What didn't wait for me there! Inevitable riches, of course, and marvellous experiences. When in a few years I return home with sacks full of dollars, the whole village will stare open-mouthed, and listen enviously to my story.[7]

Narratives like this are tragic in the sense that they basically celebrate the protagonist's capacity to endure, and comic in the sense that they aver the wealth of choices without refusing the bad ones. It has to be pointed out, though, that women's view of reality tends to be more serious and down to earth, especially in autobiographical pieces.

THE SOCIOLOGICAL PERSPECTIVE: A REALISTIC-NATURALISTIC POINT OF VIEW

Catholic missionaries, the only professional group of early Slovene emigrants, did not believe in the myth of America in the first place. They did not have to, for their task had been firmly determined by their employers[8] before actual emigration. They were to spread Catholic faith in the sparsely inhabited Indian territories, untouched as yet by white men. They had no intention whatsoever of becoming rich and they did not know when, if at all, they would return home. Their private, although published,[9] letters included detailed descriptions of the territories, the people (diverse Indian tribes), specific weather conditions, vegetation and so on. The oldest were letters written in Latin by Marko A. Kappus[10] dating back to the seventeenth century, whereas the letters written by Friedrich Baraga and Franz Pierz (both nineteenth-century) belong among the very best documents of their kind. The realistic descriptions of their hard life and work were thoroughly didactic although limited in scope; they never touched upon such secular subjects as would be of interest to most prospective emigrants. Although Baraga and Pierz wrote and published books as well and although they had many followers, their highly articulate writings reached only a limited audience. What they did, though, was of no small value: they laid bare the real cultural topology of America's Indian territories. In their anthropological approach they clearly avoided any presentation of urban America, of the millions of people buried in cities struggling for their survival. This, too, was America, the America left to be discovered by thousands of 'ordinary' emigrants who began streaming in during the last two decades of the nineteenth century.

For these later emigrants, the myth of America died almost the moment they set foot on foreign soil. Confronted with the drab reality – lack of jobs, low pay, dangerous work, hatred of aliens, ugly industrial environment arousing homesickness, linguistic and religious barriers, alien food habits, etc. – the myth of America vanished into thin air. Emigrants, the majority of them poor peasants or the proletarian underclass, hurled themselves into hard work and a merciless struggle for existence. Regardless of whether their myth of America eventually did materialise (after a lapse of time involving much hard work), their initial response to the New World was disappointment, bitterness and resignation. 'One long big grey village – that was the first devastating impression Cleveland had made on me,' says Vatro Grill in his memoir.[11] That was back in 1913 and Grill, like so many others, was confronted with an utterly alien cultural system, something neither he, a country boy, nor anyone else, had been prepared for. He was caught up in the chaos of multiple negative experiences which he eventually overcame by strong will and determination, changing his too-Catholic-sounding name in the process.[12]

The majority were less forunate than Grill. One of the characters in Irene Planinsek Odorizzi's anthology of biographical sketches remembers:

> America surprised me! I thought it would be possible to secure the same type of employment which I had enjoyed in Zagreb [she was a governess], but my dream was shattered when I realized that my unfamiliarity with the English language was a barrier to any good position. Many times amid my tears of discouragement I would say, 'If that ocean weren't there, I would walk back home.'[13]

Odorizzi's book recounts the typical yet heterogeneous stories of average immigrant experiences that in many ways equalled those of Slovene-Canadians and Slovene-Australians. In his novel *Za dolar človečnosti* (A dollar's worth of humanity), Ivan Dolenc (Canadian Slovene) presents different Slavic characters, including himself, in the process of transformation from mythical ideal to drab reality:

> In the wee hours I spread the map of the city across my bed and I copied the names of roads and streets leading from my place to the employment office . . . Soon I stood in front of my first address. A handwritten notice was posted at the entrance: 'No Help Wanted'.[14]

These simple, uneducated or poorly educated emigrants had only meagre literary resources to rely on. Hence most of what they were able to produce (and produce they did a lot!) is nowadays pushed aside by literary historians as third-class literature. Yet, from the sociological point of view, these narratives are important as well as interesting because they yield a lot of useful information, such as the motives for emigration, the administrative procedure, the frequent circumventing of the rules by males in order to avoid conscription, the journey, Ellis Island and troubles getting through it, the shock experienced upon the first encounter with the Promised Land, and (rarely!) the Dream come true.

Some talented early emigrants exhibited a wider perspective. Louis Adamic's great energy and considerable narrative skill made him one of the best-known Slovene-American authors of his day, and had he not succumbed to the urge to write so much and so fast he might well have produced a body of work of more enduring merit. Adamic's most important works reflecting his migrant experience are *Laughing in the Jungle* (1932), *My America: 1928–1938* (1938), *From Many Lands* (1940) and *A Nation of Nations* (1945), not to mention his short stories based on immigrant themes. Adamic was a revolutionary in the sense that he showed America demythologised, America hostile to foreign immigrants, America in the midst of economic crisis. Yet also America as a melting pot, as 'a nation of nations'. Before Adamic, practically every Slovene emigrant had believed in the myth of America and was sobered when faced with the negative version of his/her dream, the antimyth. Adamic, however, responded to that with humour, laughing in the jungle that was his metaphor for America. Not many, however, were of the same disposition. Anna P. Krasna, the poet and prose-writer, responded with a kind of fighting resignation. Hers was the reaction of a saddened, helpless and frequently hopeless writer, yet one who, eventually, did not give up. There were many others who occasionally felt that the American reality was too ugly, too much in contrast with the fairytale world of their dreams, to become the subject of their literary endeavours. Instead they preferred to escape into the Old World past that seemed light years removed from America.[15] Or they chose to write stories and poems for children which, although often burdened with the dark aspects of America, nevertheless sought to express their disillusionment with tact – as in the poetry and short stories of Katka Zupančič. Others, like Frank Maganja, resorted to pure humour, whilst a few openly expressed their resignation – Ivan Bukovinski for example. Disillusionment with America provided nourishment for nostalgic literature as well as for literature of an explicitly proletarian genre whose authors (e.g. Ivan Molek and Joško Oven) set about capturing a sense of the times.

Since emigration to Canada and Australia took place at a later time, the emigrants' experiences are necessarily, to some extent, different also in geographical terms.[16] The biggest difference of all, however, is in the nature of the emigrants themselves. Unlike the nineteenth-century emigrants to the USA, those who emigrated to Australia or Canada were better educated in every respect and this shows in their literature as well; their literary talent finds a home in our next category.

THE PSYCHOLOGICAL EFFECTS OF EMIGRATION

Slovenes are a people who are emotionally very strongly attached to their homeland. Consequently, emigration has always been a necessary evil, an act regrettable in itself although unavoidable. For those who emigrated legally, the moment of actual farewell represented the first bitter pill. What had so far appeared as a thrilling adventure lost a good deal of its shine when the time came to say goodbye forever.

The anonymous author of what is likely to be the oldest autobiographical narrative written by a Slovene-American[17] reminisces about the day of farewell from his parents, his brothers and sisters. The parting is rendered in a voice tinged with deep emotion. At the moment of leave-taking, the narrator is fully aware of the gravity of the situation, of the fact that it may well be that he is seeing his loved-ones for the last time in his life. In a similar manner the author of presumably the second oldest autobiographical narrative[18] is bidding farewell to his native land: 'O, glorious, romantic Bled! ... How I would like to hug you today for the last time ....'

The overwhelmingly sad occasion of departure triggered off in many emigrants the first misgivings which, more often than not, materialised more fully a few weeks later. Many found themselves in an emotional abyss, wishing that they had never been lured across the ocean. All of a sudden they felt uprooted. Homesickness is therefore one of the dominant themes of Slovene emigrant literature, particularly poetry and prose. With the passing of time, however, emphasis shifted from the Old Country toward the New World and the emigrants found themselves pushed into a state of emotional disapora: theirs was to a large degree a literature expressing the tragedy of being of two worlds. Many 'green' immigrant authors weaved their stories around the feelings of isolation and loneliness. One of the most painful experiences of the emigrants was undoubtedly the loss of old values (Slovene manners, humility, strong family ties . . .). Life in America lost its Old World spontaneity and slow pace; in many ways it lost its naturalness. Slovene-American-Australian author Frank Mlakar successfully balanced three levels of his protagonist's experience in the novel *He, the Father*: pre-emigrant, immigrant (in America), and return.[19] Once in America, Osip Princevich stops noticing the seasonal changes in nature. The ugly industrial environment – so unlike the lovely natural simplicity of his native village – kills the natural man in him and turns him into a money-making machine. Mlakar's novel is a psychological-realistic one and places strong emphasis upon his characters' psyche and their will (or lack of will) to make a living in America, which Mlakar sees as the land of opportunity for the diligent, the brave and, above all, the daring. The Old World is not idealised either: upon the protagonist's return it proves as impoverished and evil as ever. There are numerous works of literature which beautifully illustrate such emotional waste.

Several emigrant authors exposed in their writings the fairytale aspect of the New World as well as its brutal reality. And they juxtaposed them with the equally dual nature of their homeland. Many of them were never able to resolve this eternal inner conflict and felt torn between two worlds for the rest of their days. Two Slovene-American authors titled their memoirs *Med dvema svetovoma* (Between two worlds) and *Med dvema domovinama* (Between two homelands): Vatro Grill and Anna P. Krasna respectively.[20] The phrase has since turned into a key refrain of Slovene emigrant literature. Krasna herself returned to Slovenia in the 1970s but repeatedly visited the United States. Her literature of the 1970s and 1980s reflects her emotional quandary: 'You are far away now/ but you live inside me/ and will not be buried.'[21]

Another Slovene-American, Janko N. Rogelj,[22] explores in detail both the New World reality and the Old Country 'unreality'. For Rogelj there had always been a little light shining that guided him through America – his Old World memories. In his unpublished *Spomini* (Memoirs) Rogelj admitted that in moments of crisis he would mentally return to his homeland which thus became an extension of reality for the author. Paradoxically though, this only made him feel even more homeless and solitary. Both in his literature and in real life he perpetually moved to and fro. Once 'at home' (in Slovenia) he would immediately feel attracted to America and vice versa.

The identity crisis grips Slovene emigrants (i.e. the first-generation immigrants) wherever they are. Reading about it in their literary outpourings, one has the feeling that it increases proportionally with their feeling of entrapment and finality. When the chances of ever returning home are practically nil, folkloric nostalgia is likely to win over the processes of mobility and acculturation. Accordingly, individuals seek consolation in group identity, in an attempt to overcome their individual grief by creating collective consciousness. When this is not possible, when an individual 'gets lost' in the vast territory of, say, Canada or Australia, authors seek the companionship of other emigrants, regardless of nationality or race. The emotional split remains, but it is approached with less sentimentality than otherwise.

Even when emigrants were adjusting themselves to new circumstances, they would continue to dream of re-creating the past: a romantic and idealised vision of an age and place that could never be recalled. There is a lot of uncritical glorification of the old 'innocent' agrarian past and it took a long time before migrant writers became aware of the fact that all they could do was to achieve a more meaningful past or, rather, to preserve in their literature what was the best in it for the generations to come.[23]

Whereas in early emigrant literature the only form of psychological response was romantic nostalgia, it later on acquired more subtle forms, especially in poetry. With the advance of years, Slovene migrant poetry has turned away from emigrants' general, common plight and become intensely individualistic. Recent American, Australian and Canadian emigrant poetry offers numerous examples. In the poem 'Domovina' (Homeland), Ivan Kobal (Australian) anticipates a cool welcome when he returns home: 'Who will suspect that the son has come home again?' he asks himself.[24] A woman poet, Stanka Gregorič, borrows Emily Dickinson's lines 'I am nobody, and who

are you?' as a motto to her poem 'Refleksije' (Reflections),[25] whereas Bert
Pribac (Australian), alone in a church in Wales ('Valižanska cerkev' – A
church in Wales) explores the roots of his secret yearning, of his wish to
belong.[26]

The above-mentioned existential dilemma has acquired almost archetypal
proportions. It surfaces in Slovene emigrant literature all over the world. The
poet Zdravko Jelinčič (Canadian) begins his poem 'Stara vrnitev' (Old return)
with the following verses: 'He came home as a foreigner, /foreign on his own
soil.'[27] Reality seems to be a riddle and a sense of belonging is what is missing.
'What's in a name?' asks Danica Dolenc (Canadian), what difference does it
make if it is Slovene or English?[28]

CONCLUSION

My argument in this chapter has been that, on a general level, fiction proper
came off rather badly in Slovene emigrant literature. The emigrants have
continued to use realistic modes of representation to relate multicultural
experience, religious and existential dilemmas as well as social conflicts.
Literature has thus assumed the role of an informal history preserving
precious details that would otherwise be lost. Various types of autobiographical writing were obviously much in demand and in many cases –
especially Australian and Canadian – have remained so until the present day.
Everywhere, emigrants have been encouraged to integrate. Regardless of how
hard they may try to hold back this development, however, the melting pot
will eventually take its toll. One must only hope that the literature at least
will remain for future generations.

NOTES

1 See Z. Šmitek, *Klic daljnih svetov: Slovenci in neevropske kulture* (The call of far-
off worlds: Slovenes and non-European cultures), Ljubljana, Borec, 1986.
2 For more detailed information on this see my PhD thesis, 'Slovenska avto-
biografija v Združenih državah Amerike' (Slovene autobiography in the United
States of America), University of Ljubljana, Department of English, 1987.
3 W. Boelhower, *Immigrant Autobiography in the United States*, Venice, Essedue,
1982, p. 222.
4 I. Planinsek Odorizzi, *Footsteps Through Time*, Washington, Landmark Tours,
1978, p. 59.
5 L. Adamic, *Laughing in the Jungle*, New York, Harper, 1932, p. 5.
6 I. Kobal, in *Anthology of Australian Slovenes*, Ljubljana, Slovenska izseljenska
matica, 1985, pp. 8–9.
7 F. Maganja, 'In delo sem dobil' (And I did get work) in *Slovenski izseljenski
koledar*, Ljubljana, 1961, p. 246.
8 The *Leopoldinen Stiftung* in Vienna.
9 The letters were published in the official publication of the *Leopoldinen Stiftung*
named *Berichte* (reports). Some, not all, were translated into Slovene and reprinted
by local Slovene newspapers.

10 The letters have been published in their original and in Slovene translation by Professor Janez Stanonik in *Acta Neophilologica*, Ljubljana, beginning in 1987. Letters of others, written in German, were published in the *Berichte der Leopoldinen Stiftung* in the year of their writting.

11 V. Grill and J. Petrič, *Med dvema svetovoma* (Between two worlds), Ljubljana, Mladinska knjiga, 1979, p. 33.

12 His real name was Ignacij (Ignatius); he changed it to Vatroslav (Vatro for short), the Slavic version of Ignatius.

13 Planinsek Odorizzi, op. cit., p. 19.

14 This extract from his book was published as 'No help wanted' in *Slovenski koledar*, Ljubljana, 1982. p. 154.

15 These authors published their stories, poems and even dramas in Slovene-American newspapers and magazines; only a few produced Slovene-written books. So far two anthologies have attempted to collect the best authors, one in English and one in Slovene: see G.E. Gobetz and A. Donchenko, *Anthology of Slovene-American Literature*, Dover, Delaware, privately published, 1977; J. Petrič, *Naši na tujih tleh* (Our people on foreign ground), Ljubljana, Cankarjeva založba, 1982.

16 This becomes obvious especially in autobiographical or autobiographically informed literature where descriptions of landscape, sites, weather conditions and the like are rendered in a most straightforward manner.

17 The author, signed M.V.V. (unidentified as yet) entitled his narrative 'Iz Cirkuš v Ameriko' (From Cirkuše to America) – Cirkuše being a small village in central Slovenia. The first installment of what must have been a longer narrative appeared on 3 September 1891 in *Amerikanski Slovenec* in Chicago. Not all issues of this paper have been preserved.

18 Again the author is anonymous and unidentified, the title of his narrative being 'Iz Ljubljane v Ameriko' (From Ljubljana to America). The 17 January 1895 issue of *Glas Naroda* (Voice of the People), New York, brought a continuation of the story that must have begun earlier, in 1894. Unfortunately, the early issues of the paper are not available.

19 F. Mlakar, *He, the Father*, New York, Harper, 1950.

20 Anna P. Krasna (1900–88), poetess, writer, journalist, went to America in 1920 after the Italian occupation of the littoral part of Slovenia. Vatro J. Grill (1899–1976), editor, writer, translator, emigrated in 1913.

21 A.P. Krasna, 'Tujina' (Foreign country) in *Pesmi izseljenke – Poems by Immigrant Woman*, Ljubljana, Slovenska izseljenska matica, 1986, p. 86. The collection includes poems written in both Slovene and English.

22 Janko N. Rogelj (1895–1974), poet, writer, journalist. He emigrated to America in 1913.

23 Apart from the anthologies already mentioned, Slovene-Australians published another volume of *Zbornik avstralskih slovencev – Anthology of Australian Slovenes* in Ljubljana in 1988. Ivan Cimerman edited a volume of Slovene-Australian poetry entitled *Lipa šumi med evkalipti* (A linden tree rustling among the eucalyptuses), Ljubljana, Slovenska izseljenska matica, 1990.

24 *Anthology of Australian Slovenes*, vol. l, 1985, p. 12.

25 ibid., p. 78.

26 ibid., p. 44.

27 In *Dnevnik-Diary: Slovene Canadian*, vol. 2, no. 15, May 1977, p. 21.

28 J. Petrič, *Naši na tujih tleh*, op. cit., p. 369.

11

MIGRATION IN CONTEMPORARY MALTESE FICTION

Arnold Cassola

INTRODUCTION

This short chapter introduces a little-known national literature from a Mediterranean island-state which has had an intense experience with migration. In fact, during the 1950s and 1960s Malta had one of the highest rates of emigration in the world.[1] At a structural level, emigration during this period was thought to be stimulated by the archipelago's narrow natural-resource base, by its very large mean family size, and by difficulties encountered in restructuring the economy away from the country's main colonial function as British naval garrison and dockyard.[2] Most of the migration literature referred to in this chapter dates from this period of maximum mobility of the Maltese population when people were moving in large numbers not only to foreign destinations (chiefly Britain but also Australia and North America) but also internally within Malta (from rural to urban districts). The insularity of Malta – both of the island itself and of individual rural communities within it – is a characteristic which also infuses some of this literature, and conditions the experiences and perceptions of the migrants themselves. This insularity – in particular the isolation of *self* – can hardly be quantified, but is a recurring theme in Maltese literature, as we shall see.

BACKGROUND TO MALTESE LITERATURE

Although the earliest Maltese writers were the product of an Islamic culture and used Arabic as the medium of their literary production,[3] Maltese literature has, until quite recently, generally followed the prevalent cultural and literary trends of the nearby Italian peninsula. Dante's tongue played a very important part in Maltese linguistic history. As Latin started losing its multisecular importance, Italian became the regular language of Maltese culture, and remained so until the early twentieth century.[4]

As regards the traditional Maltese novel, the model has been Walter Scott's historical romance, filtered through the pen of Alessandro Manzoni. With

172

Malta having become a centre of activity for Italian intellectuals, especially after the granting of the freedom of press in 1839, Italian contemporary thought and ideals were to become more readily accessible to the Maltese. Hence the enormous impact of Manzoni's famous *I Promessi Sposi* on the Maltese intelligentsia.[5]

The Italian *romanzo storico*, which is always set in the past, is meant to be a blend of fact and fiction, with the private drama of individuals overlapping into the historical (thus public) events of a particular period. Depicting the oppression of *past* tyrants seemed to be the ideal way of highlighting the political tyranny of the various *present* dominations (the French, the Austrians, the Bourbons, the Pope) that were hindering Italy from becoming one nation. The parallel situation in nineteenth-century Malta – British domination preventing a serious constitutional emancipation of the archipelago – certainly contributed towards the success of the historical novel. This genre became very popular on the island, thanks also to, amongst others, the Italian exiles Ifigenia Zauli Sajani and Michelangelo Bottari, the Maltese philosopher Nicola Zammit, and Ramiro Barbaro di San Giorgio, a Maltese who spent much of his life abroad, mostly in Naples.[6]

Towards the latter part of the nineteenth century, this genre had definitely established itself. The time was now ripe for local writers to start producing historical novels with a local setting, and written in Maltese. A.E. Caruana's *Inez Farrug* (1889), S. Frendo de Mannarino's *Barunissa Maltija* (1893), Ġ. Muscat Azzopardi's *Censu Barbara* (1893) and *Nazju Ellul* (1909), were amongst the first of a long series of popular historical novels that continued to thrive until the 1960s. Amongst the best of the more recent novels of this genre are those written by Ġ. Galea, such as *Żmien l-Ispanjoli* (1938), *San Gwann* (1939) and *Meta Nħaraq it-Tijatru* (1946).

The changing pattern of Maltese society in the 1950s and 1960s, with the development of tourism and industry on the one hand and the creation of new urban agglomerations such as San Ġwann and Santa Lucija on the other, brought about a new kind of Maltese writer. The generation born in the 1930s and 1940s was no longer concerned with the narration of the collective saga of a people yearning for nationhood (political independence became a reality on 21 September 1964); instead the hero of the modern novel became the individual self, desperately trying to cut himself off from the rest of society and striving to distance himself from its traditional beliefs and dogmas, the fossilised remains of a rapidly disappearing past.[7]

In this battle between the values of tradition (represented by society) and the non-values of the present (personified in the figure of the ever-dissatisfied *self*), it is the latter that turns out nearly always to be the loser.[8] Friggieri and Massa have pointed out how the defeated 'self' has only two options left to wash away the wounds brought about by this defeat: suicide or emigration.[9] Eschewing the morbid, the remainder of this chapter takes a closer look at the causes and characteristics of the Maltese migratory instinct as expressed in a selection of recent novels.

THE MIGRATION THEME IN MALTESE FICTION: SOME EXAMPLES

At a psychological level, emigration is usually intended to be a definitive form of escape from a society whose values are unacceptable to the individual self. Sometimes, it is seen as the ultimate solution after various other attempts at 'escaping' from the problems of life prove unsuccessful. A good illustration of this is found in Sant's novel *L-Ewwel Weraq tal-Bajtar* (The first fig-leaves)[10] where Raymond is passing through a profound identity crisis. He first attempts to combat this crisis by trying to force himself to study and find refuge in work. This having failed, he goes a step further and isolates himself from his classmates; he also refutes the traditional values of religion and the family. At a certain point, he even tries to conform by completely divesting himself of his individuality at a disco, and abandons himself to the anonymity of the dancing crowd. But all to no avail: the only apparent solution to his *malaise* is the complete recision of his national roots and total immersion in an unlimited and anonymous new world (England).

The final scene, on the boat to England when Raymond vomits violently,[11] leaves the reader sure of one thing: Raymond has freed himself of his 'undigested' Maltese past. Yet, a stronger doubt remains: will the 'liberated' self ever adjust to, and be accepted by, the new world? Sant provides no definite solution to this query. The answer is left completely to the individual reader's interpretation and/or imagination.

Incidentally, in *L-Ewwel Weraq tal-Bajtar* we are presented with three different members of the same family who, at one time or another in their lives, have to deal with the problem of emigration. The reason behind Raymond's urge to go is his feeling of being unfulfilled as an individual. His decision to leave the island is, therefore, ultimately a conscious, responsible and *self-chosen* one. On the other hand, his uncle Peter's decision to emigrate and his father's near-emigration are *imposed* on them respectively by the ethical rules and the economic and labour policies of Maltese society. In fact, Peter had no other choice left but to abandon the Maltese shores after he had been involved in a homosexual affair with a young lad.[12] For the close-knit Catholic society of Malta, Peter could be none other than an outcast, forced to give up his past roots and to build himself a new life (and virginity) in a foreign land. As for Raymond's father, Saver, he seems to be very happy with the workings of Maltese society and in fact does his utmost to become part and parcel of it. Being elected on to the committee of the local football club is the greatest ambition of those people who, like Raymond's father, are glad to conform. This is because the local club – whether football, band, political or social – is the central hub of Maltese village or town life.[13] However, the threat of unemployment forces Raymond's father to take into serious consideration the prospect of emigrating. The reader is thus presented with a situation where Saver, totally happy with his environment, is forced to abandon it, while Raymond, totally unhappy with Malta, is obliged to stay there to support his family in his father's absence. It is only towards the end of the novel that the situation changes: Raymond's father finds alternative work and can thus stay in the land he feels at home in; Raymond gets a positive answer to his request for help from the other rebel of the family – his uncle Peter who had not conformed to the sexual norms of the community – and has the opportunity of moving away from the restricted (and restricting) Maltese sphere to embrace the open-minded way of life of an international community (the British one) that might not even be prepared to accept the provincial islander amongst its fold!

In Frans Sammut's *Il-Gaġġa* (The bird-cage), written in the form of a diaristic flashback, the hero, Fredu Gambin, comes to the same conclusion as Sant's Raymond: emigration seems to be the only apparent cure to the tormented self's *mal de vivre*. This ultimate solution is reached only after the hero's various attempts at fighting his *malaise* fail. Fredu first tries to broaden and reform the restricted mentality of the local village community by suggesting that the works of progressive writers such as Ibsen, Chekhov or Brecht be put on in the parish theatre.[14] His idea being rejected, he starts divesting himself of sincere emotions and lets himself go to the materialistic and transient pleasures of lust. In doing so, Fredu himself becomes a symbol of that hypocrisy which he had originally intended to combat: 'what might be unattainable in the future is mine today.'[15] However, he himself soon admits: 'I've become the hypocrite I used to detest so much.'[16]

Tired of village life, Fredu develops a love–hate relationship with the 'glamorous' city life: he finds fault with the people from the rich Maltese suburb of Sliema, but is at the same time attracted by their lifestyle which is something totally alien to provincial villagers: 'Leaving the village and settling down in Sliema is like settling down in another country.'[17] His move to Sliema virtually implies migrating to a totally different place where 'customs are different. Language is different. People are different too.'[18] Such was the disparity between city and village life in the Malta of the 1960s!

Yet, internal migration does not solve Fredu's problems. In Malta, it seems, one is always doomed to be marked by one's birthplace: 'I realised I was born a villager and a villager I still was.'[19] His ultimate conclusion is that he cannot do without emigrating from the island that imposes its restrictions upon him; not before, however, having toyed with the idea of committing suicide.[20] As with Raymond, emigration means 'endless stretches of open space',[21] but, again as with Raymond, whether this 'vastness' will be ready to assimilate the solitary islander within its framework remains a question. The reader is left with the impression that no matter how much and how far Fredu tries to escape, his problems and dilemmas survive and thrive within him.

Sant presented us with a hero who emigrates abroad; Sammut with one who first migrates internally and finally settles down abroad. In J.J. Camilleri's *Il-Għar tax-Xitan* (The devil's cave)[22] the peregrinations of the hero, Jumi, are even more complex: Jumi leaves his home village for the big city at the age of 20, comes back after eighteen years to reform his people and their superstitious beliefs, gives up and leaves again for the big city after having been defeated in his aims, but in the end returns a second time to his home village to regularise his position in life through conventional marriage.

In *Il-Għar tax-Xitan*, village and city life stand at the opposite ends of people's scales of values. The San Rokku parish priest, who represents tradition, believes that whoever is attracted by the glamour of city life is forever doomed. For him, any type of contact with the city is bound to become dangerously contagious. Young Jumi not only refutes this attitude but actually finds in the city his only refuge from the pettiness of the village, where everything about one's private life is general knowledge.[23] The enormous difference between the near-dormant village life and the excitement of city life is conveyed by the author's mythical description of the city, as seen through Jumi's eyes, which depicts a gap between city and village out of all proporation to geographical distances in little Malta. What Camilleri actually does is to render concrete an abstract reality: the mentality gap between city and village is so wide that the author can render it on the page only by amplifying the physical (and mythical) distance between one place and the other.[24]

Jumi's first departure from San Rokku signifies his rejection of the community's traditional beliefs, deeply embedded in past history. His return after several years (this is practically where the novel starts) is an attempt at reforming this mentality, after he has seen the light in the big city. However, the close-knit 'primitive' community is much stronger than the 'enlightened' individual, who is again forced to escape from the village in order to survive: first by finding temporary refuge from the narrow-minded villagers in the solitude of an isolated country abode; then by returning to the city he had come from. Having defied the ethical rules of the community, by living together with the village sinner Petriga, Jumi is left with no option but to abandon his native village.

The moral of the whole story is 'conform, or be defeated'. Jumi is totally defeated in his aspirations, not only because he is twice forced to flee his home village but also because in the end he actually returns to his birthplace to adjust to the norms of the local community in the most conventional of ways, that is, by getting married to Petriga. For a person who had already changed his place of abode three times in the relatively short span of eighteen years, the final settling down in his place of origin can only signify the ultimate defeat.

The normal migratory trend, from a small centre (the village) to a bigger one (city or foreign country), is reversed in V. Buhagiar's *Id-Dar f'Tarf l-Isqaq* (The house at the end of the alley).[25] In this novel, the city girl from Sliema, Helen, behaves in such a way as to frustrate what would be the aspirations and dreams of Raymond, Fredu and Jumi. In fact, Helen abandons what is supposed to be *the* 'glamorous' city of Malta for an isolated country house because in Sliema too hypocrisy manifests itself at its highest level. On the surface, everything in Sliema looks nice, but when real problems crop up (Helen is raped by her fiancé Roger and left with child) everybody abandons her, including her own parents. While Fredu and Jumi do their utmost to leave the village for the big city, Helen does exactly the opposite because it is only in a secluded country house that she can find some peace of mind.[26]

The idea of a temporary isolated place of refuge seems to be a popular theme with contemporary Maltese novelists. While Buhagiar's Helen leaves the city for a secluded country house, Camilleri's *Il-Għar tax-Xitan* introduces us to the concept of a temporary transfer of domicile from the 'crowded' village to a 'solitary' abode, nearly completely cut off from any human presence (Jumi finding serenity in the country shack). For Samwel, the hero of Sammut's *Samuraj*, the happy moments spent together with his lover Zabbett can only be appreciated in the semi-wilderness of his farmhouse. So much so that when Zabbett is taken away from him back to the 'civilised' world by the representatives of the 'evil' village (the doctor, the parish priest, the police sergeant), there is no purpose left in life and Samwel abandons himself to violent acts of self-destruction.[27]

While in *Samuraj* the isolated farmhouse becomes an indispensable accessory for the consummation of Samwel and Zabbett's love, in Friggieri's *Il-Gidba* (The lie)[28] Indri's secluded farmhouse becomes an indispensable accessory for the hero Natan to escape from a wife with whom he cannot communicate. In Indri's humble dwelling, Natan can get away from his wife Anna and what she stands for (the hypocrisy of a village community). Ironically, in a civilised society (the village) and in the presence of human beings (Anna), Natan is totally uncommunicative; it is only away from civilisation (in Indri's farmhouse) and in the presence of an irrational being (his dog Fefu) that Natan can find his powers of communication. In this case a dog really is a man's best friend: Fefu is the only living creature Natan can actually 'speak' to.[29] Friggieri seems to be implying that only a temporary escape into isolation can create the necessary conditions for real communication.

There is still one other alternative left for those who need to escape from their natural environment: leaving one's village permanently for yet another village. In the Maltese case, this means abandoning one small centre where everybody knows you for another small centre where nobody yet does. This is Natan's choice in *Il-Gidba*, when he decides to cohabit with his lover Rebekka in her home village. But the migration from one village to another does not bring about any change either in Natan or in people's attitudes. Wherever one 'flees', things remain invariably the same.[30]

These different forms of 'escape' do not solve any of the heroes' problems. Frans Ellul, the hero of Zahra's *Ħdejn in-Nixxiegħa* (By the spring), sums it all up when, having abandoned his fishing village for the town of Hamrun, he goes back to his native settlement once again because that is where he really belongs:[31] 'This is the only place I belong to. This is where I can breathe the air that purifies me.'

CONCLUSION

Ultimately, therefore, not even the most radical solution adopted – complete departure from one's homeland (emigration) or total separation from this world (suicide) – bring about any solace. Luigi Pirandello once stated that every man is an island, but that Sicilians are doubly so because of their geographical isolation. I would add that the Maltese are thrice an island, being also physically cut off from the nearest mainland, which is . . . isolated Sicily! This triple dose of insularity cannot be escaped even by leaving Malta: the individual sense of isolation caused by the physical and geographical isolation of the country is something the Maltese are destined to carry within themselves inexorably. The *toccata e fuga* theme of contemporary Maltese fiction is the literary expression of a very tangible Maltese and, at the same time, 'eternal' reality: man's inability to understand himself.

NOTES

1 H. Jones, 'Modern emigration from Malta', *Transactions of the Institute of British Geographers*, 1973, no. 60, pp. 101–20.

2 R. King, 'Recent developments in the political and economic geography of Malta', *Tijdschrift voor Economische en Sociale Geografie*, 1979, vol. 70, pp. 258–71.

3 See M. Amari, *Biblioteca Arabo-Sicula*, Turin and Rome, Loescher, 1880, vol. l, p. 241, and M. Amari, *Storia dei Mussulmani di Sicilia*, Catania, Prampaloni, 1939, vol. 3, pp. 773–4, 785.

4 On the history of the Italian language in Malta see A. Cassola, 'Malta', in F. Bruni (ed.), *L'Italiano nelle Regioni*, Turin, UTET, 1992, pp. 861–74, and A. Cassola, 'Malta', in F. Bruni (ed.), *L'Italiano nelle Regioni – Testi e Documenti*, Turin, UTET, 1994, pp. 843–59.

5 On the activity of Italian exiles in Malta during the nineteenth century, see V. Bonello, B. Fiorentini and L. Schiavone, *Echi del Risorgimento a Malta*, Milan, Cisalpino-Goliardica, 1982; G. Mangion, *Governo Inglese, Risorgimento Italiano ed Opinione Pubblica a Malta, 1848–1851*, Malta, Casa San Giuseppe, 1970; O. Friggieri, *Movimenti Letterari e Coscienza Romantica Maltese (1800–1921)*, Milan, Guido Miano, 1979.

6 Ifigenia Zauli Sajani (1810–83), the wife of a lawyer from Forlì who sought refuge in Malta after the 1831–2 uprisings in Italy, was one of the first authors to publish historical novels in Malta. Her novels include *Gli Ultimi Giorni dei Cavalieri di Malta* (1841), *Il Ritorno dell'Emigrato* (1841), and *Beatrice Alighieri* (1847). Michelangelo Bottari (1829–94) first came to Malta in 1849, launching *Il Corriere Mercantile di Malta* in 1856 (with Gugliemo Finotti). He wrote many historical

novels in Italian, some of which were translated into Maltese. Nicola Zammit (1815–99), philosopher, doctor and architect, was the editor of *La Crociata, La Fenice* and *L'Arte*. Notable amongst his innumerable publications is the historical account *Angelica o la Sposa della Musta* (1880). Ramiro Barbaro di San Giorgo, a journalist, writer and politician, was twice forced to flee the island due to his journalistic and political activity. He was the correspondent of *La Gazzetta d'Italia* in various European cities including Berlin, Athens, Paris, Budapest and Vienna. Amongst his most significant literary works is *Un Martire: Romanzo Storico Maltese del Secolo XVI* (1878).

7 I use the male gender in this sentence simply because the heroes of these Maltese works almost invariably *are* male.

8 Contemporary Maltese poetry and fiction reflect that sense of *malaise* which had gripped European intellectuals about a century earlier. Basically, the *credo* of the French *decadent* movement of the late nineteenth century, filtered through the works of Marinetti and the futurist movement, but mainly through Anglo-American writers, started bearing its Maltese fruits in the mid-1960s. *Kwartett*, the first anthology of 'modern' Maltese poetry, was published in 1966; *Aħna Sinjuri* (We're rich), a novel by J.J. Camilleri, saw the light in 1965.

9 O. Friggieri, 'L-evoluzzjoni storika tar-rumanz Malti' (The historical evolution of the Maltese novel), in *Saġġi Kritiċi*, Malta, A.C. Aquilina & Co., 1979, pp. 295–312; and D. Massa, 'The post-colonial dream', in *World Literature Written in English*, Spring 1981, pp. 135–49.

10 A. Sant, *L-Ewwel Weraq tal-Bajtar*, Malta, Union Press, 1968.

11 ibid., p. 315.

12 ibid., p. 86.

13 See J. Boissevain, *Saints and Fireworks: Religion and Politics in Rural Malta*, London, Athlone Press, 1965.

14 F. Sammut, *Il-Gaġġa*, Malta, Klabb Kotba Maltin, 1971, p. 25.

15 ibid., p. 48.

16 ibid., p. 51.

17 ibid., p. 106.

18 ibid.

19 ibid., p. 81.

20 ibid., p. 18.

21 ibid., p. 133

22 J.J. Camilleri, *Il-Għar tax-Xitan*, Malta, Klabb Kotba Maltin, 1973.

23 ibid., pp. 8, 101.

24 ibid., pp. 17–18.

25 V. Buhagiar, *Id-Dar f'Tarf l-Isqaq*, Malta, Klabb Kotba Maltin, 1975.

26 ibid., p. 52.

27 F. Sammut, *Samuraj*, Malta, Klabb Kotba Maltin, 1975, pp. 162–8.

28 O. Friggieri, *Il-Gidba*, Malta, Klabb Kotba Maltin, 1977.

29 ibid., p. 15.

30 ibid., p. 101.

31 T. Zahra, *Ħdejn in-Nixxiegħa*, Malta, Klabb Kotba Maltin, 1975.

12

LITERARY PERSPECTIVES ON JEWS IN BRITAIN IN THE EARLY TWENTIETH CENTURY

Stanley Waterman and Marlena Schmool

INTRODUCTION

In terms of time and space, this chapter deals with a very small portion of Jewish migrations over the past five centuries. It records the struggle of an immigrant generation to establish itself in a strange land, and examines the dilemmas of their sons over the linked issues of assimilation and retention of Jewish identity. It does this by comparing literary descriptions of events with situations recorded by historians and other social commentators. Whereas nineteenth-century Jews were represented by Jews in Anglo-Jewish novels of the time for a primarily non-Jewish audience, the literature of the twentieth century appears to have little of this mediating role.[1]

The Jewish experience in Britain since the middle of the nineteenth century has been one of a community built by Eastern European immigrants on earlier Sephardi (Spanish-Portuguese) and Central European foundations. This migration, part of a wider mass movement of Europeans, is well documented in both memoirs and fiction. It is interesting to note, and perhaps a reflection on the cohesiveness of the immigrant Jews, that the literary sources refer more specifically to a people and a culture than to a social geography or cultural landscape.

The stream of Jewish migration to Britain became a mass movement in May 1881, just a few weeks after the 1881 Population Census. At that time, the Jewish community of the United Kingdom comprised about 60,000 people. During the following thirty-five years nearly 150,000 immigrant Jews settled, although it may have appeared to be much larger to some.[2] The majority of the pre-1914 immigrants, coming from the Russian Pale of Settlement, settled in London's East End, although there were sizeable communities also in Manchester, Leeds, Glasgow, Liverpool and elsewhere.[3] Between 1881 and 1914, the Anglo-Jewish community increased almost five-fold through immigration and concomitant natural increase, and the descendants of this immigration gradually became the leaders of the community.[4]

ORIGINS

The traditional character and sheer numbers of the immigrants from Eastern Europe were a disconcerting challenge to the members of the established Victorian Jewish community who were in an advanced stage of adaptation into English society. Most of the 'established' Jewish community were not of the ghetto and had made great strides towards becoming Englishmen of the Jewish persuasion. The Yiddish-speaking newcomers had moved by one or two steps from a *shtetl* or small-town environment in the Pale to the slums of industrial Britain.[5] Israel's first president was part of this immigration. Not untypically, his first step was to the nearest town. He describes the country-side around his native town

> which stood on the banks of a little river in the great marsh area which occupies much of the province of Minsk ... flat, open country, mournful and monotonous but ... not wholly unpicturesque. Between the rivers the soil was sandy, covered with pine and furze; closer to the banks the soil was black, the trees were leaf-bearing ... All about, in hundreds of towns and villages, Jews lived, as ... for many generations, scattered islands in a gentile ocean.[6]

Occasionally, contact with a larger town was part of the preparations for the journey. In Anthony Sher's *Middlepost*, the central character has travelled from the *shtetl* of Plunge to Telz to have an English document drawn up to present to the authorities on arrival.

> They were grateful their journey from Plungyan to Telz was almost over ... Smous's parents were huddled on the front of the wagon ... Smous lay on the back, crammed among his father's goods ... In over thirty-five years ... [his father] had never set foot out of the Plungyan district, even when it had been easier to do so; the permits for today's trip had entailed lengthy applications and hours of queuing ... Telz was a disappointment. It was a city after all, so he had been expecting something far grander – not these dark, ragged streets clinging to the hills. ... The roads were crowded and hostile, everyone hurrying in different directions, gasping from the cold.[7]

The move often launched the immigrants into a lifelong cycle of poverty. The journey itself was arduous, involving lengthy overland and maritime sections. A train via Leipzig and Halle to Rotterdam and the Harwich boat-train to London was one of many channels by which the immigrants arrived in Britain. A more common route was by land from Lithuania or Poland to one of the great German port cities, usually Hamburg or Bremen, and thence to London, Hull or Newcastle-upon-Tyne.[8]

The experience of the immigrant en route and on arrival has been widely described.

The shore line sickened him. While out at sea, the reality of the new life that awaited him, and the remembrance of the severed ties at home in Lithuania, had both become meaningless. But now, faced with the shore, [Rabbi Abraham Zweck] experienced again the pain of departure, and the fearful anticipation of his arrival. He would never see his parents again.[9]

Where the decision to move was voluntary, anticipation of a new life came to the fore. Bernice Rubens, who expresses the poignancy of departure in *The Elected Member*, later presents a contrasting positive picture of migration in *Brothers*. Here, above all, the will to survive is paramount and prompts the move. This is encased in a chilling statement of attachment to place.

> Hardest of all was to say good-bye to one's dead . . . Now all they would have would be memories and a list of alabaster names to tell their children. Now they themselves would be pioneers, would start a history in a new country, would create their own dead for the children's memories, new monuments for their myths, new legends for their inheritance.[10]

In such circumstances, a 'document' was the only means of communication with the host society and was also important for maintaining one's identity or for providing an identity on demand for whomsoever should request it. Lacking knowledge of the language, the tiniest contact with the local Jewish or Gentile community eased the passage; the 'document' provided this service.

> His travel documents and money were kept in the inside pocket of his coat, while his most precious possession, a letter of introduction written in English, he kept in a waistcoat pocket close to his breast. It was a beautiful scroll of parchment with a yellow satin ribbon, drawn up by a linguist in Telz, and proudly described by Smous's father as his 'diploma in English.'[11]

In the same way, Rabbi Zweck, although equipped with just a slip of paper rather than a beribboned scroll,

> put his hand in his inside pocket, checking once again on the slip of paper that was his sole contact in the strange world that approached him. He looked at the crumpled sheet. The name was Rabbi Solomon, and the address, number 16 of an unreadable street followed by the letters E.2.[12]

If the paper was lost, memory could be used:

> suddenly his mind monitored Khasina's address and the vision of the small card he had left on the tavern counter. Pavel Khasina, 24 Bute Street, Cardiff, Wales, England [*sic*]. He repeated it in the phonetic

version they had taught themselves ... then screamed it aloud across the sea, as if appraising his sponsor of his arrival.[13]

This lack of language was a formidable barrier.

[He] looked up and saw the officer smiling at him. 'Another half an hour and we'll be landing', he said. Abraham Zweck raised his eyebrows. It was the safest movement to make if you didn't understand the language. It could mean 'yes' and 'no' and the nuances of 'perhaps' and 'nevertheless'.[14]

Although anti-semitism and the threat of persecution constituted important factors in the decision to migrate, economic betterment was also a strong motivating force. For many, the emigration was a deliberate process, based on some knowledge of the intended destination. It could, perhaps, be the response to a life-event. Thus, on the death of his mother, David Levinsky found

My former interest in the Talmud was gone. The spell broken irretrievably ... My surroundings had somehow lost their former meaning. Life was devoid of savour, and I was thirsting for an appetiser, as it were, for some violent change, for piquant sensations. Then it was that the word America first caught my fancy. The name was buzzing all around me.[15]

ABSORPTION

As the flood of migration became of increasing concern to the established community, the immigrants' sense of isolation was somewhat assuaged by organisations such as The Jews' Temporary Shelter, to help the newcomers.

As he reached the end of the gangway, a man approached him and asked in Yiddish if he wanted some hot soup ... 'It's free from the Jewish Board of Guardians,' the man said, and he pointed to a banner with that legend above the stand. . . . Then he heard someone call ... 'Bindel' ... and he turned around and saw ... Bok running towards him. 'They are here' he cried ecstatically, 'The Shelter have come to meet us'.[16]

Even some later migrants, arriving in Britain from Central Europe in the 1930s or after the war – educated, cultured and without a severe language handicap – encountered similar feelings of strangeness as had the *Ostjuden* three or four decades earlier.

He stood by the grey pavement while the cold-faced crowds went by. Otto wore a long black coat and a black hat. He held a small black attaché case. His shoes were black too, and even his face had a blackness in it; the eyes were black and the flesh had a thinness through which it seemed black bone was pushing. Otto Kahane had spent two years in Dachau Concentration Camp.... The shops in Cricklewood High Street were full of good things. The faces of the people were in bloom with the cold ... Otto nodded to himself. Once he had belonged to a world like this. In Poznan before the Germans came, there were shops as good as this. Better, Otto smiled to himself and nodded again, much better. Otto Kahane: Watch Repairer and Jeweller. That was nice.[17]

After the long and hazardous trip from the point of departure, maintaining an identity or setting out with a new one was often the next hurdle.

He was born Zeev Zali – that is the name on his birth-registration papers and it is written clearly, without smudge or error. However, some thirty-five years later, when he made his great journey from his native Litva to British South Africa, his document of identification then bore the name Zeev Immerman. His certificate of entry, completed upon arrival, showed a different name again – Maurice Josif Brodnik. Within a few months he was known as neither Brodnik, Immerman, nor Zali, but Smous ... which isn't a name at all, being the word in Afrikaans language for hawker or peddler, his trade at that time.... He often thought how Yiddish the word sounded, despite being Afrikaans, and of the several names at his disposal, it became his favourite.[18]

Though change of identity was symbolised in change of name, identity and sense of self could also be altered in other ways. At the hands of a mentor the newly landed David Levinsky was taken to:

a barbershop with bathrooms in the rear. 'Give him a hair-cut and a bath,' [the mentor] said to the proprietor. 'Cut off his side-locks while you are at it. One may go without them and yet be a good Jew.' ... When I took a look in the mirror I was bewildered. I scarcely recognised myself.[19]

Negotiating the geography of the new environment was often a hazardous ordeal. The journey could prove harrowing and permitted the immigrant to perceive his own strangeness in all its defenceless exposure.

He looked down at himself. His small black boots and white wool socks protruded untidily beneath his long black coat.... The shadow of his wide-brimmed black hat crossed the tip of his boots, and ... his side-locks ... cast their shadows, and he rocked his head to and fro, trying to fit the shadow hat on his boots and the curls alongside....

He followed them to London, through the customs and on to the train with an occasional raising of the eyebrow, and without uttering a single word. He couldn't remember how he'd reached the address in the east end of London. All he recalled were people staring at him, in the street and on the strange tramcars . . . Then gradually, people stared no longer.[20]

The unfamiliarity and outlandishness of the Jew was apparently greater than that of other groups which had preceded him into those neighbourhoods that served as immigrant repositories. Their alien dress and customs made it appear to observers that their numbers were greater than was actually the case.

These aliens . . . [the Jews] were not like the Germans who had been coming in considerable numbers over the years to work in the breweries or the sugar refineries, or the Irishmen who settled near the docks. . . . [E]verything about them, their language, their garb, their side-curls and beards, their eating habits and non-drinking habits, their way of life, their very manner of speaking set them apart from their neighbours, and they had no need to descend in countless hordes to make their presence felt.[21]

Although the people were familiar, the new physical environment would be vastly different from that left behind. Nothing could have been starker than the contrast in physical conditions. Compare, for instance, Weizmann's earlier description of the Pripet marshes with the scene in which Israel Zangwill introduces his ghetto:

A dead and gone wag called the street 'Fashion Street', and most of the people who live in it do not even see the joke. If it could exchange names with 'Rotten Row', both places would be more appropriately designated. It is a dull, squalid, narrow thoroughfare in the East End of London, connecting Spitalfields with Whitechapel, and branching off in blind alleys, . . . its extremities within earshot of the blasphemies from some of the vilest quarters . . . in the capital of the civilised world.[22]

The immigrants and their offspring continued to live a very cramped existence. Life in the early decades of the twentieth century is remembered with differing levels of acceptance and/or rebellion but the general picture painted by those then at the lower social levels is very similar.

Then I remember our moving to Broughton Buildings. We had taken two flats, each of two rooms. One flat rose out of the basement where all the dirt and dust and refuse of fifty flats lay after it had been shoved down slots on each of six landings of the tenement. Our other flat was . . . atop the one below. . . . Downstairs in the basement, there were rats. And bugs. Upstairs in the show rooms there were just bugs. Legions of them.[23]

Each flat had two rooms and ten people. Tenements were much the same everywhere. However, even in retrospect, Ralph Finn does not admit to childhood anger. Ralph Glasser's recollections of the Glasgow tenement where he grew up are of yet more miserable conditions: 'The Victorian building, in red sandstone blackened by smoke from Dixon's Blazes, was in decay. Splintered and broken floor boards sometimes gave way under your feet. The minimal plumbing hovered on the verge of collapse.'[24]

Both Glasser and Finn, as children of immigrants, used secular schooling as a way out of the ghetto and physical deprivation. This involved a distancing from the community and from the overarching sense of Jewish identity it came to provide. At the same time, each underlines the social and cultural distance between themselves and the host population, but each evaluates it differently. Recalling a fight at elementary school, Glasser reports:

> I was never attacked again. That perhaps proves nothing. But it must be remembered that part of the prejudice of that time was that the Jew triumphed over the guileless Christian by art and subterfuge, that he was somehow slippery, hard to pin down, a coward. I had stood and fought, and though I had lost the battle, I had done something to weaken the myth.[25]

For Finn, there was a sense of superiority related to group solidarity.

> I often went with my friends to the neighbouring districts ... to see *the slums*. We called those places slums. We had no idea that we also lived in one. Maybe the fact that most of Broughton Buildings was a Jewish territory had something to do with it. With Jews, children come before beer. ... Let's face it. Intelligent people in those days did not frequent pubs.[26]

For this reason, Finn finds the greatest social distance between immigrants and 'Choots', second- and third-generation English-born Jews of Dutch origin.

> They ate *traife*, forbidden food. They loved their jellied eels. They drank beer. They lived by gambling. Most of them were either bookmakers or employed by bookmakers. And they struggled along until greyhound racing came to Britain. Then they went to the dogs and grew rich. ... The Choots children did not go to *cheder*, Hebrew classes, in the evening, and very few went to the Jews' Free School.[27]

Cheder was continuity with the place of origin, an attempt to maintain a pattern of education which had sustained the close-knit, all-embracing, traditional communities of the Pale. When confronted with a questioning mind and an accessible outside world, the very method that had ensured a basic literacy in the Jewish world of Eastern Europe became counter-productive, often inducing ambivalence towards Jewishness in the immigrant's son. Religion and Jewish knowledge became viewed as a kind of despotism.

186

We Jewish children acknowledged the superiority of the Gentile method in one field: religion. He was practically exempt.... For four days in the week – and a good slice of Sunday – we were forced to spend the time between school and supper cooped up in the *Cheder*, the Hebrew school. Anything less like a school you never saw. It was usually a cellar-kitchen or disused workshop where the tutor, a bearded, unkempt, smelly old man in his dotage, mumbled at you for hours on end out of a large book.... We prayed for the last day of *Cheder*: the day that would leave us as free as the Gentile children. When you left the *Cheder* after your *barmitzvah* ... it took you exactly one week to forget all you had learned.[28]

Religion, while implicit to Jewish identity, was not central to it; it provided the cornerstone of a sense of peoplehood but was more important to the older immigrants.

In our neighbourhood, religion was a kind of family affair, to be treated with irony and ambiguity. People made sly jokes about rabbis and whenever things didn't work out well they addressed their asides to the *Rabboine Shel Oilem*, the Lord of the Universe, chiding him for not contriving a better fate for his Chosen People.[29]

[*Zaida* (grandfather)] was fanatically devout.... On Friday evenings he went off to synagogue to usher in the Sabbath. When we were young, my brothers and I went with him. We respected him and did it to please him.... He spent most of Saturday at the synagogue. He spent most of the holy Fast days and Festivities in the synagogue.... He did it because he believed.[30]

While sons and daughters turned away from religion, for the original immigrants the familiarity and nostalgia for food and culinary traditions frequently conjured up images of the former environment.

In the Whitechapel Road it was all bright lights and crowds of people, smart as paint, taking a Saturday night stroll after working the week as machinists and underpressers and cabinet-makers.... They crowded into restaurants for lemon tea, and swelled out of the public houses waving bottles, their arms about each other's necks, their children waiting at the doors with narrow chests.... Joe took giant strides past Russian Peter with his cropped beard and Russian peaked cap. Russian Peter usually had wreaths of garlic cloves and pyramids of home-pickled cucumbers on his barrow, a large box with handles mounted on two wheels ... Russian Peter's cucumbers were pickled by a special recipe he brought with him from Russia.[31]

Just as the character of the Anglo-Jewish community had been indelibly altered by the influx, so the character of the immigrant was changed by the move. Not least among the elements influencing this change was the response of the settled community. Financiers and professionals saw their places in British society threatened by the waves of working-class migrants, illiterate not only in English, but also in the 'civilised' tongues of central and western Europe, yet literate – and literary – in their mother-tongue, Yiddish.[32]

EMPLOYMENT

Earning a living was an immediate issue. Societies existed to aid immigrant absorption and relieve pressure on established Jewish society. A major problem was that there were few callings open to the Jewish immigrant. Nevertheless, the Jews brought with them a range of modern skills from the *shtetl* that could be adapted. There were artisans and many immigrants engaged in petty trade.[33] Above all, they were literate, and could cope with a version of democratic decision-making. Despite children's revolts against parental experience, traditional discipline had taught them how to defer gratification, preparing them for the new economic rigours as well as training them to view critically their given conditions of existence. As a community, Jews also knew how to function in a diverse and ethnically varied world. As they interacted with all kinds, they also developed their own modern culture.

Although most immigrant Jews hoped to maintain their identity, both the 'established' leadership and the newly acculturating wave of immigrants expected that the most recent arrivals would undergo transformation.

> 'A Rabbi you shouldn't bother with,' the father was saying. 'For you,' he said, weighing him up as if after years of acquaintance, 'business is better. Your own business, you marry, you have a family. No troubles . . .'. 'Forget the *Rabbonischkeit*,' he almost shouted. 'A business you should find.'[34]

In the move from the Ukraine to the Welsh mining valleys, young Aaron Bindel did not remain a tavern-keeper but was trained as a credit-draper – a tally-man for clothes – in villages like Caerphilly and Senghenydd. This occupation:

> was a trade most favoured by those immigrants who did not have a particular skill. Its great advantage was that it required little initial capital and even that could be credited. . . . The profit margin was narrow, the work-hours long and arduous but . . . some of them, after a year or two, had bought their own houses and some had used the profits to set up a shop of their own.[35]

The clash between the East European immigrant and those Jews settled prior to 1870, part of a wider societal class struggle, was augmented by fundamental intergenerational differences. Where the settled community, not itself monolithic, avowedly set out to turn the newcomers into Englishmen, these changes were bound to occur anyhow.[36] Moreover, these differences could sour relations between generations. This situation was unlikely to be resolved easily where economic necessity curtailed education and led to early employment in a sweatshop or similar dispiriting environment.

> I hadn't calculated on a sweatshop in my childish plans. To be frank, I hadn't calculated on work at all at the time I was put to it. I imagined myself safe for at least another two years at the Central School I was attending. My parents had other plans. They felt that somebody in the family besides my father ought to do some work, and as I was the eldest of six children they suddenly withdrew me from school a few weeks after I had turned fourteen. It was a long time before I forgave them that.[37]

Jewish employment patterns included a wide spectrum, from the working class, the artisans, pedlars, hawkers and small shopkeepers of the immigrant neighbourhoods to the rich merchants.[38] Nevertheless, for most immigrants, particularly in the metropolis, the sweatshop was the basic reality of their existence. In these circumstances, some Jews became active in socialist organisations and union activities, many of which were related to the cultural framework. Becoming a union member did not always emanate from socialist principles or ideology. The union rooms served as much as a convenient meeting-place for immigrant workers willing to help one another in a new environment as a centre for political organisation.

> The Union was in Whitechapel Road, and in the week there were not many tailors there, but on Sunday mornings they filled the room and spread out into the street, chatting in their long coats about this or that, small groups of them for a hundred yards up the Whitechapel Road. Sometimes a master-tailor would come up and say, 'Have you seen Chaim? I got three days' work for him.' The Union room itself was dirty, with dusty windows on which someone had written with a finger . . . . Behind the trestle Mrs. Middleton, the caretaker, stood cutting rolls, pouring tea, and talking Yiddish with some old tailor who, like, Mr. Kandinsky, looked in to hear what was happening in the world.[39]

The community feeling of the immigrants in their new environments is underlined frequently, and sometimes nostalgically. A successful child of the ghetto returns to the area of first settlement. Life has been good, but the upward social move has come at the expense of the close community life of the immigrant neighbourhoods – or at least its memories and, to an extent, its romanticisation.

> The taxi rolled down a hard street between stunted houses and this was where Hannah Adler sprang from, this was the street where she went to school, this the one where grandpa bought *chola* on a Friday night,

this where they went to synagogue: The Great Spitalfields Synagogue. . . . This was real, this was the solid reality of where life started
and what life came from and so they went to Whitechapel and down to
Bethnal Green where life ended in the old Jewish cemetery. It was a
community. Here in the old Jewish East End which now, as they sped
down the Mile End Road, seemed so empty and finished, there had been
a community, so that a child knew what he was about, born among his
own people, knew what moulded him and what he might become, born
in a community which might be frightened or intimidated but in which
people remained, not fainting against each other for love, but Jewish,
Jewish though they all die for it, together.[40]

The reality for the children of the immigrants was less homogeneous and
less community-bound. Dannie Abse, growing up in the factual equivalent
of Bernice Rubens' fictionalised Cardiff, saw himself as 'a Jew in the mirror',
having as a child 'had but the vaguest conception of the difference between
Christian and Jew'.[41]

David Daiches, in the between-wars world of Edinburgh, was doubly
marginalised – at school as a Jew, and in the small Edinburgh Jewish
community as the son of the rabbi, thus escaping the anti-semitism of the
tenement-dwellers. At the same time, he had 'no Jewish friends to whom I
could turn for real companionship. The Edinburgh Jewish community was
small – about four hundred families, and . . . though I knew the other Jewish
children in the city . . . I was intimate with no one.'[42]

While for some, the loss of this warmth was viewed with regret, others
moved willingly away from their roots.

Sarah had promised to visit her sister one Saturday and she persuaded
Alec to go along with her. It was a long ride to Barnes. . . . It took the
best part of an hour to get there. . . . Mrs. Saunders kissed Sarah
affectionately, and shook Alec's hand. She summed him up at a glance.
Not too bad. Clean and respectable, probably intelligent. And didn't
look too much like a Jew either. She was glad of that. . . . She didn't
want the whole neighbourhood to know she was Jewish.[43]

Many immigrant Jews rapidly left their marginality behind. The success of
the Jewish immigrants, their mobility upward from peddler and artisan to
store-owner and businessman, to membership of the free professions and
academia, engendered as much jealousy and hostility as their very being Jews
had caused in the more traditional world whence they came.[44] The hostility
was not just anti-semitism, but hostility from those Jews who had been
comfortably accommodating themselves to English society until the arrival
of their co-religionists.[45] Although there was a well-established communal
framework within which immigrants could be readily absorbed and turned

into productive Englishmen without being a burden on the general society, it took time to realise that both were on the same side.

THE SUBURBS AND MIDDLE-CLASS VALUES

Many Jews feared the anti-semitism that success attracted, even when they had moved out of areas of first settlement, mainly from the 1920s onwards.[46]

> As a retail shopkeeper, Isidore was especially vulnerable to mob violence. It was for fear that his family's presence would somehow draw the contagion out to Cricklewood that he had, before the war, barred them from the East End. It was a place to get out of. He regarded his days there with no nostalgia. He and his brothers had fought hard to escape the ghetto. [Isidore] ... had fled from Poland for no romantic reasons.... Isidore fled from fear and the fear had stayed with him all his life and pursued his son.... The shop was a terrible responsibility.... When Hannah insisted on opening a branch in Finchley ... Isidore went through the 30s with doubled apprehension; every onset of Fascism was a double pain to him.[47]

But these fears about Gentile reactions, however well founded historically, could be misplaced, perhaps telling as much about the Jews' apprehensions as about Gentile attitudes.

> Of course [Miss Ronald] knew Mrs. Saunders was a Jewess, but she never liked to mention it. She always avoided talking about Jews in her presence. It was tacitly understood that Jews were non-existent as far as Mrs. Saunders was concerned. She would really have liked to know more about that peculiar people, but her friend was so secretive on that subject. She always turned red, and shut up like a clam whenever Jews were mentioned.[48]

The move from the East End to the suburbs, along the Underground lines and beyond, was not always accomplished with a clean conscience. Not only was there nostalgia for the old community life, there was fear for the future. This mixture of fear of the unknown and an acute awareness of the variations in status of the various London suburbs settled by Jews emerges time and again in Jewish writing. The following exchange between Isidore Adler and his wife Hannah tells much about the perceived desirability of places such as Finchley, Cricklewood and Golders Green in north-west London.

> It just happened they were in the middle of moving and Isidore was in a bad mood.... 'I was against this from the beginning. I don't want to move. I don't see why we have to move. I was against moving from the beginning.'
> 'You know you like the house,' Hannah Adler told her husband. 'You picked it.'

'I picked it! I paid for it.' Isidore Adler leaned forward reproachfully. 'I paid for it.'

'You always wanted to live in Golders Green.'

'I never wanted to. What's wrong with where we always lived? Here.'

'You want to live in Cricklewood High Street all your life?'. . .

Woburn Road was on the Finchley side of Golders Green. It contained neat, detached houses. Number fifteen had a green, fish-scale roof. Green tile pillars supported the portico over the front door. The face of the house was made up with white stucco. The dormer windows had green tile lids. Isidore cared little about the house one way or the other, but it was certainly an improvement to be able to garage the Wolsely on the premises rather than two streets away, as he had done for years. Still, he didn't see the sense in moving. As soon as you'd saved enough money to live decently in one neighbourhood, you had to move to another you could only just afford. What was the sense in it. He had always lived above the shop. He liked to live above it. That way, you knew it'd still be there in the morning. He fumbled to find the new door key in the new leather key-ring Hannah had given him. Always finding excuses to spend money . . . Well, she had put the *mezuza* on the doorpost. That at least was something.[49]

However, movement away from one enclave did not necessarily entail a divorce from community. After all, communities were people and people were moving. The community evolving in the wake of this migration might not be exactly the same but, by the 1950s, it was nevertheless well established.

'Perhaps we'll try Golders Green,' I said without enthusiasm. Golders Green was bourgeoisie land: red roofs, green cautious hedges, cherry blossom, short-haired men polishing their Sunday cars. But is was on the Northern Line, the next station after Hampstead . . . [Following a move from NW3 to NW11], I knew many Jews chose to live in Golders Green and that was fine; but I hardly expected a fifteen-minute walk (with the wind against you) to make all that difference. After all, there were striking bearded men in both places, although beards in 1957 were out of fashion. Artistic young men in corduroys wore them in Hampstead; black-coated *schnorrers* and orthodox rabbis in Golders Green.[50]

Dannie Abse's brother Leo chided him: 'Why do you choose such an uninteresting place to settle in? It's a suburban wilderness'. But then Leo relented, 'Mother and father at least are glad you're living in an area where you can so easily buy the *Jewish Chronicle*.'[51]

Not every immigrant went through each stage of residential settlement and the difficulties that had been encountered by those arriving at the turn of the century. For many German-speaking migrants of the 1930s, the move into English society was rapid and assimilation easy.

> They were German refugees.... The flight of the Müllers had been in the early days, without panic and with all their possessions. Jack's father's business had been an export affair to England so that there had been little upheaval in their change of address. Both his father and his mother spoke English fluently, and through the business were already well connected with the upper strata of English social life. They travelled first-class from Ostend to Dover and early in the morning and when only the white cliffs were looking, they dropped the umlaut from their name, and landing as the Millar family, they spoke to the customs officer in faultless English, declaring their monogrammed silver. Upshot Rise was a natural home for them. It was almost a duplicate of the Beethovenstrasse where they had lived in Hamburg. Quiet, silent, and reliable. Like Upshot Rise, it lay in a dream suburb, a suburb of dream houses, a spotlessly clean nightmare.... [It] was not a handholding street. When you turned into it, you wiped your feet and minded your manners. Each drive sported a car or two.... Behind each set of white curtains lived people who touched each other seldom.[52]

And while this is what a considerable number of the 1930s immigrants could afford – or aspire to – from the outset, it was also the model for many of the children and grandchildren of earlier immigrants who had made the social jump upwards, and who were keen to hide their ethnic origins.

> Oakfern Drive ... came up from behind the All-England Tennis Club and joined Wimbledon Parkside. It was a quiet, dignified road with rough-set whitewashed kerbstones and white verges.... The house was everything they had planned ... and the neighbourhood was respectable without being pretentious.... Their new neighbours ... soon asked Colin and Tessa in for drinks and plainly the suspicion that they were Jewish never arose, which was an agreeable change from North London.... Equally, he was pleased that when people asked where he was living and he replied 'Wimbledon', their first thought was not, as it was when he had had to say 'Cricklewood' or 'Golders Green', that he was a Jew.[53]

These ethnic and social frustrations were apparently to remain with many who had thought that they were able to make, or had actually made, the metamorphosis from immigrant Jew (however cultured and acculturated) to Englishman. As Frederic Raphael recounts:

> When I was seven years old ... my British father was transferred by the Shell Oil Company to their London office. It was 1938.... I accepted the move without much regret: life would presumably not change that

much. When we arrived in London, my parents avoided the obviously Jewish suburbs. We found a flat in Putney. My English grandmother was not happy. 'I should say Roehampton, if I were you'.... My parents were not seeking to conceal their Jewishness, but they did not wish to immerse themselves in it either.... All the usual contradictions were in the melting pot: while my Lithuanian grandmother viewed all non-Kosher food with disgusted suspicion, her husband – who had been born ... near Munich – was not above a good old ham sandwich when we went to a ball game. My father preferred cricket ... and his accent remained decidedly British. He had been to St. Paul's and St. John's College, Oxford ... he elected to embrace ... his posting to London.... 'Well ... at least you'll be able to grow up to be an English gentleman rather than an American Jew.'[54]

CONCLUSION

The issues addressed in this chapter can be examined under the rubrics of identity and process. The Jews arriving in Britain towards the end of the nineteenth and in the early twentieth centuries had a clear religious and ethnic identity. This identity, largely provided by the social and geographical position of the Jews among the Gentile majority in the Pale of Settlement, was associated with an exceptionally strong sense of place – the *heim*. Although there was a mass exodus of Jews from Poland and the former Jewish Pale, most of the Jews remained until the Holocaust of the 1940s which ended the Jewish association with this region with sharp finality.[55]

For those who did leave, an identity that was linked with their sense of place was sharply disturbed. The process of emigration shattered their affinity with their home region, home settlement, with their home. Yet the image of their home, enveloped in the memory of the place, the culture and the social milieu which created it, is what they transported with them to their new milieu. The Jewish elements of this past included the synagogue, the *cheder*, family structure and relationships. The artisans' skills and the tradition of debate and discussion transformed into a new democratic environment allowed them to adapt to the new situation or to recreate an identity with a new sense of place.

Whereas many of the immigrant Jews *may* have wished to recreate as closely as possible their lost environment, there were several forces which acted jointly and separately to prevent this. They ensured that the identity and sense of place created at the point of entry were not simply revamped but new. One of these forces was the insistence of the *in situ* Victorian Jewish establishment that these strangers, or at the very least, their children, be metamorphosed to English men and women of the Jewish persuasion. Another was the nature of British society which, in spite of the anti-alien agitation which accompanied the arrival of the masses was, on the whole, an open society providing opportunities within the wider social fabric. Entrepreneurship was rewarded, permitting movement out of the ghetto and the opportunity for development of a new, British form of Jewishness.

Processes of cultural and social adaptation operated more or less uniformly throughout the British Jewish communities until the outbreak of the Second World War. Though some flow of immigrants occurred in the decade following 1945, immigration on a scale which could change the character of the community ceased, and the Jews stabilised and eventually declined.

The literary sources create images of an immigrant generation and its offspring adapting to a new urban and industrialised environment. Some of these literary images were produced more or less simultaneously with the process of immigration itself; others were reflective and draw on the experiences of immigration and acculturation from a distance in time of up to several decades and removed in space from immigrant neighbourhood to affluent suburb. Almost all the literary sources dwell on the issues of community and change. Even when the authors were bent on escape, they still recognised the fundamental influence of community, which was, and to a large extent still is, the overriding structural element of British Jewry.

NOTES

1 See B. Cheyette, 'The other self: Anglo-Jewish fiction and the representation of Jews in England, 1875–1905', in D. Cesarani (ed.), *The Making of Anglo-Jewry*, Oxf____ ___asil Blackwell, 1990, pp. 97–111.

2 _____, *Point of Arrival – A Study of London's East End*, London, Eyre ____ __5, p. 122; C. Jones, 'Jewish immigration c. 1870–1911', in C. Jones ____ *____tion and Social Policy in Britain*, London, Tavistock Publications, ____ _7; V.D. Lipman, *A History of the Jews in Britain since 1858*, ____ University Press, 1990.

____ ____ussian background of the Anglo-American Jewish immigra-
____ *____the Jewish Historical Society of England*, 1964, vol. 20, pp.
____ ___m Lithuania to the Ukraine encompassing parts of Russia
____ ___lbert, *The Dent Atlas of the Holocaust*, London, Dent,

____ _, East End, West End: Anglo-Jewry and the Great Immigration,'
____ *Quarterly*, 1988, vol. 35, no. 4, p. 26; I Finestein, 'Changes in authority ___o-Jewry since the 1930s: a critical view', *The Jewish Quarterly*, 1985, vol. ___, no. 2, pp. 33–7.

5 See Y. Bar-Gal, 'The *shtetl* – the Jewish small town in Eastern Europe', *Journal of Cultural Geography*, 1985, vol. 5, pp. 17–29; L. Prager, *Yiddish Culture in Britain – A Guide*, Frankfurt-am-Main, Verlag Peter Lang, 1990.

6 C. Weizmann, *Trial and Error*, London, Hamish Hamilton, 1949, p. 11.

7 A. Sher, *Middlepost*, London, Chatto and Windus, 1988, p. 24. Plunge is located about 50 km east of the Baltic sea-port of Memel in Lithuania, and Telz is a short distance further east.

8 See A. Newman, *A History of the United Synagogue*, London, Board of Deputies of British Jews, 1980; S. Waterman and B.A. Kosmin, *British Jewry in the Eighties*, London, Board of Deputies of British Jews, 1986.

9 B. Rubens, *The Elected Member*, London, Abacus, 1969, p. 48.
10 B. Rubens, *Brothers*, London, Abacus, 1984, p. 134.
11 Sher, op. cit., p. 24.
12 Rubens, *The Elected Member*, op. cit., pp. 48–9.
13 Rubens, *Brothers*, op. cit., p. 148.
14 Rubens, *The Elected Member*, op. cit. p. 49.
15 A. Cahan, *The Rise of David Levinsky*, New York, Harper Colophon, 1966, p. 59. First published 1917.
16 Rubens, *Brothers*, op. cit., pp. 152–3.
17 F. Raphael, *The Limits of Love*, London, Fontana, 1960.
18 Sher, op. cit., p. 9.
19 Cahan, op. cit., p. 101.
20 Rubens, *The Elected Member*, op. cit., pp. 49–50.
21 Bermant, op. cit., p. 123.
22 I. Zangwill, *Children of the Ghetto*, Leicester, Leicester University Press, 1977, p. 9. Originally published 1892.
23 R. Finn, *Time Remembered*, London, Future Editions, 1985, pp. 11–12. Originally published 1963.
24 R. Glasser, *Growing Up in the Gorbals*, London, Pan Books, 1987, p. 7.
25 ibid., p. 22.
26 Finn, op. cit., pp. 13–14.
27 ibid., p. 17.
28 W. Goldman, *East End My Cradle*, London, Robson Books, 1988, pp. 23–6. First published 1940.
29 E. Litvonoff, 'The Battle for Mendel Schaffer', in *Journey Through a Small Planet*, Harmondsworth, Penguin, 1979, p. 50.
30 Finn, op. cit., pp. 50–1.
31 W. Mankowitz, *A Kid for Two Farthings*, London, André Deutsch, 1953, pp. 98–9.
32 B. Williams, 'East and West: class and community in Manchester Jewry, 1850–1914', in D. Cesarani, op. cit., pp. 15–33.
33 Bermant, op. cit., p. 125.
34 Rubens, *The Elected Member*, op. cit., p. 51.
35 Rubens, *Brothers*, op. cit., p. 157.
36 R. Livshin, 'The acculturation of the children of the immigrant Jews in Manchester, 1890–1930', in Cesarani, op. cit., pp. 79–96.
37 Goldman, op. cit., p. 86.
38 Lipman, op. cit., p. 52.
39 Mankowitz, op. cit., pp. 81–2.
40 Raphael, op. cit., p. 443.
41 D. Abse, *A Poet in the Family*, London, Robson Books, 1974, p. 15.
42 D. Daiches, *Two Worlds*, Edinburgh, Canongate Publishing, 1987, p. 11. Originally published 1956.
43 S. Blumenfeld, *Jew Boy*, London, Lawrence and Wishart, 1986, p. 131. Originally published by Cape, 1935.
44 C. Holmes, 'J.A. Hobson and the Jews', in C. Holmes (ed.), *Immigrants and Minorities in British Society*, London, Allen and Unwin, 1978, pp. 125–57.
45 Williams, op. cit.
46 See V.D. Lipman, 'The rise of Jewish suburbia', *Transactions of the Jewish Historical Society of England*, 1965, vol. 21, pp. 78–103; Lipman, *A History of the Jews in Britain since 1858*, op. cit., ch. 9.
47 Raphael, op. cit., pp. 157–9.
48 Blumenfeld, op. cit., p. 139.
49 Raphael, op. cit., pp. 15, 20.

50 Abse, op. cit., p. 169. On the social geography of Jews in London, see S. Waterman, *Jews in an Outer London Borough – Barnet*, University of London, Queen Mary College, Research Papers in Geography, no. l, 1989; S. Waterman and B.A. Kosmin, 'Mapping an unenumerated population – Jews in London', *Ethnic and Racial Studies*, 1986, vol. 9, pp. 484–501; S. Waterman and B.A. Kosmin, 'Ethnic identity, residential concentration and social welfare: Jews in London', in P. Jackson (ed.), *Race and Racism*, London, Allen and Unwin, 1987, pp. 254–71; S. Waterman and B.A. Kosmin, 'Residential patterns and processes: a study of Jews in three London boroughs', *Transactions of the Institute of British Geographers*, 1988, vol. 13, pp. 75–91. See also J. Connell, 'The gilded ghetto: Jewish suburbanisation in Leeds', *Bloomsbury Geographer*, 1970, vol. 3, pp. 50–9.
51 Abse, op. cit., pp. 162, 166.
52 B. Rubens, *Mate in Three*, London, Abacus, 1966, pp. 9–10.
53 Raphael, op. cit., p. 124.
54 F. Raphael, 'The curiousness of Anglo-Jews', *The Jewish Quarterly*, 1984, vol. 31, p. 13.
55 B.-C. Pinchuk, *Shtetl Jews Under Soviet Rule*, Oxford, Basil Blackwell, 1990.

13

RETURN MIGRATION IN AMERICAN NOVELS OF THE 1920s AND 1930s

Neil Larry Shumsky

INTRODUCTION

Throughout the twentieth century, American novelists have used the experiences of migrants and ethnic-group members as an important element in their depictions of life in the United States. They have frequently explored questions about the adjustment, adaptation and assimilation of migrants as a way of probing culture, inter-group relations and individual psychology in American society. This is especially true of writers who themselves were migrants.[1]

During the 1920s, some of these authors began to create characters who initially migrate to the United States but later consider going back to their homelands and sometimes actually do so. This rather restricted literature therefore allows us to glimpse a rarely recognised and even less frequently understood phenomenon, return migration. Although it is estimated that more than one-third of the migrants to the United States between 1870 and 1930 returned to their homelands (a return migration of more than 11 million people), it is widely believed (especially by Americans) that the United States was the final destination of most migrants and that, if given their choice, a substantial part of Europe's population would have chosen to live in America. Because of this belief, return migration from the United States is rarely acknowledged, much less studied.[2] Consequently, its novelistic representation, particularly in the years after the end of essentially unrestricted migration to the United States, is extremely important.

Although every novelist's treatment of return is idiosyncratic and reflects the individual temper of their characters, these fictional portrayals, taken together, explore the reasons why some migrants chose to stay in the United States and others chose to go back. In particular, they investigate the personal dimensions of that choice and the considerations that swayed individuals as they decided where they wanted to live. They also highlight the differences between those who remained and those who returned. In these novels, the possibility of return almost always exists for their protagonists, regardless of where they come from. The characters recognise the option of going back and contemplate return at critical moments, especially family crises. Although economics sometimes creates a context that influences staying or going back, decisions usually result from characters' realisations about their own personal identity and values within the context of the socio-economic institutions of both the United States and the homeland. Although social and economic institutions clearly influence the lives of these fictional characters, authors writing about return migrants focused not on institutional restraints and determinants but rather on how their characters personalise and internalise their condition.

In some cases, writers overtly contrast the United States and the homeland and force their characters to choose between them. Sometimes they pit two characters against each other, with one advocating departure and the other persistence, and this disagreement reflects divergent values and priorities. Sometimes this conflict divides husband and wife, with him representing one kind of temperament and her another. Other times, novelists fashion an internal struggle that wrenches a single individual trying to balance clashing emotions. But regardless of how authors people these contests, their fictional characters consciously choose America or the homeland depending on personal assessments of life in the two societies and their own priorities.

To these fictional characters, the United States represents the new and the modern; the homeland stands for the old and the traditional. In the former, they can make more money, but in the latter they can live more peacefully. In America, they can change, but in the homeland, they can abide. Whether these fictional characters decide to stay in America or return, the novelistic depiction of their decision-making process reveals much about the difference between migrants to and return migrants from the United States.

THE UNBIDDEN GUEST – UNWILLING RETURN

One of the earliest works to use return migration to compare the United States and another country is *The Unbidden Guest* by Silvio Villa.[3] In telling the story of a man who returns to Italy unwillingly, it begins to contrast the personalities of those who prefer the United States and those who prefer their homelands.

When Benjamin, the narrator's younger brother, begins to experience severe back pains a few years after coming to the United States, his physician diagnoses nephritis and presses him to go back to Italy. He needs a special diet, but more than anything else, he requires home surroundings and family care. That night, alone in his room, the young man bursts into tears.[4]

He reacts so strongly because of his comparison of Italy and the United States. To him, Italy exemplifies dream and leisure; America, work and action. Benjamin prefers the latter, and he has shaped his life accordingly. Before he became sick, he had told his brother that 'work and honesty' were the bywords of America and that Italy lacked 'the same spirit of work and activity'. He told Carletto that they should 'be modern' and 'react against our national tendency to forget life's realities, to lose ourselves in dreams'. He criticised Italians for their 'tendency to give a sterile admiration to the great without a corresponding positive effort to emulate them', and he exhorted his brother to be different. The two of them must 'stick to the positive, to act and not to dream'. They should 'accept the teaching of these people here, who do not boast of half-divine deeds, but can show day by day the fruit of their labour', and he exhorted, 'Let us work hard!' Carletto agreed, and they resolved to 'forget dreams'.[5]

Four years after Benjamin has returned home, Carletto visits him in Turin. Now paralysed, Benjamin laments his departure from America. He cannot forget 'the great hopes' they had built there; and memories of America's vigour continue to 'haunt' him. That night, Carletto realises as never before, 'the immense gift that it is to act'. And though he considers remaining in Italy to be with Benjamin, he cannot abandon America. In the following years, back in the United States, he is 'reborn into a man'. Every experience opens his mind and stimulates his will, 'developing gradually [in him] a new spirit – the spirit of America'.[6]

MOON HARVEST – RETURN AND FAMILY CONFLICT

The Unbidden Guest is unusual in portraying an unwilling return. Most of the novels show migrants going home because they prefer life there. And yet, their characters perceive the United States and the mother country in much the same way as Benjamin. To them, the new land embodies change, progress and materialism; the homeland, stability, tradition and spirituality. Migrants who prefer the former remain; those who value the latter return. Another novel of the Italian-American experience, *Moon Harvest* by Giuseppe Cautela,[7] uses a disintegrating marriage and the wife's subsequent death to develop this contrast and explain the differing values of those who stayed and those who departed. It also introduces the significance of family relationships as a determining factor in the decision.

In *Moon Harvest*, a young schoolteacher, Romualdo, decides to leave the village of Ortonova because of its hidebound traditions. 'He wanted to breathe freely, he wished to expand his ideas of human progress and emancipation.' In Ortonova, 'an iron circle of opponents' was 'biting his heels like a pack of hungry hounds'. As he once explained, every time he crossed the square he hated having to salute the very people who made his life uncomfortable. In contrast to this static world, he imagines the marvels of America and sometimes feels that no place could contain his dreams. 'He had to go, [and] he had to see.' So strongly does he feel that he insists he will never return.[8]

Romualdo's wife, Maria, differs. 'Modest and silent . . . meek and tender towards all', she tells her mother that they will come back to Italy. Later, when her mother writes and asks when to expect them, Maria asks Romualdo. And when he answers, 'God knows,' she replies to her mother 'some day'.[9]

Romualdo finds all that he had imagined in America and savours the opportunities it presents him, especially the chance for new experiences. Most important, he falls in love with the vivacious, affluent, demanding and Americanised Vicenza Di Dedda who wears morning gowns of diaphonous fabrics and delicate colours. As he gradually recognises how he and Maria have changed in America, Romualdo realises that 'the difference between her old life and life in this new and terrible land . . . had changed all his views, all his conceptions of society.'[10]

While Romualdo flourishes in the United States, and glories in its opportunities, Maria withers. When she realises that he has fallen in love with Vicenza, she feels 'a homesick longing to see all that plain again under the full eye of the moon', and she wants 'to escape from here, from this unfriendly atmosphere'. She yearns 'to return again to the simplicity of her native place, where she could move about serene and happy. . . . Her roof called her, with its nests of swallows, with their whistle-like calls in the morning air.'[11]

Ultimately, Maria decides to return to Italy where life possesses a 'depth' and a 'rhythm of eternity. Things there seemed made to live forever with the illusion of a dream.' In contrast, 'America was the land of rude reality, and she was too fine a plant to resist the blasts of the northern wind. Even her hair was losing its lustre and softness of ebony.' Maria cannot survive without 'her native sunshine and the perfumed air of acacia'. However the time for return had passed, and it is too late. On the day of her departure, Maria collapses and dies. Distraught over her death, Romualdo himself decides to return to Italy temporarily, and he asks Vicenza to visit him, hoping that 'in that atmosphere' she will absorb Maria's 'simple, deep spiritual force'.[12]

Both Villa and Cautela portray an America suited for people with speculative, expansive temperaments who relish change and growth. Benjamin, Carletto and Romualdo all prefer to remain in America; Maria prefers to return. Benjamin, Carletto and Romualdo all thrive there and savour the opportunity to develop in new and unexpected ways; Maria dies.

GREENHORN – UNREQUITED RETURN

Paul King, the pen name of Paul Kiralyhegyi, a Hungarian-American author, develops similar ideas and perceptions in *Greenhorn*, published a few years later.[13] Throughout this book, the narrator (identified only as Paul) vacillates between staying in the United States and returning to Budapest, and his attitude toward the United States mirrors his attitude toward himself at particular moments. When he is successful, he feels positive, enjoys the United States, values growth and intends to remain. When he is failing, he questions his own ability, craves stability and wonders if America is right for him. Ultimately, he remains in California because opportunity of every variety arrests him.

Paul originally migrated to America for complicated reasons. When his middle-class father discovered him working as a waiter and in love with a waitress, he publicly humiliated the boy, and Paul decided to kill himself. When he tried to borrow a gun to carry out his plan, his friend Charley suggested going to America instead. As they talked about the future, their minds filled with images of 'freedom, liberty, wealth, happiness'. Charley summed up their thoughts, 'we don't want to be slaves of tradition, slaves of our parents, slaves of stupidity any more. We must go to America! ... Tall and impossible buildings, beautiful girls, interesting people, big money.'[14] And the two of them decided that the future, at least for now, lay across the ocean. They intended to remain only temporarily, and Paul told his girlfriend that he would come back with money.[15]

When Paul arrives in New York, the contrast to Budapest overwhelms him, and he wonders if he can adjust to all the changes. Would he ever succeed or would he 'write home for money, and return, beaten ... in a few months?' These questions 'disturbed and bothered' Paul, but they also make him 'happy' because he is 'in the centre of life's vortex'. He desires 'change, wide perspectives, possibilities, unlimited possibilities'.[16]

As Paul's experiences in the United States fluctuate between positive and negative, his feelings about being there fluctuate between exhilaration and anxiety. After he gets one job, his optimism soars. 'There were unlimited possibilities.... Prosperity must be on the way.' Suddenly, 'life ... appeared beautiful.' He is 'on the road to success'. And when a friend presents a different opportunity, Paul jumps at the new chance. Suddenly, his current job 'looked grey and dismal'. 'Who wanted to be a clerk for a lifetime?'[17]

This 'great' opportunity sputters, fizzles, and collapses; and failure generates thoughts of return. Although Paul and his friend find other jobs, they feel no interest in their work. They have 'caught a dangerous sickness', homesickness, and 'the desire to return to Budapest grew and grew'. They want to see 'the houses, the signs, the familiar faces; and above all, to find [their] younger selves'. They will be going home 'beaten and humble, but home'. Paul discovers that he 'could not keep on struggling any more. Life and success in America were for strong men, ... for better men than [he].' So, he and his friend dream of saving a thousand dollars to finance their return.[18]

And then the extraordinary occurs. Paul wins nearly that amount shooting dice. 'An indescribable and unbelievable happiness' overcomes him. 'Europe! Budapest!' But as he walks home, exulting in his good fortune and dreaming of Hungary, two gunmen rob him and take every cent, leaving him more dejected than ever – and stranded in Detroit.[19]

Paul now becomes a bus-boy, and goes through the motions of life, continuing to despair, until he describes himself to a friendly patron as a failure who can succeed at nothing. Immediately, the customer chides him.

'How can you talk like that? . . . You can do great things in America still. . . . You can make a man out of yourself if you will only try.' More than that, acting like a character straight out of a novel by Horatio Alger, he offers Paul a job in a bank. The result? Paul's homesickness disappears, replaced once again by optimism and confidence. 'It had been shameful to be ready to admit defeat. [He] would have to try again.' He will 'not be going home for a while yet. Later on, [he] would be a successful banker, a financier; and then, maybe, a great limousine would be a reality after all.' He is 'young, and determined to try, try, again'. He 'would go higher and higher'.[20]

Paul's fortunes progress rapidly, and then he receives a letter from a friend in Hollywood. Her descriptions of life there renew his restlessness and his urge for fresh experiences. Suddenly, working in the bank seems 'like a jail' whereas 'everybody who amounts to anybody in this world lives in Holly-wood. One could do big things there.' The more he thinks, the more boring Detroit becomes. 'Always the same . . . always the same places . . . the same people.' The upshot? Instead of returning to Budapest, he boards the next train for Hollywood, prepared to be a 'greenhorn' all over again. His craving for change and new experiences have once again overwhelmed his desire for stability; opportunity matters more than security.[21] To someone with these predilections, Hollywood seems preferable to Detroit just as the United States is more desirable than Hungary.

Greenhorn was only the first of half a dozen novels written during the Great Depression that considered return migration. In each of them, eco-nomics affects characters' decisions about coming to the United States, remaining in the United States or returning to the homeland. However, economics affects return in unexpected ways. For one thing, the Depression *per se* does not play a major role in the novels or receive much acknowledge-ment; characters who are unsuccessful attach no blame to American economic institutions or economic conditions generally. More than that, characters do not return because they have failed and given up hope. Nor do they return because they have succeeded and can go home triumphantly. Most of the characters who consider return have prospered at least moderately in the United States, and their prosperity enables them to decide where to live on the basis of personal factors and preferences, especially family circumstances, rather than on the basis of economic necessity. In this sense, they resemble Romualdo and Maria of *Moon Harvest* in which Maria wants to return because her marriage to Romualdo is disintegrating and Romualdo returns because of Maria's death.

BELLY FULLA STRAW – RETURN AND
THE FAMILY LIFE-CYCLE

Belly Fulla Straw, by the Dutch author David Cornel deJong,[22] tells the story of Harmen and Detjen Idema and their four children, and its comparison of the United States and Holland replicates that found in *The Unbidden Guest*, *Moon Harvest* and *Greenhorn*. Although Harmen remains in the United States for about fifteen years, succeeds in business, and rarely thinks about going back, he returns to Holland when his personal ties in America have been snapped.

Like Paul in *Greenhorn*, the Idemas came to the United States for a complex of economic and personal reasons. Harmen and Detjen told people over and over again that they were migrating for the well-being of their children, so that they would 'have the chance'. But it was 'a really insignificant quarrel' with his family that had 'swollen into the desire to go away, to America, a desire which had become stronger when people told of untold wealth and great freedom'. With 'their eye already on America', they also considered Detjen's dislike for the town where they lived, her quarrels with his sisters, and the lack of opportunity for their older son. Harmen's education and his life with Detjen had extinguished the 'deep and elemental simplicity' necessary for him to stay happily in Meerdum, and so they left, 'forever'.[23]

In America, they discover a land of economic opportunity, but also a land of materialistic louts who demand strict conformity to their own standards of behaviour; and neither Harmen nor his wife ever feel completely at home. After three years in the United States, Harmen recognises that materially he is substantially better off, and that 'there were many compensations for the things he had been giving up.' But he also 'realise[s] how painful the readjustment had really been, and how disconcerting several more years of it would be'. Were it 'not for the family, he would return at once'.[24]

Years later, when the family no longer exists, he feels no reason to remain in America. The first break occurs with the death of Detjen, closely followed by increased acrimony with his two oldest children. At about the same time, Harmen receives a letter from his father suggesting that he return to Holland and take over the family business. The invitation appeals, and Harmen's reaction emphasises the familiar contrast between the United States and the homeland. 'No more going to the new, but finding the lost again. No more pains and emptiness' caused by children. 'No more derision and hatred and mistrust, no more walking alone in a cold and unfriendly city.' Instead, he will have 'the familiar, the past, the peaceful. . . . Return. Why not, why not?' His father's letter 'held out the possibilities of peace and rest'.[25]

Idema is not yet ready to break from the two younger children with whom he still has close ties, and neither of them will accompany him. However, steadily mounting conflict with the older children makes him realise that he cannot live through any of his offspring, and he tells his favourite daughter that 'it must end, and very soon', to which she sadly agrees.[26]

The final breach with America accompanies the death of his long-time friend and partner. Another tie is broken. When Idema pays a condolence call on the widow, she dejectedly urges him to return. He has put his hopes on his children, and to no avail. He is 'trying to do the good, and trying to be honest', and 'there was nothing there.' She asks if he has been lonely, and he replies, 'Uprooted'. She asks him,

> 'Why don't you pack up your things and go away? Why don't you go back, as you used to think you would like to? There is nothing to hold you here, nothing very strong. And now that he's [her husband] gone . . . there's another thread broken.'

He replies that he has decided to go back to the things he had left when he was young. A few months later he departs.[27]

KINGDOM OF NECESSITY – RETURN AND RE-EMIGRATION

Characters who consider return place a high value on what they have left behind. Those who actually return value what they have lost more than what they have found. In balancing this equation, the nature of family relationships is perhaps the most important element. Characters' decisions show that migratory chains could pull people back to Europe as well as impel them to America. Sometimes characters have trouble determining their priorities and conclude that they have erred, first by migrating to America and then by returning. As a result, they travel to the United States a second time and become 'repeat' migrants. Morris Hyman in Isidor Schneider's *From the Kingdom of Necessity* is such a figure.[28]

Although the characters in this novel are unusual in some respects, its comparison of life in the United States and life in the Polish village of Horodemal strongly resembles the analysis of earlier novels. As the novel begins, Morris Hyman, a Jewish tailor, has just returned from the United States, bringing with him the first sewing machine ever seen in the region. Visitors come from great distances to see this marvel and to find out why someone has foresaken the prosperity of America.

Morris provides two answers. Although he concedes that he could live better in America than in Poland, he argues that a man without his family is 'in a wilderness'. Then he shocks his audience by claiming that America has killed religion. When asked to explain how a country without laws discriminating against Judaism could kill religion, Hyman announces that laws and police and pogroms do not kill religion; riches do. In America, Jews sell their religion for money. They work during hours that they should devote to their families, and then they sell God's time. In the beginning, they work during morning prayers. Later, in exchange for overtime, they work during evening prayers. Ultimately, they work on the Sabbath itself.

> Time is money! they say, and when they cash in time, no one strictly inquires whether it is an hour taken from comforting the sickbed of a beloved one, from the companionship of husband and wife, or from God. They cash in God, and without God, what are they? Hollow eggs!

Hyman tells his incredulous audience that it is easier to be a Jew in Russia. He lived well in America, and he made more money there, but it tormented him to see 'the sight of Jews selling their Jewishness'. So he came back to the village where a Jew could live like a Jew.[29]

Hyman has not divulged the complete story. He has not disclosed his economic failure, nor has he revealed that he failed because of his temperament. America, constantly changing, required 'alert, ingenious, decisive people, who could quickly adjust themselves to changes and take advantage of them'. It was not suited 'to men with passive virtues, to the humble, self-denying and peaceful'. It had no room for the Morris Hymans, who wanted 'to live each day exactly as . . . the day before, in the same rooms, among the same people, at the same work benches'. Less than two years after Hyman arrived in America, the garment industry began manufacturing ready-made clothes, and the 'old, quiet ways of the tailor's life changed accordingly'. The rhythm of machines replaced the rhythm of his body, and they had 'a scream and a pace like a tornado'. These machines had frightened Morris out of America, had 'harassed and driven him until he had to run away'. He could not face the new conditions and decided that, 'rather than risk his life in this terrifying land with its haste and change, he would buy himself a machine and return to the old country to settle down to a peaceful, slow, stable life.'[30]

Morris Hyman's story does not conclude with his return to Poland. No more successful in Horodemal than New York, he packs up his family and returns to the United States, thus emphasising two other aspects of the return migrant experience – repeat as opposed to return migration and the ongoing conflict of values. In the novels on this theme, some migrants come to the United States, feel dissatisfied with what they discover, think that they preferred life back home, and return, only to discover that America looks better from a distance than from close up. Although they long for the tranquillity of home, they still value the opportunity available in the United States so they travel west again. These repeated moves result from the difficulty of balancing the competing attractions of the two countries, weighing change and progress against stability and tradition. Even many of those who make a decision and stick with it continue to wonder if they have chosen correctly.

THE GRAND GENNARO – RETURN AS WEAKNESS

The depiction of return migration in Garibaldi Lapolla's *The Grand Gennaro*,[31] a classic novel of the Italian-American experience, parallels other fictional portrayals from the 1920s and 1930s. The United States appeals to people with expansive, aggressive temperaments, the homeland to those with restrained and yielding personalities. Like other authors, Lapolla symbolises this contrast by devising conflicts between closely related characters, and, as in other novels, death could occur because a character was in the wrong environment. Lapolla also identifies thinking about return with the death of a close relative and the snapping of family ties.

At the beginning of the novel, a conflict is developing between two business partners, Rocco and Gennaro. Rocco possesses a relaxed and pleasing nature, 'given to laughter and cheer and taking things as they came. He would rather be friends than drive a bargain; he would rather remain friendly than gain an advantage.' His business is making him wealthy and will soon enable him to return to Italy and marry Elvira. He plans to take over her father's farm, 'rich in green grass, with luscious grapes big as plums on the slopes, a cow, and sheep, and real horses'. But Gennaro differs. Determined 'to pile up a stack in a short time no matter how he did it', he has stolen half of Rocco's business. Later, after Gennaro has taken complete control of the business, he tells Rocco that they will build the company together. But they will not 'make a few dollars . . . and go back to the old country with a feather stuck in our hats and a few gold chains hanging across our bellies'. They will bring their families to America and enjoy a new life. Rocco wants none of it, and he tells Gennaro,

> 'The American fever is in you. The touch of gold in your hands has been too much. We peasants are not used to it. It does queer things to us – the good it makes mean, the mean brutal. Give me the money that's mine, and you can have the business. You stay here, I'll go back to the old country.'

Rocco's words bounce off Gennaro who continues his ruthlessness. He rapes his landlady and thinks afterward, 'I'll be boss here . . . peasant once, but no more . . . I know now . . . take what you want when you want it. That's the trick!'[32]

An exaggerated, more cruel and vicious version of Romualdo in *Moon Harvest*, Gennaro, too, hated the narrow-mindedness of Italian society. He had despised being deferential to his social betters, and he loves being in America where all that has ended. Here, he is afraid of nothing, and he is going to stay because he is quick, shrewd and has no fear; the strong man is bound to win. Let others go back, those 'ninnies and fools and the stuff of dunghills'. They could 'grovel at the feet' of their superiors.[33]

Just as Gennaro resembles Romualdo, his wife Rosaria resembles Maria, and the similarities between *The Grand Gennaro* and *Moon Harvest* are striking. Like Maria, Rosaria never feels comfortable in America, and her unhappiness is compounded by a family tragedy that she blames on their migration. Her oldest son Domenico rapes a neighbour's daughter, enters the army, and dies in the First World War. As Rosaria ages, she flees from her unhappiness by escaping into the past and romanticising life in Italy where her family had been honoured. Gennaro brought her to America, and what did she get?

The new coal stove, the electric lights, the oil-cloth on the floor, the icebox on one side of the kitchen, a table with a marble top. But as she said, 'this,' 'this,' her voice became sadder and sadder, tinged with more contempt. 'These things are not for me.... I have nothing. I want nothing. Domenico is gone . . . that is all I know.'

As Rosaria slips into her own private world, Gennaro consults physicians who advise that she return to Italy 'where the change of air and environment might recall her from the deep darkness into which she had fallen'. Several weeks after her departure, Gennaro learns that she did not arrive. One night aboard ship, she felt chilled, went to the hospital, and, also like Maria, died before she reached her destination. Gennaro understands precisely why she died and told a close friend that 'some of us can't be just pulled out of our nests. We die like the birds. Some, they become strong and fly high in the wind. Like me. Poor Rosaria.'[34]

Rosaria is not the only character in *The Grand Gennaro* who contemplates returning to Italy. Although the others only consider going back, the timing of their thoughts is significant. Each follows a family crisis and suggests a need to find security by returning to the familiar. After Domenico rapes Carmela, her father says that the family will go back to Italy, a plan which his wife calls a way of 'escaping your trouble'. As Donna Maria lies dying on the day her husband opens a new store, he implores her to live, telling her that the shop is 'going to make us rich . . . and we'll go back to Castello-a-Mare and live the way we should, the way people like us should'. And when Davido Monterano dies, his teenage son's initial response is to take his younger brother and sister back to their relatives in Italy.[35] Although none of these returns actually takes place, the timing of their contemplation reveals once again the meaning of the homeland. It signifies shelter and moorings in a strange and sometimes stormy world. Those who most feel the need for refuge return.

A uniform set of ideas about return migration pervades these novels of the early twentieth century. Some people had the right temperament for migration, others did not. People with one kind of personality should stay in the United States, those with another should return. The destruction of family relationships in America, or the need to renew relationships in the homeland, frequently precipitated the return of migrants who did not belong in America. Chains of migration could keep people in America or pull them back to Europe. Many of these conceptions reappear in a pair of novels written about Scandinavian-Americans in the late 1930s.

TAKE ALL TO NEBRASKA – RETURN AND INTERNAL CONFLICT

Take All to Nebraska by Sophus Keith Winther tells the story of Peter and Meta Grimsen and their children.[36] These Danish migrants arrived in

Massachusetts three years before the novel begins, and, although they never return, the agony of deciding to stay reveals how difficult the decision could be. In addition, their personal interactions show how conflict could occur within a single individual as well as between individuals; and the basis of their decision to remain strengthens the interpretation that personal relationships played a crucial role in causing people to stay or go back.

The Grimsens have done poorly in Massachusetts, largely because they lack adequate knowledge. Peter brought his family to America because he believed 'that this was the land of opportunity and plenty, that this was the poor man's haven'. He had heard that he could acquire rich land without a stone on it, and 'the promise of large acres of free land was the lodestar' that called him across the Atlantic 'in the hope of gaining independence, in the faith of a greater destiny'. He had not known that there was no free land on the east coast, nor had he expected 'shallow soil and more stones than he had ever encountered in Denmark'.[37]

After three years of struggle, he admits defeat and is now taking his family to Nebraska, but as they ride the train west, he keeps wondering if he has made the right decision. Should they be 'journeying still farther into the vast stretches of this unfriendly land', or should they be returning to Denmark? He is especially troubled because of the difference he perceives between himself and his wife. He knows that she would have preferred going back to Denmark and her mother, her only living relative. He believes that she knows she will never see her mother again because people who go to America do not return. On the other hand, he himself knows that 'he could not have gone back.' After all, could 'a man live on companionship? he asked himself with scorn'. As he puts it to Meta, 'we couldn't go back even if we wanted to.'[38]

In the opening segments of the novel, Winther portrays Peter and Meta having the same relationship as Romualdo and Maria, Gennaro and Rosaria. He belongs in America, she does not. He wants to stay, she wants to return. 'Secretly' she 'had hoped at times that they might, possibly, go back to Denmark'. But as the narrative develops, attitudes toward return become more complex, and Winther reveals that both husband and wife have divided hearts. The conflict is as much internal within each of them as it is between them, and expected roles are reversed. He argues for returning, she for staying. One summer day, after hot windstorms have destroyed their corn crop, Peter tells Meta that they have reached the end; that they cannot continue. She admits that she has thought many times of returning to Denmark, but she cannot see how it would help 'to come there with empty hands'. How would they get back, and where would they find money for the tickets? There is nothing they could do. 'Oh, if only we had stayed in Denmark!' But, after saying that, she concludes the conversation by expressing just the opposite feeling. 'I am glad we are here in America. Our children will not have to be farmers. They can rise to better things in this democratic country.' Peter experiences the same inner struggle. At one moment, 'the possibilities for the future in this new country fired his imagination', and he feels 'a challenge in this land that appealed to his courage as well as his industry'. And yet, 'in the back of his mind he had never given up the idea of some day returning to the old country. *Like all emigrants* he secretly treasured the hope that by some means which he did not understand, he would again be back home and a well-to-do farmer.'[39]

Although both feel conflict, only after some time do they realise the extent of each other's ambivalence, come to terms with their conflicts and achieve some resolution. After they have finally overcome a long string of adversity and appear to be on the road to success, their landlord leases the farm that they have steadily improved to another tenant. With this seemingly final blow, Meta tells Peter that they should 'forget it all and go back to Denmark'. Her words 'stirred a deep and fearful thing in him. For seven years he had buried those words in his own heart.' And now, 'almost fiercely', Meta pours out her bitterness and cries that she 'couldn't stand it any longer'. Her outpouring begins many further discussions between them. Meta insists that they return, and Peter recognises that 'Denmark was still home, but whenever he thought of leaving America his heart sank.' He 'could not make up his mind'. His loyalty to his wife and children pull him to Denmark, but the 'plans that stole into his mind, catching him unawares', keep him in America. Then, one day, he visits a neighbour who implores him to remain in Nebraska and suggests a different farm that they could rent. When Peter goes home and hesitantly tells Meta that maybe he should look it over, that there 'really [is] no hurry about going back to Denmark', and that they 'could always do that', her response shocks him. She announces that they have to stay in America, 'not next year only, but always'. Their oldest son is refusing to go back with them. And she realises that if he stays, 'his brothers would come over here to him one after another as they grew up, and we would sit alone in our old age. America is our home now for good or ill. Today I discovered that for myself, and I don't know but that I am glad.'[40]

THE EMIGRANTS – TWO HOMES OR NONE?

The Grimsens stay in America because of the priority they give family relationships in the larger context of their ambivalence about whether they prefer Denmark and its familiarity or the United States and its opportunity. Likewise, Morten Kvidal, the hero of Per Bojer's *The Emigrants*,[41] decides to remain in the United States, but only after he has first gone back to Norway, as Morris Hyman returned to Poland. But whereas family and opportunity combine to keep Peter and Meta in the United States, Morten chooses American opportunity over his Norwegian mother and the family farm about which he had dreamed for years.

Like other fictional migrants, Morten came to the United States planning to remain a few years, make some money and return to Norway. 'Poverty on a small [Norwegian] farm was little better than slow torture', and he intends to 'come back as a rich man'. When he first arrives in the Dakotas, his resolve never wavers. He has gone, 'but not for good. . . . He could feel the old country inside him all the time.' He would acquire hundreds of acres, 'cultivate them, build farmhouses . . . sell the whole lot for thousands of dollars . . . hurry home and throw himself into the work awaiting him there.' He has extraordinary visions of what he will do when he returns. 'He would make his mark in the parish and in the country.' But before that, 'he must make good out here on the prairie. . . . make money, pile up the dollars quickly, make his fortune in a few years – and then he could start in earnest, for the real work was to be at home.'[42]

After five years, Morten's resolve to return seems to be weakening. Since he learned that his old girlfriend has married another man, 'the old country seemed to be farther away than ever', and the breaking of this bond undermines his tie to Norway. Relationships to friends on the prairie suddenly mean as much as those to his family back home. 'His mind was becoming divided in two. He had countrymen there and countrymen here, but it seemed as if the latter needed him most.' The prairie seems to sing about the future 'and the opportunities it presented to a man who had the will to work and get on in life. . . . If he really meant to climb the ladder – high, high up – America was the very place for him.' And yet, he feels homesick and 'grew sick with longing to see it all again'. Sometimes, 'he got down to a deeper self, beyond the struggle for money, horses, more land, more power, and position – to something calm and holy.' In America, 'he was a slave to the land and the almighty dollar', but back home, 'he would become his true self again. He must go back.' And yet, 'how could he leave all this? But what about . . . [the farm] in the old country? Who was looking after it now? Hadn't he promised to return soon? to build a big farm there?'[43]

When Morten learns that his old girlfriend has died, he finally decides to go back. In some strange way, her death seems to give her back to him, and he no longer feels estranged from Norway. 'After this he could not wait any longer; he must tear himself away and hurry home.' That spring, he leaves.

At first, he responds to home as he had expected. 'To lay one's head on Mother's pillow again, to feel oneself at home, to be lulled to sleep by the roar of [the] waterfall – what more could one wish? He slept soundly and well, with a sense of well-being he had not known for many years.' But within a few days, he begins to experience the same tensions he had felt on the prairies. His head feels 'queer' when he looks at the mountains, the valley and the fjord. This landscape is 'so unrestful' and 'made him so dizzy. . . . Was it because he was so used to the prairie?' And yet, he 'had all that he had longed and pined for in a strange land! Every day the landscape and its memories seemed to be getting him more and more into its power, drawing him closer and closer to itself.' He has no clear sense of where he belongs and what he intends. 'Had he come home for good, or had he not?' Part of him says that his ancestors struggled to tame this farm, fertilised it with their sweat. He himself has spent eight years in America collecting the money to build and improve it. Another part tells him that in America he was a leader, in Norway he is nothing. The Norwegian

> landscape was weaving a spell about him, ... but was he going to sacrifice his life for it? Or should he tear himself away and escape? His country was very dear to him, but what about all that he had on the other side? Did that count for nothing?[44]

Three events settle his mind to return to the United States, and each of them resembles the bases of decisions made by other fictional characters. First, he meets a young woman whom he finds attractive. He had known her before he left Norway, she had been a close friend of the woman he had left behind, and her sister lived near Morten in America. Now he finds himself wishing that he had met her on the prairie. Her appeal intensifies when he discovers that she is considering joining her sister in North Dakota, and he finds himself thinking that if he advised her to go, he would have to go too. This developing relationship (one that would culminate in marriage) clearly plays an important role in drawing him back to America. So does his perception of traditional ideas that he now rejects. After he has been in Norway for several months, he is appointed a poor-law guardian, and at the first meeting he attends, he disagrees with the chairman. Morten thinks the committee is spending too much money and that people should earn their own livelihood. When the chair replies that Norway 'had outgrown the idea that the poor ought to suffer', Morten 'asked whether the poor were not supposed to work either'. After being outvoted,

> he went home swearing under his breath; they'd all be paupers in this country sooner or later. On the other hand, he could see the prairie, lying there in the spring breeze, and calling for young men and women who were willing to work and trust to the strength of their own hands.

Norway and America suddenly seem to have different values, and he realises that he has to 'choose, Morten, choose'. The final event that pulls him back to America is a letter describing the changes that had occurred during his absence. North Dakota has become a state, construction of a new road has started, plans for a railway are in the air. Things 'were beginning with a vengeance'.[45]

The night that letter arrives, Morten lies awake thinking about the future and deciding what to do. As he views Norway's magnificent countryside, 'he felt that his soul was spellbound here forever.' The scenic features chant, 'You are we. We are you. Wherever else you go, you will always be an alien.' And yet, 'another wave had caught him and was carrying him farther and farther away: the more he loved all this that was his own, the more he longed to leave it.' In fact, he has already made up his mind to leave. 'He would go back. He must. But he shrank from admitting to himself that it would be for good this time.'[46] Although he returns to North Dakota, he visits Norway seven times more, and each time he thinks he is going back permanently. On one of his trips, he tells a shipmate,

'When you're here, you feel you can't be happy out of the old country, and when you've been there a little while, you begin to look out across the ocean, and you find that you're happier over here after all!'

Years later, he answers a close friend who asks if he was still homesick, 'Can you ask? Do you suppose there has been a single day, in all these long years, when I didn't say to myself, I'm going home, soon?'

Perhaps Augustine, a character in *The Coming of Fabrizze* by Raymond de Capite, expresses Morten's feelings as well as anyone. On one occasion, when Augustine announces that he is returning to Italy once again, a friend remonstrates with him. 'First you were there . . . and then you were here. And then you went back. And now you come and go again. Where is the end of it?' To which Augustine replies simply but knowingly, 'The end of it? In the end, I'll have no home at all.'[47]

Despite Augustine's plaintive cry, he does have a home – or perhaps he has two. If we think of him, and other return migrants, only from the perspective of where they lived last, they become either immigrants (if they remain in the country to which they migrated) or return migrants (if, after some point, they remain in their homeland and never again leave). Those who become return migrants seem to have decided, despite whatever doubts and uncertainties they may have had, that they belong in their native lands. They prefer them because tradition, stability, the old and the past matter more than progress, change, the new and the future. More than that, their personal ties and relationships to people 'back home' are stronger and more important than their ties and relationships to people in America. Chains bind them to their homelands as well as to America. Those who become immigrants and remain in the United States have the opposite feelings and preferences.

But the fictional depiction of return migrants suggests another interpretation, that they have two homes, that Augustine does not have to choose one or the other, either the United States or Italy, that he can partake of both, that he can move back and forth (perhaps literally, certainly mentally) between the two countries. The novelistic portrayal of return migrants, and potential return migrants, suggests that a substantial proportion of all migrants were in a constant quandary about where to go and where to live and that they wondered constantly where they could make the best lives for themselves. They saw the advantages and disadvantages of both societies, and they wanted, to the greatest extent possible, to enjoy the advantages of both. They wanted the opportunities of the future but the traditions of the past, the stability of the known but the excitement of the strange, the love of those back home but also the love of those here at home. In creating these

213

characters, the novelists who depicted return migration showed how difficult it was either to live in both of these worlds simultaneously or to choose between them. And in painting these portraits, they have enriched our understanding of the migrant experience.

NOTES

1 See, for example, W. Brown and A. Ling (eds), *Imagining America; Stories from the Promised Land*, New York, Persea Books, 1991; C. Fanning (ed.), *The Exiles of Erin: Nineteenth-Century Irish-American Fiction*, Notre Dame, Ind., University of Notre Dame Press, 1983; D.M. Fine, *The City, The Immigrant, and American Fiction, 1880–1920*, Metuchen, NJ, Scarecrow Press, 1977.
2 The few works that have given substantial attention to return migration include H. Jerome, *Migration and Business Cycles*, New York, National Bureau of Economic Research, Inc., 1926; B.B. Caroli, *Italian Repatriation from the United States, 1900–1914*, New York, Center for Migration Studies, 1973; K. Virtanen, *Settlement or Return: Finnish Emigrants (1860–1930) in the International Overseas Return Migration Movement*, Turku, The Migration Institute, 1979; W.S. Shepperson, *Emigration and Disenchantment: Portraits of Englishmen Repatriated from the United States*, Norman, Oklahoma, University of Oklahoma Press, 1965; K. Hvidt, *Flight to America. The Social Background of 300,000 Danish Emigrants*, New York, Academic Press, 1975; L.-G. Tedebrand, 'Remigration from America to Sweden', in H. Runblom and H. Norman (eds), *From Sweden to America. A History of the Migration*, Minneapolis, University of Minnesota Press, 1976; F. Kraljic, *Croatian Migration to and from the United States, 1900–1914*, Palo Alto, Calif., Ragusan, 1978; G. Moltmann, 'American-German return migration in the nineteenth and early twentieth centuries', *Central European History*, 1980, vol. 13, pp. 378–92; D. Cinel, *The National Integration of Italian Return Migration, 1870–1929*, New York, Cambridge University Press, 1991; N.L. Shumsky, 'Let No Man Stop to Plunder! American hostility to return migration, 1890–1924', *Journal of American Ethnic History*, 1992, vol. 11, pp. 56–75.
3 S. Villa, *The Unbidden Guest*, New York, Macmillan, 1923.
4 ibid., pp. 178–9.
5 ibid., pp. 170–6.
6 ibid., pp. 185–7, 194.
7 G. Cautela, *Moon Harvest*, New York, Dial Press, 1925.
8 ibid., pp. 1–4.
9 ibid., pp. 26, 94.
10 ibid., p. 145.
11 ibid., pp. 192–3.
12 ibid., pp. 145, 191, 193, 214–15, 253.
13 P. King, *Greenhorn*, New York, Macaulay, 1932.
14 ibid., pp. 17–18.
15 ibid., p. 42.
16 ibid., pp. 103–4; see also pp. 203, 252.
17 ibid., pp. 282–3.
18 ibid., pp. 286–7.
19 ibid., pp. 292–3.
20 ibid., pp. 296–9.
21 ibid., pp. 302–7.
22 D.C. deJong, *Belly Fulla Straw*, New York, Alfred A. Knopf, 1934.

23 ibid., pp. 6, 23.
24 ibid., pp. 113–14.
25 ibid., pp. 258–60.
26 ibid., p. 296.
27 ibid., pp. 306–7, 314–15.
28 I. Schneider, *From the Kingdom of Necessity*, New York, G.P. Putnam's Sons, 1935.
29 ibid., pp. 7–8.
30 ibid., pp. 7–10.
31 G.M. Lapolla, *The Grand Gennaro*, New York, Vanguard Press, 1935.
32 ibid., pp. 14–15, 17–18, 22.
33 ibid., p. 38; see also pp. 61–2.
34 ibid., p. 200.
35 ibid., pp. 134, 253, 279.
36 S.K. Winther, *Take All to Nebraska*, New York, Macmillan, 1936.
37 ibid., pp. 2–3.
38 ibid., pp. 4–5, 23.
39 ibid., pp. 279, 291–2 [my emphasis].
40 ibid., pp. 301–5.
41 P. Bojer, *The Emigrants*, New York and London, D. Appleton-Century, 1938.
42 ibid., pp. 52, 114–15, 150–1.
43 ibid., pp. 205–7, 226–8.
44 ibid., pp. 265–7, 273, 274–5.
45 ibid., pp. 277–9.
46 ibid., p. 279.
47 R. de Capite, *The Coming of Fabrizze*, New York, McKay, 1960, pp. 41–2.

14

BIRDS OF PASSAGE OR SQUAWKING DUCKS

Writing across generations of Japanese-Canadian literature

Audrey Kobayashi[1]

'One lives in sound, the other in stone.'
(Joy Kogawa, *Obasan*)[2]

INTRODUCTION

Like most cultural communities, Japanese Canadians have a shared mythology, through which they imagine their history, define their geography and probe the limits of their community.[3] One of the myths concerns the 'silence' of the Issei (first generation) with respect to their political and human rights, as opposed to the activism and anger expressed by the Sansei (third generation), and the ambivalence of the Nisei (second generation) between the two other worlds. The Issei are viewed as stoic, constrained by a culture of complex codes and interdictions to persevere (*gaman*) in silence in the face of what would for others be intolerable oppression. Reference to their supposed attitude of *shikata ga nai* (it can't be helped) has become a cliché for the 'old values'.

My purpose here is not to break down the myths, to transform them into 'reality', or to expose beneath them an essential, putative 'truth' but, rather, to engage the 'myth of myth'[4] to show how the practice of myth-making structures, contains and codifies the terms of shared existence for cultural groups. There is nothing beneath the myths but myths. My purpose therefore is to write between the myths, to set up the countertextuality of writings of different generations of Japanese Canadians as expressions of a dynamic and variable collectivity of expression. These writings are a succession of journeys between. They represent not static expressions of generational difference, but creative transformations of intergenerational tension, occurring within the context of 'real' journeys that took place from Japan to Canada and from emigrants' Canadian homes to exile during the 1940s.

The myth of generational difference is synonymous with the myth of the

216

Issei as 'other', with their foreign ways and unassailable silences. They are referred to metaphorically as *watari dori* ('birds of passage'), those who have undertaken the journey from Japan to Canada. Paradoxically, they are usually signified as cranes, free and graceful, an image that belies its opposite image of voices turned to stone, hardened against the effects of modernity and change, and the sheer injustice of their constant uprooting, first from Japan, then in the passage from immigrant to permanent resident of Canada, then – the final injustice – in their forced removal from British Columbia during the 1940s.

The myth of generational difference tends to focus on the different reactions of the three generations to the injustices suffered at that time: the stoicism of the 'birds of passage' against the squawking of the politically activist younger generations. To understand this tension, it is necessary to understand how the myths have been constructed and reconstructed within community discourse, as well as within the broader social discourse that has defined all Japanese Canadians as other.

SPEAKING THE SILENCE: THE ISSEI

The Issei literary legacy is far from silent. It is a rich tradition of vernacular poetry engaged in not only by a creative elite but by a large proportion of Issei society. Poetry 'circles', or clubs, have represented one of the most important social institutions among early immigrants since the turn of the twentieth century, and have become even more pervasive in the post-Second World War era as ageing immigrants have turned to poetry as a major retirement pastime. That this poetry is not more widely appreciated is due in part to its being written in Japanese, which most Nisei understand only poorly and most Sansei not at all, and in part to its ephemeral nature. The poems are composed orally at regular meetings; if written records are kept they are not always accessible. None the less, there are some anthologies in translation,[5] and several in Japanese.[6]

Poetry is an ancient Japanese art form, adapted during the latter part of the nineteenth century (the Meiji Period) from a pastime of the social and cultural elite to become a pastime of the general population. By the turn of the century, when Japan had achieved the highest literacy rate of any country in the world, no village was without at least one poetry circle. Virtually all Japanese school children were well versed in the classic forms of poetic expression: the *tanka*, a verse of five lines and thirty-one syllables arranged in a sequence of 5–7–5–7–7; and the *haiku*, a verse of three lines and seventeen syllables arranged in a sequence of 5–7–5.

The best Japanese poets work within these syllabic structures to create profound and inspired images, expressive of the ascetic mysticism that characterises much of Japanese art and philosophy. None the less, the highly structured nature of Japanese poetry, combined with its widespread practice, ensures that much if not most of its vernacular expression is trite and

217

repetitive. But the significance of this writing is not in the quality of its product, but in the fact that its practice as a shared activity, a social act, provides a basis for common expression of the aesthetic and the emotional. Each poem is thus a mimetic reconstruction of feelings whose terms are publicly acknowledged. More importantly, each reading/writing is similarly a mimetic act; it is impossible to speak of poems in this context, but only of poet-ry. Poetry is a normative expression, then, that carries the restrictions as well as the comforts of all norms.[7] Elsewhere, I have described the result as 'structured feeling', in an adaptation of Raymond Williams' concept of 'structures of feeling'.[8] Such 'structures' provide not only a means of continuing past traditions but also a channel for the remaking of tradition under conditions that may otherwise prove traumatic because of the rapidity or the severity of change. Japanese poetry, in other words, is more than an everyday expression of common values; it is a means of normalising extra-ordinary events.

As such, it was used by the Issei both as a means of maintaining their cultural heritage and as a means of adjusting to the difficult circumstances of life in Canada. Extant sources, mainly newspapers and private papers, provide an understanding of the many contexts in which poetry was used as group cathartic for the Issei. Ito describes the social organization of poetry circles in British Columbia and Washington State starting in the 1890s, and provides examples of poetic responses to events as they occurred within the community.[9] For example, the following are some of the poems submitted after twenty-three Japanese immigrant men were killed in an accident while working on the Great Northern Railroad line near British Columbia:

Here the tombstones speak:
A record the endurance –
Sad and plaintive tale.[10]

He who lost life's game
Then buys one lone plot of ground –
For his burial.[11]

All the way over
To America, and then
Life came to nothing.[12]

Ito's massive collection chronicles other occasions for poetry, composed at regular meetings of poetry circles or other social groups, seasonal celebrations or gatherings to commemorate people and events. Whatever the circumstances, however, the poems serve to transform the trauma of passage, reconstituting and structuring the experience along normative cultural lines, and thereby initiating a new passage through immigrant life. Themes included nostalgia for home:

With tears in my eyes
I turn back to my homeland,
Taking one last look.[13]

No hope to go back..
Yet I dream of my home place,
In sleep returning.[14]

resolve to adapt to a new life:

Resolved to become
The soil of the foreign land,
I settle down.[15]

resignation:

Illusion and I
Travelled over the ocean
Hunting money-trees.[16]

Here in this country
My traditions and customs
Escape my embrace.[17]

perseverance and stoicism:

Vexed beyond my strength,
I wept. And then the wind came
Drying up all tears.[18]

or the particular concerns of Issei women:

Hope for my children
Helps me endure much from it,
This alien land.[19]

New Years – and mother
Today, too, wears her apron.
No holidays for her.[20]

As if the trauma of migration were not enough, in early 1942, all Japanese Canadians were forcibly removed from their homes, stripped of human rights and most of their possessions and confined to internment camps or forced to work as indentured farm labourers. The exile imposed by the War Measures Act, which allowed such extraordinary treatment of Canadian citizens as well as landed immigrants and Japanese nationals, lasted until 1949 when, for the first time, Japanese Canadians began to enjoy the same civil rights as other Canadians.

Reactions to this treatment varied among the 23,000 Japanese Canadians affected but, perhaps in response to the degree of trauma experienced, the poetry written by interned Issei went far beyond previous verse in its attempt to normalise the event. The following lines were written during a poetry competition held to 'commemorate' three years of incarceration near Slocan Lake in British Columbia:

> Spring water, warming
> I am happy after three years,
> Sitting at lake's edge.[21]

BREAKING THE SILENCE: THE NISEI

According to Ken Adachi, the attitudes of the Issei conformed to the stereotypes of stoicism found in their poetry:

> The Japanese [sic] were inclined to follow lines of least resistance since their cultural norms emphasized duty and obligation as well as the values of conformity and obedience.... The lack of aggressive behaviour and high dependency was part of the *enryo* (restraint) or *gaman* (forbearance) syndrome which explains much of Japanese behaviour. Not to conform was unthinkable ... Disruptive behaviour was censured, discipline and obedience were mandatory so that self-control, resignation and gratitude were highly desirable. Issei felt that suffering and hard work were necessary ingredients of character-building.

In contrast:

> Nisei had, on the whole, been denied the acceptance necessary for wellbeing and full personality development by both their Issei parents and white society. Society was again rejecting them, and many reacted with passive withdrawal and intense anger which was masochistically concealed and passively expressed.[22]

If the myths of Issei character were internalised poetically, the work of Adachi, a Nisei, serves to reconstitute them for subsequent generations. His scholarly interpretation is also an expression of the differences that were acted out in intergenerational relations. The Nisei, unable because of language barriers to appreciate the complexity of the Issei's 'structures of feeling',[23] cast the Isseis' stoicism within the stones of silence, all the while searching for their own passage toward acceptance in Canadian society. It was not an easy journey.

That journey has been immortalised by Joy Kogawa, a Nisei, whose *Obasan* is one of the most critically acclaimed English-language novels to be published in Canada in recent decades.[24] It is widely viewed as a pre-eminent example of the 'postmodern' genre in Canadian literature, both because of its political project of resisting the oppression of the government, and because it gives voice to previously silenced, and marginalised, members of Canadian society. Kogawa places those voices in the mainstream. But Kogawa's journey is still one that moves between silences, as she too enters the myth:

> There is a silence that cannot speak.
> There is a silence that will not speak.[25]

The mainstream has written back to Kogawa, further mainstreaming the experiences that Kogawa documents by 'interpreting' those experiences in ways that re-invent Japanese Canadians as 'other' and confound the silence. A few examples should suffice to show that mythic representation occurs at yet another level. For Harris,[26] 'in all *immigrant* communities the first, second and third generations represent crucial stages in adjustment to the adopted culture' (emphasis added). This positioning of the minority not only seals the entrance to the dominant culture, but also – and this despite some very perceptive comments on generational difference in the rest of the review – shows the extent to which Kogawa's book has been misunderstood; it makes Kogawa's not only an 'other' voice, but an exotic and irredeemable other.

The silencing occurs even more forcefully by Rose, who uses the theoretical work of Edward Said to reinvent the Issei:

> Clearly *Obasan* is historical fiction which is minority fiction, but as sophisticated minority fiction it offers not only a critique of injustice to a racial minority, but a critique of values within that minority which have tended to reinforce abusive majority attitudes. Kogawa through Naomi makes it clear that Japanese-Canadians are silent largely because it is the 'Oriental' way to be silent. In other words, what Edward Said calls 'Orientalism' has been so internalized by the Oriental minority, that their silence is an inadvertent bow to the occidental hegemony which legitimizes their abuse.[27]

Rose's objective is a theoretical one, to develop a method of critical 'inter-textuality' whereby the 'text' and 'the critic' can be caused to intersect meaningfully with 'the world'.[28] This convoluted appropriation of Kogawa's text, rather than celebrating minority expression, denies them a voice as anything but other, while making a mockery of their 'fictionalised' history. If Kogawa uses silence as trope in order to give her characters voice (and it is partly this paradoxical quality that makes *Obasan* such a fine novel), Rose thus uses silence as a means for denying voice.

In yet another hegemonic theoretical appropriation, Potter uses Kristevean theories of M/mother to translate the silences of *Obasan* along yet another mythic journey. Rejecting the surface story 'told by a woman about her childhood as a Japanese Canadian during World War II', Potter finds a subtext of 'the tale of a totem, a tribe, and the sacrificial victims – and the dialectic is one of abjection'.[29] While many of Potter's subsequent

221

observations enrich the reading of the text, that richness is recast in the final observation that what Kogawa achieves is:

> a coming to terms with one's race and gender (body as a sign). These words for me contain faith in the power of the feminine, and her saving grace, despite the continuous ebb and flow. The semiotic that Kristeva hypothesizes can only be attained by reaching below the surface of our learned feelings of repulsion for the M/mother(s) body.[30]

Is it possible to reappropriate this text, pry it from the grasp of such hegemonic discourse? If not, then neither is it possible to repossess any literature that is uniquely Japanese Canadian and expressive of their experience. If so, however, it will not be by any simple refutation of theoretical perspectives that do not seem to be, somehow, appropriate. For the critical interpretations are also myths of myths, and cannot be discounted as 'untrue' if we are unwilling to accept an essentialised notion of the 'real' meaning of Japanese-Canadian literature. It will be, rather, in distinguishing between those myths that turn to stone and those to sound, those that silence and those that, as Willis notes, *speak* the silence.[31] Or, in recognising, as Simon and Garfunkel once put it:

> the vision
> that was planted in my brain
> still remains
> within the sound
> of silence.[32]

Obasan seems to have taken Kogawa on a journey far from the stilted *haiku* of her ancestors. The structured feeling of neo-classical Japanese verse, and the stoic responses of its poets, has given way to what Willis calls 'an imaginative triumph over the forces that militate against expression of our inmost feelings'.[33] And yet, for all the passion of Kogawa's prose as she documents Naomi's very un-Issei-like journey for a sense of belonging, the silence is given sound in the end, and the connections are made: 'My loved ones, rest in your world of stone. Around you flows the underground stream. How bright in the darkness the brooding light. How gentle the colours of rain.'[34] And the final lines of the book evoke the mystical one-ness of the haiku:

> Up at the top of the slope, I can see the spot where Uncle sat last month looking out over the landscape.
> 'Umi no yo,' he always said. 'It's like the sea.'
> Between the river and Uncle's spot are the wild roses and the tiny wildflowers that grow along the trickling stream. The perfume in the air is sweet and faint. If I hold my head a certain way, I can smell them from where I am.[35]

Much as the water of Slocan Lake washed solace over Issei poets stranded upon its shores, the landscape of southern Alberta instils in Naomi a sense of peace. But it is a particular kind of peace, not derived from the landscape itself in any naturalistic manner, but from the interaction of human beings within a landscape that has particular aesthetic features constructed in the likeness of human emotion. The voices of Naomi's family come together, finally, in all the richness of their difference but also in their peculiarly geographical association with place. Their journeys take on new meaning in this final passage, as actual journeys between physical places. For all three generations, expression needs this grounding.

There is more than a slight touch of romanticism, of course, in Kogawa's conclusion. Few ever achieve Naomi's sense of a journey completed. But this novel, like most, is archetypical, a myth of myths; as such it re-creates common vision and should be understood above all as representative of a community for whom shared space, and the structured aesthetic responses that such space engenders, is indeed common ground. If there is a lesson to be learned from intertextuality, therefore, perhaps it is not in the somewhat arrogant tones of grand theory but in the generational intertextuality that allows (and of course always has allowed irrespective of Kogawa's work) communication between Issei and Nisei despite formal differences of language and style.

But there is yet another level at which these common experiences must be read. A fact seldom noted in the critical interpretations of *Obasan* is that the character of Naomi's Aunt Emily, the aunt who lives in sound, the 'word warrior . . . a crusader, a little old grey-haired Mighty Mouse, a Bachelor of Advanced Activists and General Practitioner of Just Causes'[36] is modelled upon a real-life woman named Muriel Kitagawa (1912–74) whose papers, now held in the National Archives of Canada, provided Kogawa with inspiration for her book. Those papers, and Muriel Kitagawa's story, were revealed recently by Roy Miki,[37] who writes:

> For Muriel Kitagawa, the writer, the journey to Toronto prompted her to think and write about her Nisei past. The abrogation of her rights in 1942 was infuriating, but the disappointment in her country struck much deeper. All the hopes and aspirations of her generation, all the struggles they had waged to break down the discriminatory barriers excluding them from full participation in the life of their society – these appeared to be lost in the maelstrom caused by the very forces of racism they had been trying to eradicate. Her Nisei generation had endured a whole spectrum of social, political and legal restrictions because of their ancestry. Now, with the uprooting, they were thrown into another crisis, even more trying. 'This is the time,' Muriel wrote in *A Record of Dignity*, 'to tighten our belts, to gird our courage, to condition ourselves to every kind of privation and endure . . . endure . . . and still endure.'[38]

The attitude of *gaman*, so strongly associated with the Issei and with their sense of dignity, traverses the generations, stripped of its stoicism, to become a model of perseverance rather than endurance, and it is this difference that animates the Nisei desire to engage rather than withdraw from the processes of oppression. For Muriel Kitagawa, that engagement was expressed in an eloquent testimonial to her place in Canadian society and to the importance of coming to terms with the uprooting:

> At first I was rather shy about it, though very proud, because in spite of hardships, of hunger too, there was this feeling of *belonging*:
> 'This is my *own*, my native land!'
> Then as I grew older and joined the Nisei group taking a leading part in the struggle for political liberty, for economic equality, I waved those lines around like a banner in the wind:
> 'This is my own, my *native* land.'
> Ten years later, when our struggle hadn't brought any tangible results I wept those lines with the slow grief of knowing that the ten years would soon be twenty:
> '*This* is my own, my native land!'
> When war struck this country, when neither pride nor belligerence nor grief had availed us anything, when we were uprooted, despoiled, and scattered to the four winds, I clung desperately to those immortal lines:
> 'This *is* my own, *my* native land.'[39]

Those words became an inspiration, and a kind of battle cry, for the Japanese-Canadian redress movement, which culminated in a settlement on 22 September 1988.[40] Many of those most actively involved in the movement were Sansei, few of whom had experienced the uprooting, few of whom had ever seen Japan. But their journey too has been in quest of a place, and if their means of expression evokes squawking ducks rather than birds of passage, their expressions show a deep sense of generational continuity.

THE SOUND OF SILENCE: THE SANSEI

The Sansei generation grew up inspired by Simon and Garfunkel rather than Bassho (a classical *haiku* poet) or Scott (whose ballad, as part of the canon of English Canadian education, had inspired Muriel Kitagawa). Their tone is modern, their verse unstructured. They reinvent the myths of their community with impatience rather than perseverance, but not without a strong reverence towards past journeys and roads taken. I am one of them, a fact that cannot be ignored if generational intertextuality is to have any meaning. For my own words must be interpreted against the nudgings of a grandmother and a grandfather who were poets, and whose words speak directly into my personal silence. I once wrote back to them:

224

Senbei and sushi giver
Surrounded by miso aura
You maintained dignity.
You brought life to coloured paper
And in its shapes we saw
the world you cared about. . . .

Did you know
When your hopes came eastward
That your hands would become thick with calluses
So the paper wouldn't fold any more?
Did you know what they would do to you?[41]

These words are clearly those of a Sansei. To generalise further about the common nature of Sansei experience is difficult; none the less, among the growing body of Sansei literature, most of it poetry, we can hear a common if not singular voice. The journey continues, a passage between the lines of racialised identity. For Gerry Shikatani, the ghost of the *haiku* poets transforms the Canadian landscape:

a trickling
of a hidden stream/
some place
in the ear
where it appears
this idea heard
dances where
it is.[42]

Haruko Okano intertextualises more directly:

My syntax is Japanese-Canadian,
formed by generations of hushed voices.
Pressed white,
like manju. Pulling away softly
from lips barely parted.
A sweetness of language lost.[43]

But despite the loss:

The time of silence passes.
We grow unobtrusively, insistently.
A maple tree, a Japanese maple tree –
symbolic of our presence.[44]

The passing of silence, however, has occurred within a social context that is intensely political. If in their poetic expressions these Sansei create a some-what romanticised linkage to their past, that linkage only serves as an

225

inspiration for a much more bitter engagement with the present. In *Bitter-sweet Passage*, Maryka Omatsu writes to her dead father:

> Now, every time I am in Vancouver and I pass your corner I am filled with an anger so palpable that my mouth starts tasting of dry sawdust. I knew that you missed British Columbia with the longing of a lost love. After all, how could you forget the rain-drenched lushness and the staggering beauty of the mountains backdropping the Pacific Ocean? Still, I could understand why you never returned there. Just turning a corner would have unleashed memories that would have eventually consumed you.[45]

Finally, Roy Miki's journey between places and generations evokes the full sense of writing between two worlds, of generations rooted and uprooted and rerooted/routed through the spaces of our cultural myths:

> & to be among   there are
> no   roots
> there   ships wait
> to be moved   and they
> too cross   the pacific
> once   we said
> we say   the world lay
> a mixture   the sun
> the sea   our children
> like marigolds   our boats
> our nets   we
> filled our houses   now
> thin & brittle   the cold clings
> now   the inland sea[46]

CONCLUSION

'Umi no yo,' said Naomi's uncle, it's like the sea. There is a common landscape for Japanese Canadians, reconstructed from place to geographic place and from time to time and generation to generation, mythologised to create a sense of connection, rootedness and a community of experience. If that landscape has meaning as an inland sea through which surges our spoken silence, however, it is not only because that meaning is woven as a common thread in the themes and metaphors that our words produce. It is also because our journeys have involved the negotiation of cultural practices and the myth-making through which we define ourselves and justify our lives. That is more than a romantic gesture to trite poetic phrases; it is the struggle to give voice to an identity that is not only other than the dominant society, but seeking its own dominance. That identity shifts frequently between stereotypes and innovation, between graceful aesthetic expressions that immortalise tradition and strident squawkings that seek to silence those expressions but end up remaking them in new image. The resulting intertextuality lands us in what Gillian Rose calls 'paradoxical space',[47] where we are both centred and marginalised, racialised and gendered in ways that sometimes emancipate and sometimes control, and where the precise contours of place are founded on the myth of myths. Our words may be therein destabilised; they are not silent.

NOTES

1 I wish to thank Katie Pickles for her assistance in gathering the materials for this chapter, and Mark Rosenberg for his inspiration.
2 J. Kogawa, *Obasan*, Markham, Ont., Penguin, 1981, p. 32.
3 The reference to 'imagined communities' comes from Benedict Anderson. His analysis, however, is based on 'large cultural communities' whose nationalism is the antithesis of the small and marginalised communities I am addressing here. B. Anderson, *Imagined Communities: Reflections on the Origin and Spread of Nationalism*, London, Verso, 1983.
4 P. Fitzpatrick, *The Myth of Modern Law*, London, Routledge, 1992, p. 17.
5 Kisaragi Poem Study Group, *Maple: Tanka Poem by Japanese Canadians*, Toronto, Continental Times, 1975; G. Shikatani and D. Aylward, *Paper Doors*, Toronto, Coach House Press, 1981.
6 D. Kobayashi, *Kobayashi Denbei to sono ku*, Tokyo, Denbei Kobayashi, 1963; Kisaragi Tankakai, *Kaede*, Toronto, Kisaragi Tankakai, 1972; Nikkeijin hyakunen hatsu iinkai, *Nikkeijin hyakunen hatsu kinen kushi*, Toronto, Continental Times, 1979.
7 An analogy might be drawn between Japanese poetry and contemporary rock music. It is experienced as an expression of freedom, yet acts as a normative agent of conformity.
8 A. Kobayashi, 'Structured feeling: Japanese-Canadian poetry and landscape', in P. Simpson-Housley and G. Norcliffe (eds), *A Few Acres of Snow: Literary and Artistic Images of Canada*, Toronto, Dundern Press, 1992, pp. 243–57; R. Williams, *Marxism and Literature*, Oxford, Oxford University Press, 1977, p. 31.
9 K. Ito, *Issei*, Seattle, Executive Committee for Publication of Issei, 1973. Ito's translation style maintains the original 5–7–5 syllabic cadence of the *haiku*, thus capturing the original sense of rhythm. Despite the 'creativity' of freer forms of *haiku* translation, I also prefer this format. This issue is of considerable concern to *haiku* translators, or to writers of *haiku* in English; it is not an issue for those who write in Japanese, for whom the strict form defines the genre.
10 Henko, translated by Ito, op. cit., p. 883. Most poets wrote under a pen name. The further identity or place of residence of the poets is unknown, but they were a mixture of Canadian and American immigrants, and poems were freely exchanged across the border. The distinction is not relevant for the poems cited here.
11 Ryuka, translated by Ito, op. cit., p. 883.
12 Daikichi Nishio, translated by Ito, op. cit., p. 884.
13 Seijin, translated by Ito, op. cit., p. 34.
14 Mutsuko, translated by Ito, op. cit., p. 51.
15 Ryufu, translated by Ito, op. cit., p. 11.
16 Kijo, translated by Ito, op. cit., p. 32.

17 Seijin, translated by Ito, op. cit., p. 349.
18 Yoshie, translated by Ito, op. cit., p. 255.
19 Katsuko, translated by Ito, op. cit., p. 602.
20 Tamu, translated by Ito, op. cit., p. 274.
21 Kobayashi, 'Structured feeling', op. cit., p. 254.
22 K. Adachi, *The Enemy that Never Was: A History of Japanese Canadians*, Toronto, McClelland and Stewart, 1976, p. 225.
23 It is significant that for Ito, a postwar Issei, a history of Japanese immigrants in North America would be incomplete unless interpreted through poetry; Adachi's history of Japanese Canadians, written from a Nisei perspective, makes no mention of poetry. This is so despite the fact that he used community newspapers, which virtually never left the press without the inclusion of poetry.
24 It won the Books in Canada First Novel Award in 1981 and the Canadian Authors' Association Book of the Year Award in 1982. It has since been translated into more than a dozen languages.
25 J. Kogawa, *Obasan*, op. cit., Preface.
26 M. Harris, 'Broken generations in *Obasan*', *Canadian Literature*, 1990, vol. 127, pp. 41–57.
27 M.R. Rose, 'Hawthorne's *Custom House*, Said's *Orientalism* and Kogawa's *Obasan*: an intertextual reading of an historical fiction', *Dalhousie Review*, 1987, vol. 67, pp. 286–96.
28 ibid., p. 295.
29 R. Potter, 'Moral – in whose sense? Joy Kogawa's *Obasan* and Julia Kristeva's *Powers of Horror*', *Studies in Canadian Literature*, 1990, vol. 15, p. 117.
30 ibid., p. 138.
31 G. Willis, 'Speaking the silence: Joy Kogawa's *Obasan*', *Studies in Canadian Literature*, 1987, vol. 12, pp. 239–49.
32 Simon and Garfunkel, *The Sound of Silence*. © Paul Simon/Pattern Music Limited 1957. With kind permission of Pattern Music Limited.
33 Willis, op. cit., p. 249.
34 Kogawa, op. cit., p. 246.
35 ibid.
36 ibid., p. 32.
37 M. Kitagawa, edited by R. Miki, *This Is My Own: Letters to Wes and Other Writings on Japanese Canadians, 1941–1948*, Vancouver, Talonbooks, 1985.
38 R. Miki, in the introduction to Kitagawa, op. cit.
39 Kitagawa, op. cit.
40 M. Omatsu, *Bittersweet Passage: Redress and the Japanese Canadian Experience*, Toronto, Between the Lines, 1992.
41 A. Kobayashi, 'Obasan', in The Powell Street Revue and the Chinese Canadian Writers' Workshop (eds), *Inalienable Rice: A Chinese and Japanese Canadian Anthology*, Vancouver, 1979, p. 12.
42 G. Shikatani, 'The stream', in Shikatani and Aylward, op. cit, p. 121.
43 H. Okano, 'Sansei', in *Come Spring*, North Vancouver, Gallerie, 1992.
44 H. Okano, 'Canada, neh?' in *Come Spring*, op. cit.
45 Omatsu, op. cit., p. 37.
46 R. Miki, 'five', in *Saving Face*, Winnipeg, Turnstone Press,1991, p. 17.
47 G. Rose, *Feminism and Geography*, Cambridge, Polity, 1993.

15

VULCAN'S BROOD

Spatial narratives of migration in Southern Africa

*Jonathan Crush**

AWKWARD INTERVALS

Gold mining, and the system of oscillating labour migration upon which it is based, has acquired a paradigmatic significance for understanding larger issues of race, class and politics in South Africa.[1] The result is a plethora of academic writing on the origins, entrenchment and reluctant transformation of the modern world's most enduring, organised and rapacious system of international labour movement.[2] Gold also has a special kind of symbolism for the literature of colonial and post-colonial resistance in South Africa. Any work of fiction set in Johannesburg must inevitably engage at some level with the foundation of gold, for without it the city could not have come into existence.[3] Travellers exploring past and present South Africa inevitably pass through Johannesburg and they too write the city and its mines as emblematic of the country as a whole.[4]

In academic writing, there has been a complementary effort to grasp the interior worlds of the migrant experience.[5] Historically, most migrant workers were poorly educated, even illiterate. If they had not been, they would probably have worked elsewhere. In the archival record, there are only the barest written traces of the migrant voice – in court transcripts, missionary records and commission proceedings. Oral history promises an entrée into the world of the migrant.[6] But there is sometimes a disarming naïveté about the interpretive challenges posed by oral texts.[7] Life-history is the dominant technique and scholarly narrative the dominant mode of representation. Both are deeply problematical vehicles for understanding subjectivity and experience. Migrants rarely remember in neat chronology. Scholarly narratives filter, combine and recombine the rough contending voices of oral histories, smoothing them into a coherent whole designed to illustrate some larger truth. Original transcripts, when accessible, speak tantalisingly of a world outside and beyond the intrusions of the interviewer and the measured prose of the scholarly construction.

Fiction, unencumbered by scholarly imperative, offers an alternative way of constructing the experience of migration. In South Africa, works of fiction that take migration as their theme are generally forms of outsider representation and many are lodged within the deeply entrenched conventions of colonial discourse.[8] Probably the best contemporary example is the prolific popular novelist Wilbur Smith. Learned critics agree that Smith's corpus has little redeeming merit. Yet, outside South Africa, Smith is probably better known and certainly more widely read than 'literary' figures such as Brink, Breytenbach, Coetzee and even Gordimer. Smith's writing is of interest for what it reveals about the mind of (white) South Africa and its attempts to control the language and terms in which the black experience is spoken.[9]

In the case of the South African mining industry, Smith's 1970 novel *Gold Mine* provides a powerful example of closure premised on a particular kind of geographical imaginary.[10] The novel, for all its pretensions to show the life of the mine, shows minimal comprehension of the life of the miner. Its constructions are driven by a mechanical and technological determinism in which all behaviours and beliefs are a natural extension of the physical and geological environment of gold mining. Physical processes become, in turn, metaphors for migration itself, silencing and effacing a far more complex reality.

Geography can be as much an enabling as a silencing device in the literary text, as various post-colonial representations of migration make clear. Paul Carter argues that an authentic spatial rendition of the migrant experience would begin by regarding movement 'not as an awkward interval between fixed points of arrival, but as a mode of being in the world'.[11] The black South African playwright, Zakes Mda, chooses precisely this vantage point from which to explore the experience of migration. Mda's plays are set not at the fixed points of departure and arrival but at the transitional places in-between – 'The Hill', 'The Road', 'The Dead End'.[12] By situating migration in this way Mda attempts to visualise the homestead and the mine as they are mapped in the consciousness of the migrants, unencumbered by the brute presence of these places.

In the colonial and post-colonial representations of migration to be explored in this chapter, the language of space is central. But these are ultimately and always the constructions of outsiders moving, in the case of post-colonial writing, progressively closer to the migrant experience but never being of that experience. Is there any way of moving across the boundary between these rival kingdoms and determining how the spatiality of migration is represented from within migration itself? Cultural historians have recently begun to pay greater attention to the oral texts generated by migrants themselves.[13] In the performance arts there are voices which are not themselves constructions of the interaction between narrator and scribe. These deeply poetic texts are from an entirely different structure and realm of meaning. What is striking about many of the texts that have been recorded, none the less, is the centrality of spatial imagery to the articulation of the migrant experience. This is not a spatiality that unfolds in the unilinear, ordered and purposeful fashion of the outsider narrative, however. It is a vivid spatiality of unruly fragments and disconnections.

CHEAP BLOOD

Every year over 500 migrant South African miners are killed in rockfalls, rockbursts and underground explosions. Between 1945 and 1984, over 50,000 miners died while at work, 20,000 of these in accidents.[14] Thousands more are crippled and maimed each year – some 110,000 between 1972 and 1976 alone. Occupational disease exacts a heavy toll on those who survive.[15] In mining-industry discourse the accident rate is an unfortunate but inevitable by-product of the unstable geology of deep-level mining.[16] Since the early 1980s the National Union of Mineworkers (NUM) and progressive researchers have seriously undermined this view, locating the accident rate as much in the social as the physical conditions of production.[17] Sociologist Jean Leger has demonstrated that black miners have an elaborate 'pit-sense', that alerts them to dangerous working conditions. Ignoring the evidence of the 'talking rocks', white miners in search of production bonuses consistently put themselves and their black gangs in dangerous work situations, often with devastating consequences.[18] The idea that unsafe, and avoidable, mining practices are at least partially responsible for the catastrophic mortality rates in the mines has been vigorously contested by the mining industry.

In Wilbur Smith's *Gold Mine*, a tedious and predictable plot is interspersed with graphic and lurid depictions of the bruising and brutal conditions of underground gold mining. The story has barely begun, for example, when the character known as 'the squealer' is introduced: 'the sound of a grown man, with his legs crushed under hundreds of tons of rock, perhaps his spine broken, dust suffocating him, his mind unhinged by the mortal horror of the situation.' A rescue party finds the squealer, 'his face running with the sweat of terror and insanity . . . He began to scream, but suddenly the sound was drowned by a great red-black gout of blood that came gushing up his throat, and spurted from his mouth.'[19] This kind of blood-soaked language suffuses all of Smith's many subsequent depictions of underground accidents. Lethal geology is also colour-blind in its devastating effects as black and white miners are maimed and destroyed by the power of the rock. Smith's narrative builds predictably to the final cataclysm in which hundreds of workers die in an unlikely underground rockburst and flood. The representations quickly take on a formulaic and voyeuristic quality and lose their impact precisely because they leave nothing to the imagination.[20]

In the late 1960s, when he conducted the research for his novel, the mining industry's views on the causes of underground mayhem were virtually hegemonic and Smith accepts them without question. The novel begins with a detailed geological description of the formation of South Africa's deep-lying

231

gold reefs and makes an immediate didactic link with the human cost of gold mining:

> When you go down into the ultra-deep levels below eight thousand feet and from those depths you remove a quarter of a million tons of rock each month, mining on an inclined sheet of reef that leaves a vast low-roofed chamber thousands of feet across, then you must pay, for the stress builds up in the rock and the focal points of pressure change until the moment when it reaches breaking point and she bumps. That is when men die.[21]

The conclusion is obvious: men die because of dangerous geology. Why then is gold mined at all? There is no explanation other than the 'greed and lust it conjures up in the hearts of men'.[22] Warped geology and generic greed, rather than the specific conditions of gold markets, capitalist production and exploitative labouring, account for the danger and death that sooner or later colours gold red.

The physical environment of gold mining supplies Smith with an inexhaustible fund of metaphors with which to write the human experience of migration. Take, for example, the geographical imagery used to describe the migrant labour system at work:

> From the swamps and fever lagoons along the great Zambezi, from among the palm groves fringing the Indian Ocean, out of those simmering plains that the bushmen called 'the big dry', down from the mountains of Basutoland and the grasslands of Swaziland and Zululand they [the recruiters] gathered the Bantu, the men themselves completing the first fifty or sixty miles of the journey on foot. Individuals meeting on a footpath to become pairs, arriving at a little general dealer's store in the bleak scrub desert to find three or four others already waiting, the arrival of the recruiting truck with a dozen men and their luggage aboard, the long bumping grinding progress through the bush. The stops at which more men scrambled aboard, until a full truck load of fifty or sixty disembarked at a railway siding in the wilderness. Here the tiny trickle of humanity joined a stream, and at the first major centre they trans-shipped and became part of the great flood that washed towards 'Goldi'.[23]

The imagery is hackneyed and laboured but the intent is clear – to portray migration as a natural phenomenon, as inevitable as a river rising in several different watersheds and moving inexorably to the sea. Here the certainty of the process (and the writing out of all coercion, agency and resistance) is achieved by analogy with the physical world. Geography becomes a means to silence a more complex and nuanced reality.

Primordial racial characteristics are determined by the physical environment – simmering plains, mountains, grasslands, the Tropics all produce their own particular 'tribal' identities. These identities are partly overcome by the still more powerful determinism of gold mining. The terrors and challenges of the physical environment forge a kind of deracialised 'Vulcan's brood', men bonded together by a common enemy. Racial difference is erased and masculine 'virtues' of toughness, brute strength and bravery lauded. The two major characters – the black Big King and the white Rodney Ironsides – are even named for their physical prowess. Even as Smith gestures at a complicated set of social relations between white and black miners, he successfully erases the racism, domination and gratuitous brutality of the underground work station that the NUM has so successfully contested since the early 1980s. While the reader is privy to the intimate social and sexual life of the white miner, the black miner remains beyond Smith's representational powers. The result is a highly stereotyped set of images. Ironically, in trying to erase the odour of racism from his main white character, Smith reveals the depth of his own.

Black miners are recruited for work throughout the sub-continent by a centralised recruiting organisation. For Smith, the agency's role is entirely benign, to 'make certain that a man taken from an environment that had not changed in a thousand years' would retain his health, happiness and sanity and return with positive tales of his experiences. Smith has little to say about the rural end of migration except in terms like these. Radical historians of the 1970s and 1980s were to completely demolish the kind of colonial mythology that assigned timeless, static and backward characteristics to the source regions of migration, showing instead how deeply they had been marked and transformed by labour migration.[24] When Smith researched and wrote, the mythology of pristine tribalism remained strong.

Smith's constant uncritical usage of apartheid terminology – such as 'girl' to describe black women and 'the Bantu' to describe male miners – coalesce into depictions of the inevitability of ethnic conflict:

> Basuto is [sic] one of the fighting tribes of the N'guni group. These wiry mountaineers rushed into the conflict with the same savage joy as the Shangaans, a conflict that raged and roared out of the single room to engulf the entire population of Dump City. One of the girls . . . shrilled that peculiar ululation that Bantu women used to goad their menfolk into battle frenzy.[25]

Ethnic violence is certainly a long-standing plague on the mines. The origins, according to Dunbar Moodie, lie not in primordial ethnicities but in specific mine conditions and practices.[26] Smith, like most other colonial writers in the Robert Ruark mould, can conceive of violence only in terms of primitive, atavistic impulses. Violence is a pleasure, a recreation, a source of 'savage joy'. Black women are virtually invisible throughout Smith's account. Here, and in the incident that follows, they surface in the colonial imagery of animalism and unbridled sexuality.

There is little in Smith's depiction of African atavism that is not echoed elsewhere in the Chamber of Mines' official submission to the 1974 Du Randt Commission of enquiry into mine unrest or in the everyday discourses of white miners and mine managers.[27] These colonial myths, as Maughan-Brown points out, 'serve to stress the essential "otherness" of blacks, and thereby provide an implicit rationale for political strategies which seek to enforce separateness and deny blacks the right to political and social equality'.[28] Smith's maps of the migrant labour system are meant to rationalise the heinous geography of apartheid itself.

OUTSIDERS

In the colonial discourse of Wilbur Smith, geographical imagery completely erases any space for articulating the migrant experience. But this does not mean that the spatial inevitably plays a reactionary role. In the plays of Zakes Mda, for example, alternative possibilities emerge.[29] Mda attempts to define a distinctive migrant space within which those voices can be heard. To illustrate this point, I focus on his 1979 play about migrant Basotho mineworkers, *The Hill*.[30] The play is set on a rocky granite outcropping known as Qoatseneng Hill, a prominent feature of the landscape of the Lesotho capital, Maseru.

When I first saw *The Hill* in Lesotho in 1980 it was an extremely ironic experience. The play was performed at the Maseru Holiday Inn (also the purveyor of pornography, gambling and illicit sex to white South Africans).[31] From the Inn one could see Qoatseneng Hill; and the Holiday Inn, with its security guards and high fence, was clearly visible from the hill. On one side of the hill itself was the opulent Hilton Hotel, protected from the hill's cave-dwelling other residents by screeds of barbed wire. A social-worker friend tried to persuade the manager of the Hilton Hotel to donate the daily leftovers from the five-star restaurant to a feeding programme for unemployed miners. The manager had a deal with a local pig farmer who paid well for the scraps and the request was turned down. An attempt to buy a tent for hill residents to sleep in was stopped by the Lesotho government who openly chastised the plan for implying that the state was doing nothing to help. Scores of unemployed migrants slept in the open every night outside the mine recruiting office at the Maseru traffic circle, opposite the Catholic Cathedral. Many more took shelter among the caves and rocks of Qoatseneng Hill. In its stark geography, the Hill juxtaposed the extremes of poverty and wealth which are endemic to mining and migration. Mining executives trying to ingratiate themselves with the Lesotho government stayed in the Hilton on the hill; mineworkers who mined the gold that made them wealthy lived in caves over the ridge.

By confining the play to this one circumscribed space Mda can attempt to unravel the conflicting currents, characters and contradictions of the migrant labour system as they ebb and flow through one small part of the overall system. Through the symbolic value of the hill, Mda is able, in Carter's words, to explore migration as a 'mode of being in the world'.[32] The Hill is a truly migrant space, unlike any other. One of Mda's characters refers to it possessively as his 'house and home' though both are really elsewhere. It is neither the mine (where management's power and surveillance severely constrain the activities, words and relationships of the miner),[33] nor is it the rural home, where the family and the community are paramount, where miners are expected to be 'fathers' and 'husbands' and are judged in terms of how well they provide for their families. The Hill is a travelled space where migrants are truly migrant and can relate to each other in ways that are impossible anywhere else.

The Hill is also fundamentally a space of exclusion, a new location in a system which used to be defined in terms of the binary opposition between mine and home. This is an intermediate, twilight zone, a zone of deprivation and uncertainty, but one which is none the less experienced by increasing numbers of migrants and for growing lengths of time.[34] Before the mid-1970s, any Basotho migrant who wanted a mine job was virtually assured of getting one. This came to an abrupt end after 1975 when rising mine wages and growing regional unemployment dramatically expanded the pool of workers willing to accept minework. The mines became an employment centre of first recourse rather than last resort.[35] Miners with jobs feared losing them and tended to stay on. Fewer recruits were now needed. New workseekers and experienced miners not in possession of re-employment guarantees found it increasingly difficult to get a mine contract. They could not stay at home and wait, nor could they travel to the mines. All contracting was through the Chamber of Mines' far-flung network of rural and urban recruiting stations. The only way to get a contract was to hang around one of these offices. Many took up semi-permanent residence, living a precarious hand-to-mouth existence while they waited.

The Hill is the place where a migrant culture of exclusion and desire takes root. An employed miner who stops over at the hill for a night is ostracised by the unemployed hill-dwellers – 'why should he intrude on our peace and happiness?' demands the young workseeker. The main characters are unable to secure the mine contracts that will take them away from this zone of waiting. The poverty and degradation of their existence is portrayed in frank, yet sardonic, language. Miners seek casual labour but not for more than a day or two otherwise they might miss a mine contract; they beg, borrow and steal any food they can; they scavenge through the dustbins of Maseru West where the discarded food is richer; they compete furiously for everything including, ultimately, the contract; they sell their blood to the Hospital in exchange for cash; and they compare the size of their faeces in a desperate game of one-upmanship to demonstrate that they are eating better than their peers and therefore experiencing the indignities of their situation less acutely.[36] They refuse to admit defeat and go to the coal mines recruiter – 'I am a man of gold', declares the experienced miner to his younger accomplice.[37]

Yet, for Mda, their whole existence is ultimately premised on the possibility of leaving. Thus they speak continuously of other places rather than of where they are. They also enact a series of plays-within-the-play to bring those other worlds to the Hill. The homestead, the recruiting office, the mine church, the search for temporary work in Maseru West are all recalled with venomous parody. The technique decentres the stable worlds of the migrant-labour system, viewing them as they are refracted through the consciousness and desire of the migrant traveller. The Hill is the place where the migrant looks backwards and forwards, backwards to where he has come from, forwards to where he hopes to go. It is also where imposed identities and daily humiliations can be angrily contested, well beyond the reach of management's disciplinary power.

The migrant-labour system is premised upon the erasure of individual identity. Management's fundamental aim is to enumerate, aggregate and homogenise: 'I am from the place', observes the veteran miner, 'where men are but numbers engraved on plastic bracelets.'[38] Mda restores a semblance of individuality through three miners whose station, ambition and experience are radically different: the young novice, the older miner waiting for work and the veteran on leave from the mine. The older miner needs work to rebuild his shattered rural homestead and to provide for his impoverished family. The novice, in a rather overdrawn portrait of the target worker, wants only the 'good things' that mine money can buy – clothes, a car, a stereo, a reputation. The veteran miner circulates continuously between the mines and Maseru, without ever going home, articulating a new kind of truncated and unfulfilled migration: 'I have not seen them for four years. Every year things follow the same pattern. I come back loaded with clothes and money. The next day my suitcase and pockets are empty. I can't go home to my home village naked and empty-handed. What else is there to do but join again?'[39] The different characters are certainly stylised (in an ironic twist none is actually named) but they do establish the fact of difference and individuality. Mda demonstrates that there is no single miner – but men with different expectations and experiences of migration.

Mda's migrants do not see the central problem of migration in terms of the threat to life and limb in a hazardous industry. Indeed, his characters barely mention the dangers of minework. Rather, the greater danger, in almost Foucauldian terms, is what it does to the bodies that are classified, dehumanised and disciplined:

> Privacy! You will learn soon enough that from now on privacy is a thing of the past. Your shit will have privacy at your home where you are a man. Where you are the father of your children and the husband of your wife. The mines will teach you a different lesson, and you better start

learning it now. We all shit in open lavatories there. Father and son together. We all wash in communal shower rooms. There is no privacy in nakedness.[40]

Migration strips away all normal bodily functions, beginning with the fortnightly sale of blood to the Hospital – 'the day we eat and drink our blood' – for 'crisp rands, tins of fish and milk'. It continues in the many humiliations of the recruiting process and is completed on the mine itself:

> What is not degrading in the land of gold? The medical examinations through which you'll go, are they not degrading? When all the recruits stand naked irrespective of age and relationship, only to have the heartbeat examined, is that not degrading? Is it not degrading to sleep in the bug-infested hostel of the Native Recruiting Corporation while waiting for the train to take us across the Mohokare River ... away from our beloved Lesotho? Is it not degrading to be packed like stinking sardines in the train? What do you say about Mzilikazi, the mine labour hospital in Welkom, where in our nakedness we are publicly x-rayed, and have to raise our arms and legs while we are being inspected like cattle for sale? Is that not degradation? Are all these things not meant to humiliate us, to make us inadequate as men and fathers to our children, and to deprive us of human dignity, so that we may dig the gold of the white man with utmost submission? You talk of degradation. You have not seen anything yet. You have not felt the painful injection they give you without changing or cleaning the needle after injecting a hundred recruits. You do not know the House of Satan where acclimatization is done. Where you run and do all sorts of exercises in a very hot room – until you shit and collapse if you are a weakling.[41]

For Mda the migrant-labour system also emasculates by depriving men of their sexual identity and gender roles, 'to make us inadequate as men and fathers to our children'. The system 'strives to castrate us'.

The migrant-labour system is also deeply gendered, providing very different constraints and opportunities for men and women. The three women who (endlessly) rob the returning migrant are themselves victims of an abbreviated form of migration which stops them short in Maseru and prevents them from going on to South Africa to seek work. The Hill also provides a space for these other migrant voices, the voices of women who have been marginalised by the deprivations of migration, and who are excluded from the gendered and patriarchal spaces of the compound and the rural homestead.

In his commentary on Mda's work, Horn argues that the women 'seem to have lost all sense of community obligation' and are 'untroubled by conscience', having robbed the man and therefore his family of all his

structure is relatively straightforward – a motive for migration, a journey from home to the recruiting office, and an impending sense of doom and loss of personal autonomy and control. Biblical allusions to South Africa as satanic and devilish enhance this sense, as does the contrast between a humane rural existence and the impersonal recruiting process.

In many *lifela*, the journey of embarkation to the mines is of secondary significance; what it permits is a broader poetic exploration of the phenomenon of travel itself. The generic name for *lifela* does not designate them as miners' songs but rather as 'songs of the travellers or ramblers'.[54] Poets identify themselves as vagabonds, orphans and wanderers. One notes poignantly 'a man's home is everywhere. Take up your stick and ramble.'[55] A *sefela* rarely unfolds in linear time. Rather, it consists of a collage of names, peoples, places, events and conversational fragments. Poetic symbols of travel, movement and mobility link together the disparate images. Travel sustains the narrative, but the narrative is itself a means of social commentary on the character and consequences of travel. As Coplan points out, 'the poet transcends boundaries in space, time, and narrative structure'.[56]

Within Lesotho, most migrants travel from home to the mine recruiting office by bus, taxi or plane. From there, miners traditionally travelled onwards to the mines by train (though in the 1980s buses replaced trains as the dominant mode). Within most *lifela*, there is a curious silence about the means of travel inside Lesotho. Many give the distinct impression that migrants journey alone and on foot: in Coplan's vivid phrase 'the homeboy with the homeless mind'.[57] The solitary journey allows the poet to centre himself in the narrative. By travelling on foot he can demonstrate his intimate knowledge of the physical and human landscape as it is journeyed across. Colourful landscape images are interspersed with depictions of the social topography of each village, its inhabitants and memorable events associated with each place.

Across the boundary, within 'the wilderness' of South Africa, travel is again more functional and purposeful, a means to an end. Yet, the imagery used to describe the miners' train is of extraordinary range and complex allusion:

> The train is a mythical, devouring watersnake called Khanyapa, tutelary diet of diviners and spirit mediums that travels in a cyclone . . .; it shake-dances like a Xhosa initiate or an entranced diviner; it's an adventurer, a warrior, a madman, a raging prairie fire, a whole menagerie of wild, swift animals, a centipede, a millipede, a swallower and a disgorger.[58]

In another *sefela*, the train is a favourite cow, a hyena, the horse of the ancestors, a performer of miracles, a wanderer, a madman with iron legs. The train, according to Coplan, is a simulacrum for the migrant himself; migrants become the train they ride – fierce, ruthless, intimidating, frightening, all-knowing. The taming of the train – 'It's whitemen, you will see they put these iron blinkers on its cheeks; it's so that it gets used to looking down the road' – represents the domestication of the migrant, his transition from wanderer to worker.

Many *lifela* construct a particularly stark set of binary oppositions between the landscape of town and countryside. Perhaps the best-known *sefela* – 'Another Blanket' – is a particularly vivid evocation of the transition between two discrete worlds.[59] The dangerous and harsh conditions of the minescape are set against an idyllic, and highly romanticised, rural existence. The dominant motif is that of power – the loss of power and control of the body when the river is crossed; its retention and celebration when at home. When migrants cross over the river of baptism that separates Lesotho and South Africa, there is a knowing adoption of new modes of consciousness and forms of behaviour better suited to the world of work. The rhythms and relationships of underground work are sharply visualised in many *lifela* with constant biblical allusion. The horrors of death and dismemberment are often unspoken, even unspeakable.[60]

The dominant spatial subject of the *lifela* is not, however, the mine environment but Lesotho. To a degree, this reflects the fact that poets tend to compose and extemporise about the places they are absent from. For most mine migrants (particularly since the 1970s when migrants were forced to work continuously or forfeit their jobs) the mine is home for the greater part of the time.[61] Nevertheless, as Colin Murray points out, the paradigm of a successful migrant career is to build up sufficient rural resources to be able to opt out of the system at as early an age as possible.[62] The rural-centrism of many migrants ensures that images of home remain central to the poetic and prosaic discourses of the mine. The images of home are fluid and contradictory. Romantic longing for the freedom of home contrasts with fear of the disintegration of family and the unfaithfulness of spouses. Soaring images of the stark beauty of the physical landscape contrast with harrowing depictions of rural poverty and privation.

Within the intermediate spaces of the migrant-labour system other performance styles and languages are vented. The *lifela*, performed without musical accompaniment before rapt audiences, are rarely voiced in the raucous bustle of the wayside *shebeen* (bar). *Lifela* poets are generally, though not exclusively, male.[63] Women are also migrants, though not necessarily to the mines and certainly not to work in the mines. Are they forever resigned to being the subjects of the male poet, deprived of a voice of their own? The film *Songs of the Adventurers* provides a stunning visual evocation of dance and song in a *shebeen* in northern Lesotho. The men sit rather sullenly around the perimeter of the performance. Women patrons take centre stage reciting, in song, a powerful set of travelling images of their own. Male Basotho stoutly resisted Coplan's attempts to equate these extemporaneous bar-songs with the *lifela*. Yet, as he points out, there are marked textual similarities between the two forms. The *shebeen* is a space invented and controlled by women, a 'rare public forum for women's discourse on problems and disabilities that are unacknowledged in other contexts'.[64] It is also a place for women to present their own interpretation of the migrant experience.

The history and contemporary forms of Basotho women's migration, particularly its interpenetration with male mine migration, are well documented.[65] The experience of female migration is refracted through *shebeen* oral culture in ways which are quite distinctive from cultural expressions of male migration. The heroic and self-centred cast of the *sefela* is absent, though the songs are no less self-reflexive or individualistic. Displacement is the basic spatial trope in women's oral culture. Women travel not to seek knowledge or to demonstrate their prowess, but as a form of personal escape – from poverty, abusive relationships or demanding relatives. Their migration is purposeful and directed – 'They call me a vagabond, but I am not a vagabond; I am taking care of business.'[66] In so doing, they are forced to abandon what they value most, including their children. The rural home is a space of longing and regret – 'I have left my poor child Thabang, yes I have left my sweetheart behind crying.' The places that women escape to are, in theory, sites of freedom. South Africa is not, as in the male *sefela*, perceived as a devilish place but often one of autonomy and independence. In practice, it often becomes a place of suffering and sorrow – 'Women on the Rand are vagrants, they wear shoes without stockings.'[67]

In the songs, women are forced into survival roles that bring further hardships not freedoms – 'For indeed I'm living in hardships . . . I'm so poor that you could give me your last garment.' They demean themselves to survive and depend on men whom they despise – 'They slander me as a prostitute, I am not a prostitute; I tell the causes: at my home, girls, at my home eating is difficult.'[68] Their songs articulate a fierce critique of male attitudes and actions. Male neglect and violence is a constant refrain. The gratuitous violence of Basotho urban youth gangs such as the Ma-Rashea is castigated.[69] There is little appropriation of or empathy with the male migrant's experience on the mines. Rather, the mines and the migrant labour system are blamed for the profound disruption they bring to rural life – 'Men died; with whom do we remain? Men departed, only the worthless remain. Real men departed, with whom do we remain? I am a wailing fool who remains among ruins.'[70] *Shebeen* culture may open up a space of defiance for the articulation of other voices but there is little celebration of that freedom. Rather, women's songs speak loudly and eloquently of what Coplan calls 'the catharsis of affliction' and displacement.[71]

CONCLUSION

A chapter such as this must, by definition, decontextualise and displace the migrant voice from the original sites of its production. Yet these voices, in the very act of displacement, can speak to other audiences than the immediate circle of *lifela* devotees or *shebeen* patrons. The poetic imaginary of migrancy immediately decentres the dominant narratives of colonial discourse associated with writers such as Wilbur Smith; a discourse that acts with such devastating effect on the bodies and lives of the migrants. Central to these narratives is a geography of word and practice that marginalises, silences and excludes. This chapter has attempted to show that it may be possible to create oppositional literatures that more closely mirror or capture the migrant experience. But there is another alternative – engagement with the arresting power of oral performance. For here there are already alternative visions in existence, imaginary geographies lodged within the everyday experience and subterranean spaces of the migrant world. The poverty of dominant geographies is exposed at the very moment these voices are vented.

NOTES

* I am particularly grateful to David Coplan for access to a pre-publication copy of his manuscript *In the Time of the Cannibals* and for his comments on this chapter. My thanks also to Rosemary Jolly for discussions on the themes of this chapter and to the Social Sciences and Humanities Research Council of Canada for its support.

1 F. Johnstone, 'Mines of gold, moons of Jupiter', *Queens Quarterly* 1993, vol. 100, p. 591.

2 For example, F. Wilson, *Labour in the South African Gold Mines, 1911–1969*, Cambridge, Cambridge University Press, 1972; F. Johnstone, *Class, Race and Gold: A Study of Class Relations and Racial Discrimination in South Africa*, London, Routledge and Kegan Paul, 1976; N. Levy, *The Foundations of the South African Cheap Labour System*, London, Routledge and Kegan Paul, 1982; A. Jeeves, *Migrant Labour in South Africa's Mining Economy*, Montreal and Kingston, McGill-Queen's Press, 1985; J. Crush, A. Jeeves and D. Yudelman, *South Africa's Labor Empire: A History of Black Migrancy to the Gold Mines*, Boulder, Colo., and Cape Town, Westview and David Philip, 1991; W. James, *Our Precious Metal: African Labour in South Africa's Gold Industry*, London, James Currey, 1992; J. Crush, W. James and A. Jeeves (eds), *Transformation on the South African Gold Mines*, special issue of *Labour, Capital and Society*, 1992, vol. 25; J. Crush and W. James (eds), *Crossing Boundaries: Mine Migrancy in a Democratic South Africa*, Cape Town, Institute for Democracy in South Africa, 1995.

3 P. Abrahams, *Return to Goli*, London, Faber and Faber, 1953; *Mine Boy*, New York, Collier, 1970. See also M. Manaka, *Egoli: City of Gold*, Johannesburg, Ravan Press, 1981; I. Steadman, 'Alternative theatre: fifty years of performance in Johannesburg', in L. White and T. Cozens (eds), *Literature and Society in South Africa*, Cape Town, Maskew Miller Longman, 1984, pp. 138–46; D. Ricci, *Reef of Time: Johannesburg in Writing*, Johannesburg, Ad Donker, 1986.

4 J. Crush, 'Gazing on apartheid: postcolonial travel narratives of the golden city', in P. Preston and P. Simpson-Housley (eds), *Writing the City*, London, Routledge, 1994, pp. 257–84.

5 See, for example, H. Alverson, *Mind in the Heart of Darkness: Value and Self-Identity among the Tswana of Southern Africa*, New Haven, Yale University Press, 1978; D. Moodie, *Going for Gold: Miners' Lives on the South African Mines*, Berkeley and Los Angeles, University of California Press, 1994; P. Harries, *Work, Culture and Identity: Migrant Workers in Mozambique and South Africa*, Portsmouth, NH, Heinemann, 1994.

6 See T. Keegan, *Facing the Storm: Portraits of Black Lives in Rural South Africa*, Cape Town, David Philip, 1988; B. Bozzoli, *Women of Phokeng: Consciousness, Life Strategy and Migrancy in South Africa*, Portsmouth, NH, Heinemann, 1991; J. Guy and M.Thabane, 'Basotho miners, oral history and workers' strategies', in P. Kaarsholm (ed.), *Cultural Struggles and Development in Southern Africa*, Portsmouth, NH, Heinemann, 1992, pp. 239–58.

7 B. Bozzoli, 'Migrant women and South African social change: biographical approaches to social analysis', *African Studies*, 1985, vol. 44, pp. 87–96; P. la Hausse, 'Oral history and South African historians', *Radical History Review*, 1990, no. 46–7, pp. 346–56; M. Miles and J. Crush, 'Personal narratives as interactive texts: collecting and interpreting life-histories', *Professional Geographer*, 1993, vol. 45, pp. 84–94.

8 On colonial discourse see E. Said, *Orientalism*, Harmondsworth, Penguin, 1978; P. Brantlinger, *Rule of Darkness: British Literature and Imperialism, 1830–1914*, Ithaca, NY, Cornell University Press, 1988; L. Lowe, *Critical Terrains: French and British Orientalisms*, Ithaca, NY, Cornell University Press, 1991; M. Pratt, *Imperial Eyes: Travel Writing and Transculturation*, London, Routledge, 1992; D. Spurr, *The Rhetoric of Empire: Colonial Discourse in Journalism, Travel Writing, and Imperial Administration*, Durham, Duke University Press, 1993.

9 D. Maughan-Brown, 'Raising goose-pimples: Wilbur Smith and the politics of Rage', in M. Trump (ed.), *Rendering Things Visible*, Johannesburg, Ravan Press, 1990, pp. 134–60; see also A. Chennells, 'Just a story: Wilbur Smith's Ballantyne trilogy and the problem of Rhodesian historical romance', *Social Dynamics*, 1984, vol. 10, pp. 38–45.

10 W. Smith, *Gold Mine*, London, Pan Books, 1970.

11 P. Carter, *Living in a New Country: History, Travelling and Language*, London, Faber and Faber, 1992, p. 101. See also E. Said, 'Identity, authority and freedom: the potentate and the traveler', *Transition*, 1991, vol. 54, pp. 4–19.

12 A. Horn (ed.), *The Plays of Zakes Mda*, Johannesburg, Ravan Press, 1990.

13 L. Gunner, 'A dying tradition? African oral literature in a contemporary context', *Social Dynamics*, 1986, vol. 12, pp. 31–8; P. McAllister, 'Beer drinking and labor migration in the Transkei: the invention of a ritual tradition', in J. Crush and C. Ambler (eds), *Liquor and Labor in Southern Africa*, Athens, Ohio University Press, 1992, pp. 252–68; P. Berliner, *The Soul of Mbirwa: Music and Traditions of the Shona People of Zimbabwe*, Berkeley and Los Angeles, University of California Press, 1988; M. Drewal, 'The state of research on performance in Africa', *African Studies Review*, 1991, vol. 34, pp. 1–64; L. Vail and L. White, *Power and the Praise Poem: Southern African Voices in History*, London, James Currey, 1992; I. Hofmeyr, *'We Spend our Years as a Tale that is Told': Oral Storytelling, Literacy and Historical Narratives in the Changing Context of a Transvaal Kingdom*, London, James Currey, 1994.

14 Crush, Jeeves and Yedelman, op. cit., pp. 192–4.

15 R. Packard, *White Plague, Black Labor: Tuberculosis and the Political Economy of Health and Disease in South Africa*, Berkeley and Los Angeles, University of California Press, 1990.

16 J. Lang, *Bullion Johannesburg: Men, Mines and the Challenge of Conflict*, Johannesburg, Jonathan Ball, 1986.

17 J. Leger, 'Safety and the organisation of work in South African gold mines: a crisis of control', *International Labour Review*, 1986, vol. 125, pp. 591–603; J. Leger, 'From fatalism to mass action: the South African National Union of Mineworkers' struggle for safety and health', *Labour, Capital and Society*, 1988, vol. 21, pp. 270–93.

18 J. Leger and M. Mothibeli, 'Talking rocks: pit sense amongst South African miners', *Labour, Capital and Society*, 1988, vol. 21, pp. 222–37; J. Leger, 'Talking

rocks: an investigation of the pit sense of rockfall accidents amongst underground gold miners', PhD thesis, University of Witwatersrand, 1992 (this important thesis is currently being prepared for publication).

19 Smith, op. cit., pp. 17–19.
20 Unlike, for example, Emile Zola's classic mine novel *Germinal*, Harmondsworth, Penguin, 1976 edn.
21 Smith, op. cit., p. 14.
22 ibid., p. 158.
23 ibid., p. 83.
24 R. Palmer and N. Parsons (eds), *The Roots of Rural Poverty in Central and Southern Africa*, London, Heinemann, 1977; C. Bundy, *The Rise and Fall of the South African Peasantry*, London, Heinemann, 1979; W. Beinart and C. Bundy, *Hidden Struggles in Rural South Africa*, London, Heinemann, 1987; J. Crush, *The Struggle for Swazi Labour, 1890–1920*, Montreal and Kingston, McGill-Queen's Press, 1987.
25 Smith, op. cit., pp. 104–5.
26 D. Moodie, 'Ethnic violence on the South African gold mines', *Journal of Southern African Studies*, 1992, vol. 18, pp. 548–613.
27 Crush, Jeeves and Yudelman, op. cit., pp. 182–3.
28 Maughan-Brown, op. cit., p. 151.
29 For a selection of Mda's plays see Horn, op. cit.; and Z. Mda, *And the Girls in their Sunday Dresses*, Johannesburg, Witwatersrand University Press, 1993. Andrew Horn calls Mda 'the most imaginative and articulate voice' to have emerged in black South African theatre: see A. Horn, 'South African theater: ideology and rebellion', *Research in African Literatures*, 1986, vol. 17, p. 222. Mda's own reflections on political theatre appear in Z. Mda, 'Marotholi travelling theatre: towards an alternative perspective of development', *Journal of Southern African Studies*, 1990, vol. 16, pp. 352–8; and *When People Play: Development Communication through Theatre*, London, Zed Books, 1993; see also R. Kavanagh, *Theatre and Cultural Struggle in South Africa*, London, Zed Books, 1985.
30 Z. Mda, *The Hill*, in Horn, *Plays of Zakes Mda*, op. cit., pp. 67–116.
31 J. Crush and P. Wellings, 'The Southern African pleasure periphery, 1966–83', *Journal of Modern African Studies*, 1983, vol. 21, pp. 673–98.
32 Carter, op. cit., p. 101.
33 J. Crush, 'Power and surveillance on the South African gold mines', *Journal of Southern African Studies*, 1992, vol. 18, pp. 825–44; J. Crush 'Scripting the compound: power and space in the South African mining industry', *Society and Space*, 1994, vol. 12, pp. 301–24.
34 Crush, Jeeves and Yudelman, op. cit., pp. 162–6.
35 J. Crush, 'Inflexible migrancy: new forms of migrant labour on the South African gold mines', *Labour, Capital and Society*, 1992, vol. 25, pp. 46–71.
36 Mda, *The Hill*, op. cit., pp. 71–8.
37 ibid., p. 74.
38 ibid., p. 87.
39 ibid., p. 89.
40 ibid., p. 77.
41 ibid., pp. 95–6.
42 Horn, *Plays of Zakes Mda*, op. cit., p. xxxi.
43 Mda's point is echoed by David Coplan who argues that 'the self-assertive, independent attitude ever more prevalent among Basotho women is rooted in their growing reliance on networks composed of other women rather than . . . of men': D. Coplan, *In the Time of the Cannibals: Word Music of South Africa's Basotho Migrants*, Chicago, University of Chicago Press, 1995, ch. 6.

44 Mda's view of women's roles within the migrant labour system is simultaneously very constrained. Women are typecast, in this male discourse, either as virginal (the nun) or predatory and immoral (the prostitutes). There appears to be no room for the more complex roles and identities actually taken by women in the countryside, in Maseru and in South Africa itself. I am grateful to Rosemary Jolly for discussion on this point.

45 Mda, *The Hill*, op. cit., p. 99.

46 See, for example, Crush and Ambler, op. cit.

47 L. Vail and L. White, 'Forms of resistance: songs and perceptions of power in colonial Mozambique', *American Historical Review*, 1983, vol. 88, pp. 883–919; K. Sole, 'Oral performance and social struggle in contemporary Black South African literature', *TriQuarterly*, 1987, vol. 69, pp. 254–71; A. Sitas, 'Traditions of poetry in Natal', *Journal of Southern African Studies*, 1990, vol. 16, pp. 307–28; V. Erlmann, '"The past is far and the future is far": power and performance among Zulu migrant workers', *American Ethnologist*, 1992, vol. 19, pp. 688–709.

48 D. Coplan, 'Performance, self-definition, and social experience in the oral poetry of Sotho migrant mineworkers', *African Studies Review*, 1986, vol. 29, pp. 29–40; 'The power of oral poetry: narrative songs of the Basotho migrants', *Research in African Literatures*, 1987, vol. 18, pp. 1–35; 'Eloquent knowledge: Lesotho migrants' songs and the anthropology of experience', *American Ethnologist*, 1987, vol. 14, pp. 413–33; 'Fictions that save: migrants' performance and Basotho national culture', in G. Marcus (ed.), *Rereading Cultural Anthropology*, Durham, NC, Duke University Press, 1992, pp. 267–95; Coplan, *In the Time of Cannibals*, op. ·cit. The making of the film *Songs of the Adventurers* is described in *In the Time of Cannibals*, ch. 6.

49 Coplan, 'The power of oral poetry', op. cit., p. 14.

50 J. Kimble, 'Labour migration in Basutoland, c. 1870–1885', in S. Marks and R. Rathbone (eds), *Industrialisation and Social Change in South Africa*, London, Longman, 1982, pp. 119–41; W. Worger, *South Africa's City of Diamonds: Mine Workers and Monopoly Capitalism in Kimberley, 1867–1895*, New Haven, Yale University Press, 1987.

51 Coplan, 'The power of oral poetry', op. cit., p. 5.

52 ibid., p. 13.

53 ibid., p. 12.

54 ibid., p. 21.

55 ibid., p. 22.

56 ibid., p. 21.

57 Coplan, *In the Time of Cannibals*, op. cit., ch. 7.

58 ibid.

59 Agency for Industrial Mission, *Another Blanket: Report on an Investigation into the Migrant Situation*, Horizon, A.I.M., 1976.

60 Coplan, *In the Time of Cannibals*, op. cit.

61 Crush, 'Inflexible migrancy', op. cit.

62 C. Murray, 'Migrant labour and changing family structure in the rural periphery of Southern Africa', *Journal of Southern African Studies*, 1980, vol. 6, pp. 141–56; *Families Divided: The Impact of Migrant Labour in Lesotho*, Cambridge, Cambridge University Press, 1981.

63 Coplan, *In the Time of Cannibals*, op. cit., ch. 6.

64 ibid.

65 P. Bonner, 'Desirable or undesirable Basotho women: liquor, prostitution and the migration of Basotho women to the Rand, 1920–1945', in C. Walker (ed.), *Women and Gender in Southern Africa to 1945*, Cape Town, David Philip, 1990, pp. 75–113.

66 Coplan, *In the Time of Cannibals*, op. cit., ch. 6.
67 ibid.
68 ibid., ch. 8.
69 J. Guy and M. Thabane, 'The Ma-Rashea: a participant's perspective', in B. Bozzoli (ed.), *Class, Community and Conflict: South African Perspectives*, Johannesburg, Ravan Press, 1987, pp. 436–56.
70 Coplan, *In the Time of Cannibals*, op. cit., ch. 6.
71 ibid.

16

FAR CITIES AND SILVER COUNTRIES

Migration to Australia in fiction and film

Roy Jones

INTRODUCTION

The classic studies of literary geography were concerned with the equally classic rural and, often, spatially circumscribed novels of region and of landscape, such as Thomas Hardy's Wessex or Mary Webb's Shropshire.[1] More recently, a number of geographers, notably those who have themselves moved from one location to a distant and more or less alien 'other', have sought to widen the engagement between literature and geography to include a range of human experience which extends beyond attachment to and description of a single and/or singular place. In this connection Porteous[2] has argued that 'Human experience of place is one major dimension, involving the funda-mental distinction of existential insider:existential outsider ... Location of experience is the second major dimension, the significant antinomy being home:away.' The matrix (Figure 2) put forward by Porteous encapsulates comparable distinctions, such as native:non-native[3] and roots:rootlessness,[4] which are particularly appropriate to the study of mobile, rather than static, individuals, groups and societies.

	INSIDER	OUTSIDER
HOME	'sense of place' (often rural)	entrapment
AWAY	the traveller	journey, exile, yearning (often urban)

Figure 2 A conceptual framework for humanist literary geography (After J.D. Porteous, 'Literature and humanist geography', *Area*, 1985, vol. 17, pp. 117–22)

The presence of 'uprooted away-outsiders' in societies of 'rooted home-insiders' is frequently acknowledged in demographic and political terms within social science. But an appreciation of the lives and experiences of such individuals and groups is, very often, only attainable through fiction.[5] Teather's personal account of 'the expatriate experience' is very much the exception within the professional literature.[6]

If geographers have been traditionally concerned with the literature of place, rather than of movement, they have also concentrated their energies on the study of the production of these fictions, rather than on their reception. Osborne has warned us that 'we must not lose sight of the role of literature as part of a mass communication system concerned with the creation and maintenance of shared consciousness.' He also reminds us that 'the effect of literature on the receiving society is much influenced by such prosaic dimensions as volume of output, distribution, readership and reception.' It is the 'mass-produced historical fabrications of . . . popular writers' which have the most important cultural (if not high cultural) impact.[7]

The phenomenon of early postwar European migration to Australia provides an ideal opportunity for the consideration of several of these issues. The influx of large numbers of more or less willing 'New Australians', from a variety of political, cultural and ethnic backgrounds, provided a significant pool of foreign 'outsiders', leaving one home and aspiring, in varying degrees, to create a new one and to become Australian 'insiders'. The sheer size of this population movement also obliged both the Australian-born and the immigrants to confront the expatriate experience, both as a personal or familial rite of passage for a large proportion of the society and as a national question. What did it mean to be Australian in the mid-twentieth century and what did, and do, Australians want it to mean in the years to come?

The literary starting-point for this chapter is Nevil Shute's novel *The Far Country*, first published in 1952.[8] According to his obituary in *Time Magazine* (25 January 1960), Shute was 'the top bestseller of all contemporary British authors'. While he did not write 'classic' novels, he was tremendously successful as a commercial author, so much so that he was able to put forward a set of proposals for the support of the Arts in Australia, these proposals being comfortably financed from his personal income tax contributions.

Shute left Britain for Australia in 1950, at a time when he was middle-aged, famous, financially successful and when he and his family were well established in their current home. The translocation of this rather unusual migrant can be appreciated in terms of conventional push and pull factors. Shute was leaving a country where he chafed under both financial and physical government restrictions, notably the high rates of income tax (one of his enduring preoccupations) and the limitations on his ability to carry out fieldwork for his novels as a result of petrol rationing. He was attracted to a country where, prosaically, both these restrictions were considerably less onerous, and where, less prosaically, 'there was all that white space on the maps', as Shute put it in an interview in 1959.[9] Perhaps, between these two extremes, 'Australia also recommended itself as a country that the English-speaking world would like to read about.'[10]

The Far Country was the first novel that Shute wrote after he settled in Australia. In it he addressed the issues of not only British, but also continental European, immigration. Although it was not one of Shute's most famous works, the novel has enjoyed good sales for much of the last forty years. No doubt that has been aided, in the recent past, by the book's adaptation as a television mini-series in 1986. This adaptation, by Peter Yeldham, an Australian playwright and screenwriter who has also worked in Britain, has been shown on Australian television on several occasions and will be discussed in its own right later in this chapter.

Given both its subject-matter and its subsequent televising, *The Far Country* inevitably invites comparison with *Silver City*, another study of the Australian migrant experience in the immediate postwar years. *Silver City* has appeared as both a book and a film, though in this case the film was produced first. The screenplay of *Silver City* was commenced in 1978 by the film's director, Sophia Turkiewicz, an Australian of Polish ancestry, who had already made two short films on postwar immigration. She started the *Silver City* project during a six-month stay in Poland which was funded by a Polish government travel grant. The final version of the screenplay was produced in 1983 as a co-operative venture between Turkiewicz and the Australian novelist Thomas Keneally, who had also spent some time in Poland. The film was released in 1984. In the same year *Silver City* appeared in the form of a novel, having been adapted from the screenplay by Sara Dowse, a writer and, formerly, a civil servant who had arrived in Australia as an 18-year-old American migrant in 1956.[11]

Although the film was only a moderate commercial success, it has been shown repeatedly on Australian television in recent years, both on the commercial networks and on the so-called 'ethnic channel', SBS.

While both novels and both films explore the migrant experience in terms of the home:away and insider:outsider dichotomies, an important distinction must be made between Shute's novel and the other three works, namely the gap of more than thirty years that exists between the dates of their creation. Shute's novel was designed to appeal to the popular market of the 1950s, the other works to that of the 1980s. The differences between Shute's novel and the other works provide us with a valuable index of what these novelists and screenwriters saw as credible, saleable or, even, worth using as propaganda at these two different periods. It should also be noted that Shute's pro-monarchist and anti-socialist propaganda was very different from the views of some of the other authors. Keneally became the Founding Chair of the Australian Republican Movement in 1991. Dowse had been a senior public servant during the reformist Whitlam Labor government but left her post shortly after the election of a Liberal-National government in 1975.

THE FAR COUNTRY

It is clear from virtually all of Shute's writing on Australia that he saw himself, and indeed all Britons, as Australian insiders. At the time there was a degree of mutuality in this view. A review of *The Far Country* in the *Adelaide Advertiser* (23 August 1952) referred to 'English' and 'foreign' migrants as two separate categories. Shute may well have both made and seen himself as even more of an insider: one of his Australian neighbours, on the Mornington Peninsula near Melbourne, was later quoted in the Canadian press as saying that there was 'nothing uppish about him like other pommies around here, [he] fights the bushfires and comes to the agricultural shows just like the other fellows'.[12]

The fact that Shute regarded himself as an insider because he was English, rather than in spite of it, is reflected in the encompassing imperial scope of the family whose characters and actions are central to the novel. The heroine, Jennifer Morton, is a young English woman visiting Australia. Her hosts are her English aunt, Jane Dorman, and Jane's Australian husband, Jack. Jennifer had been encouraged to travel to Australia by her grandmother, the impoverished widow of an Indian Civil Servant, who deplored (as did Shute himself) the condition of postwar Britain but who, as a distressed gentlewoman, believed that Australia would offer her granddaughter 'everything I had at your age'.[13] Indeed, one of the central messages of *The Far Country* is that Australia was 'in many ways like England must have been a hundred years ago'.[14] This state of affairs would certainly encompass material wealth (including low taxes and limited government restrictions) but, perhaps more importantly, a genteel, rural lifestyle: in short 'the England [of] seventy years ago when everything was prosperous and secure'.[15]

The notion that Australia represents an England from a golden past has a long pedigree in English literature.[16] Dickens, the most popular English novelist of the mid-nineteenth century, portrayed Australia as a Pickwickian Arcady untouched by the evils of the Industrial Revolution. Shute, as the most popular English novelist of the mid-twentieth century, did likewise. The Australian settings of *The Far Country* are the forests and farmlands of Northern Victoria, 'a gracious, pleasant country, well-watered and friendly'. The trees in the Dandenong Ranges were 'finer and taller than any Jennifer had seen in England'. Although 'England might have been like that once', that was in the good old days before the First World War.[17] It is more than a coincidence that both the novel's title and its epigraph are by A.E. Housman, one of the most famous celebrants of English rural nostalgia.

Shute tempered his own nostalgia with some detailed accounting. He calculated the likely labour and capital inputs and profits from a moderately sized sheep property in the district in which the story is set.[18] But, by doing so at the time of the Korean War wool boom, he hypothesised some atypically high financial returns, so high that Jennifer's aunt could expect the wool cheques to provide a 'gracious' English country house, with English plants and an English lawn. 'After all, the English country houses came when agriculture was doing well, and agriculture's doing well here now. We all came out from England and we've got the English way of doing things.'[19] Even those who did not come out from England were assumed to share the same ideology. Carl Zlinter, a Czech refugee, could feel 'What a good country this was! It had all the charm of the Bohemian forests he had loved as a young man, plus the advantage of being English!'[20] However, whatever Shute may have assumed about their aspirations, the continental Europeans in the book are portrayed as outsiders, not to be trusted in local dance halls, unwilling to 'talk English, like a Christian'[21] and sometimes guilty of rowdy behaviour in the pubs. Shute was aware of the existence of anti-migrant prejudice, but considered this to be a minor problem. When Zlinter encountered a waitress who was rude and unco-operative to him, she could be dismissed as someone who

> came from a family of Australians who had been casual labourers for generation after generation, bad stock and mentally subnormal. She and her family were bitterly hostile to all immigrants, especially the European ones who worked too hard and were guilty of the social crime of saving, thereby threatening the Australian way of life.[22]

It is implied that people holding such views are as rare as genetic freaks and, in the novel, Zlinter is portrayed as being able to solve this problem, even in a small country town, by going to another cafe.

The practical and social struggles of continental European migrants are acknowledged in the novel to a certain extent. Zlinter had to work as a labourer and undertake a protracted and expensive (and, it is implied, unnecessary) retraining programme before he could resume his medical career. For two years he lived in camps and had not set foot in an Australian home. But this period as an outsider was definitely viewed by the author as temporary. It is clear at the book's end that Zlinter's two years of labouring and a similar period of retraining will be followed by requalification as a doctor and that this, together with marriage to Jennifer, which will render him effectively British, will make him, to all intents and purposes, an Australian insider. Even if the European migrants do not achieve material or marital success, they are portrayed as happy to be in Australia. The former 'Professor of Artistic Studies at the University of Kaunas' views his life as a platelayer on the Victorian Railways as far superior to any alternative in his home country.

Overall, Australia is portrayed as a land of plenty where success, par-ticularly material success, is possible for those who work for it. Shute is clearly aware of the polemical nature of his book, a point illustrated by a scene in which Jennifer studies immigration brochures with a friend: 'It looks all right in these things, doesn't it? But then, they wouldn't tell you the bad bits, like half the houses in Brisbane having no sewerage system . . . it's probably all desert and black people behind the camera.'[23] But this is a somewhat tokenistic piece of balance. One source of information is Jennifer's cousin, a final-year Social Studies student at Melbourne University who rationalises her desire to go to London to work as a social worker by claiming that 'there's not the scope' for such employment in Australia because 'there aren't any poor people here'.[24] Most of the characters in the book, when given the choice between 'the hard, bleak winter of England' (or continental Europe) and the 'heat, ease and beauty of the Australian summer',[25] would wish to exchange their European outsider identity for that of an Australian insider and make their home in that far country. Shute does not merely imply that they all will. His view is that they will do so relatively quickly and relatively painlessly. In the process of becoming insiders, they will also become virtually indistinguishable from the monarchist English Australians who are already at home there.[26]

SILVER CITY

Judith Wright discerns a 'double aspect' in the (white) Australian vision of nature – its dualistic ability to represent, simultaneously, 'the reality of newness and freedom' and 'the reality of exile'.[27] If *The Far Country* depicts Australia, both natural and unnatural, in the former manner – albeit with a nod towards the good old days – it is the latter aspect that *Silver City* throws into sharp relief.

None of the migrants portrayed in *Silver City* is English, or even British. Australia is far more alien to them and they to it:

> They agreed . . . that, despite what they left behind, this Australia would take some getting used to. A lonely place with its weird vegetation, low wide-apart houses. The broad open sky. . . . Even the people, what they had seen of them, seemed raw . . . not started, not finished.[28]

For Nevil Shute's characters, Australia was a land flowing with steak and wool cheques, but for Sara Dowse's heroine, Nina Majowska from Poland, and her fellow refugees, 'the country was short of everything as well as people – houses, schools, goods.'[29] The housing shortage was so acute that characters in the novel are shown as living in garages and even packing cases in order to gain a roof over their heads in Sydney.[30] The relative abundance of meat, fresh fruit and vegetables in Australia was acknowledged in both novels, but the first reaction of the Polish migrants to food in the film version of *Silver City* was 'they call this bread?' When the characters were transferred from Silver City – a rural reception camp for 'displaced persons' – to a hostel in Sydney

253

they considered that it was 'worth it' to travel on two buses and a train in order to buy Polish sausage.[31] Even when the migrant hostel was left behind and Australia's bounty was more readily available, cooking was a question of 'what did one *do* with sheep meat, how could it be disguised?'[32]

The pleasant picture of well-paid work in the scenic outdoors as painted by Carl Zlinter contrasts with Nina's workplace in a hospital laundry ('laundries must have been a prototype of hell') and with that of her lover, Julian Marczewski, on a production line in a car factory where safety standards are inadequate and migrants are allocated the worst jobs.[33] Admittedly, neither the safety standards nor the quality of the work at Zlinter's timber camp were of a high order, but the ready and, by and large, grateful acceptance of these conditions by the migrants in *The Far Country* is certainly not shared by their counterparts in *Silver City*.

This would seem to imply that the characters in the two novels possess rather different views of both 'home' and Australia. The view of Poland in the immediate pre-war period – 'A Sunday in June at the Lazienski, the water a mellow blue flecked with gold, with deep reflections of the place and the overhanging trees . . . the outdoor concert. Chopin and precelki by the lake'[34] is very different from the bleak picture of immediate postwar Europe (continental and insular) painted by Shute. As a vision of 'home' it is more akin to his evocation of the Victorian-Edwardian era in Britain.

Even with regard to the landscape, *Silver City* presents a view of anti-Arcadia.

> 'These trees, they are not healthy.'
> 'Wait till you see the rest of it. The whole country looks like that.'
> 'What could you grow here?'[35]

Though parts of the countryside and the country towns reminded Nina of the Polish plain, both the natural and the cultural landscape were alienating. 'She felt the absence of the Romanesque churches, the Baroque halls as one misses the imprint of character on a self-centred face.'[36] In direct contrast to the message of *The Far Country*, the migrants in *Silver City* associated (admittedly pre-war) Europe with ease and elegance, and Australia with harshness and deprivation. To imagine Poland while in Australia was to transform the 'ragged trees' to beeches and ducks to swans.[37]

A further contrast to *The Far Country* is offered in the differing treatments of the 'fact, theme and metaphor of imprisonment', a topic which Turner sees 'in widespread use' in Australian fiction.[38] It is also, of course, a situation in which it is the incarcerated who are the outsiders. Carl Zlinter finds his 'imprisonment' in the timber camp largely congenial. Although he is counting off the days until he is 'a free man at last', he is already 'a man on his way out from years of life in camps (mainly in Europe), a man beginning to enjoy life who was unused to joy'.[39] Nina and the Marczewski family, by contrast, see Silver City and the hostel in Sydney as far more of a continuation of their imprisonment in European refugee camps. Families are separated by gender into different huts. Petty discipline is enforced by 'the *gauleiter*' and other officials who use German as a common language and are, on occasion, suspected of being Nazis in hiding. Bribery is necessary in order to obtain minor favours.

Even formal release from the camps does not necessarily bring any improvement. Nina's placement as a 'domestic' in a country town hospital was 'worse than Silver City because there was no-one to talk to'. For Julian, the whole of Australia was 'a funny place, the entire country like one big camp. The men here, the women there'.[40] For those who live in a camp, rather than a house, it is hard to create a home.

For the migrants of *Silver City* the problems of being outsiders also loomed far larger they did for Carl Zlinter and his workmates. Prejudice against migrants was considered to be the rule, not the exception. Sympathetic Australians, such as the neighbours who helped Nina and Julian to build an extension on to their garage home, stand out.

Shute acknowledged that prejudice existed, but considered that severe examples were exceptional and unlikely to give rise to major problems. The Australian farmer, Jack Dorman, in *The Far Country* was indeed concerned that his Italian labourer – but not his Australian or British employees – would become too familiar with Australian girls at a local dance. He does not appear to acknowledge the possibility of a scene such as that in *Silver City* depicting Nina's near-rape by three Australian men – including her supervisor's son – in a country town.

Moreover, for Europeans with intellectual (or, indeed, any non-manual) aspirations, Australia is portrayed as an even more unwelcoming and exclusionary place. The new migrants in *Silver City* were warned on their first day: 'No tell them you are educated – Vill be vorse for you if you do . . . This is a country where peasants do vell. You have to learn to think like a Polish peasant . . . America vant brains. Australia vant muscle.'[41] In Sydney, 'the library had for them a sanctified atmosphere, a link with their European roots.'[42]

Nevertheless, the remote prospect of a move from outsider to insider, from away to home, remained. 'It may be that . . . the best I can hope for is that my children will feel that Australia is their real home. And maybe this is what millions of emigrant mothers have felt before me',[43] as one academic writer has put it. This personal view is echoed on several occasions in *Silver City*. Nina was advised by camp officials to abandon her aspiration to become a teacher, to 'wake up and find a husband. Dream for your children.'[44] For Julian, who failed in his attempt to requalify as a lawyer, the move to insider status was being made by his son, a medical student at the time that the book ends. But *Silver City* does offer some more personal hope.[45] One minor character, Viktor, is depicted as beginning a business with one utility (or pick-up truck) and building up a fleet of vehicles during the 1950s.

Predictably, 'brains' were seen as a disadvantage in the assimilation process.[46] The 'peasants' acclimatised first. Nina's friend Helena, a girl from a Polish farm, soon became a 'kangaroo', first working for and then marrying the Australian farmer who had been Nina's penfriend. But, over a period of years, Nina was gradually Australianised also.

> Strangely, my nostalgia was not for the Poland I had scarcely known, or the one I had known too well, so abruptly immersed in war, but the place of my maturing . . . the empty power of sea and land had immersed itself in me. I came to cry at the sight of an ugly red-green tree.[47]

For the more intellectual or, perhaps, more inflexible Julian, however,

> hardly anything had changed. Ten long years. Oh, they ate a bit better and perhaps there was hope for the children, but even that was no certainty. He cast his eyes over the factory canteen, the Australians with skilled jobs up at one end, the men like him with the lousy ones at the other. A few of them at night school, struggling to compete with children. He saw years of it ahead. Free from fears – maybe. No round-ups, no beatings, no starvation. No bombs. As for the rest they might have been under occupation.[48]

But, overall, the message of *Silver City* is also one of hope. It does not offer the promised land and the get-inside-quick-fix portrayed in *The Far Country*. The delays, displacements, difficulties and sacrifices experienced by Dowse, Kenneally and Turciewicz's fictional characters are significant and enduring. They strike a number of resonating chords from Teather's account of her own progress through the protracted and problematical process of 'the expatriate experience' and add to this the dimensions of greater material deprivation and cultural difference.[49]

SILVER SCREEN

The book and the film versions of *Silver City* appeared simultaneously and were produced at least to complement, if not to replicate, each other. Production of the televised version of *The Far Country* postdated that novel's appearance by more than three decades and occurred long after the novelist's death. Although in both cases the screenplay was 'based on' the novel (or vice versa), a sufficient number of discrepancies exist between the two versions of *The Far Country* for the television mini-series to be considered in its own right.

Peter Yeldham's screenplay, like both versions of *Silver City*, is an historical piece of writing, in contrast to Nevil Shute's contemporary novel. Inevitably, therefore, latter-day concerns and perceptions of the early postwar era colour the work. In many ways, the 1980s version of *The Far Country* presents views far more in tune with those found in *Silver City* than those in its own 1952 progenitor.

The negative picture of Britain and Europe presented by the born-again Australian, Nevil Shute, is projected far less stridently in the television version. The sun shines on scenes located in London and Leicester. Some Britons of the 1940s are depicted as cheerful, well dressed and actually enjoying themselves. One television scene, allegedly depicting a classic English country churchyard but, like many of the 'British' scenes, probably filmed in Victoria, goes some way towards acknowledging the nostalgia for a European 'home', which was frequently expressed in *Silver City*. In the novel such nostalgia was seen as futile. The literary Jennifer Morton's attitude, on viewing London's architectural treasures before departing for Australia was 'so what?' Her cousin, the aspiring social worker from Melbourne University, appeared repeatedly in the novel to praise everything British and to claim that everything Australian was second-rate, including the career opportunities for social workers. Her role in the novel was to assist in making Shute's polemical points by providing a ludicrous caricature of any of the counter-arguments. This character, for a variety of reasons, would not have been very believable for an Australian audience in the 1980s and was totally omitted from the television version.[50]

If Europe receives a more generous treatment thirty years on, Australia, by contrast, is not depicted in quite such glowing terms as those of the original novel. The natural landscape of the Victorian High Country is used to considerable visual effect. This was hardly surprising so soon after the national and international success of a previous High Country film, *The Man from Snowy River*. But the Australian characters are shown as no more than comfortably-off. The emphasis on material success, if not excess, and an account – in both senses of the word – of a spending spree in Melbourne after Jack Dorman receives his massive wool cheque are absent. The mini-series does not highlight the postwar shortages depicted in *Silver City* but, since it is set in a well-established farming area at the time of the Korean War wool boom, this merely indicates the differences, in time and space, between the books' settings.

With regard to the topic of anti-migrant prejudice, however, the television version takes a radically different line from the book. In both the novel and the mini-series Carl Zlinter is called upon to carry out unauthorised first-aid work and, eventually, illegal major surgery under emergency conditions. In the novel these activities are looked on positively by most members of the Australian community. In the television version powerful local individuals, such as the police sergeant and the doctor (neutral and supportive, respectively, in the novel) are shown to be suspicious, resentful and xenophobic. Shute's Australia was largely free of arrogant bureaucrats who labelled the migrants as 'whingers', 'stirrers', 'commos' and 'troublemakers' (all terms from the mini-series). For Yeldham, as for Dowse, such figures were commonplace. The Jennifer Morton character makes repeated reference to 'small town bigots'. Yeldham's screenplay criticises the Australian Medical

Association and the Australian medical and bureaucratic establishment in general for putting so many obstacles in the way of migrant doctors at a time of chronic shortage of qualified Australian personnel. Shute is relatively generous in his treatment of the Australian powers that be, reserving his spleen for the growing public bureaucracy and fledgling National Health Service of his former home – a home which was, for him, an extremely alien place.

Clearly there are dramatic and/or televisual reasons for some of the differences in the presentation of this story in two very different media. The anti-migrant prejudice of the police sergeant and the *dénouement* of Zlinter's army service (a small issue in the novel) both heightened the drama of a classic courtroom scene. The resentments of the local doctor could be overcome by Zlinter's heroism and medical usefulness during a visually spectacular bush-fire. (The mini-series was filmed shortly after the 'Ash Wednesday' bushfires had devastated large areas of Victoria.) The anger of the migrant workers, in the mini-series, when they discovered that Zlinter was a former *Wehrmacht* medical officer was reminiscent of some of the scenes in the camps of *Silver City*. Their subsequent treatment of this 'Nazi', again in a number of scenes totally absent from the novel, provides several minutes of dramatic and action-filled television.

But it is hard to escape the conclusion that Yeldham, more than thirty years on, was obliged to make several subtle and not-so-subtle changes to the text in order to make the story credible to his contemporary audience. Like Dowse, Keneally and Turkiewicz, he was writing for people whose views of Australia, of Europe and even of the 1940s and 1950s differed significantly from those of Shute's original readers.

CONCLUSION: REFLECTIONS, REPRESENTATIONS AND RETROSPECTIONS

In both *The Far Country* and *Silver City*, the major European characters are shown to be in the process of becoming – more or less enthusiastically – Australian. Certainly this would seem to be a reasonable goal for freedom- and security-starved Czechs and Poles and steak- and sunshine-starved Britons in the immediate postwar period. Given that these were reasonable hopes, the sales success of *The Far Country* at the time of its first release would undoubtedly bear out Cockburn's contention that 'of all indices to moods, attitudes and above all aspirations, the bestseller list is one of the most reliable. There is no way of fudging it.'[51]

These moods, attitudes and aspirations of a people, if they be widely shared, are seen by Walter as 'dominant myths'. He contends that, since people are active interlocutors of their past, such dominant myths are transformed through a discourse which can be followed, *inter alia*, in the newspapers, books, films, art and television of the time. It is then, through this discourse over what we are to take the past to mean and what our hopes for the future are, that a public culture and an idea of a nation are constructed.[52]

The Far Country contains one very obvious mythic figure. Jack Dorman was commissioned from the ranks of the Anzacs on the Western Front during the First World War. He charmed a high-born English lady, who defied her parents and accompanied him to Australia. After years of struggle on the land during a depression – which Shute apparently considered as atypical in the Australian context – he became successful and prosperous during the – according to Shute, more typical – wool boom. (The appreciably more recent sales success of Albert Facey's[53] autobiographical work *A Fortunate Life*, which embodies many similar elements in its story, emphasises the continuing power of such myths.) Although Shute's novel contained its mythic figure, *The Far Country* was, perhaps, defying a dominant 'Aussie battler' myth by downplaying the degree of struggle that many new Australians, especially migrants from the non-English speaking world, were likely to face.

More importantly, Shute did not see the non-British migrants as entering into Walter's discourse and changing both Australia and its myths in the process. By the 1980s Britain had joined the European Community and its links with Australia, in terms of both trade and the significance of a shared head of state, had been greatly reduced. While these changes effectively destroyed Shute's visions of what Australia was, as portrayed in *The Far Country*, and what Australia might become, as portrayed in *In the Wet*, they had brought about, by the mid-1980s, a new collection of 'dominant myths' for a multicultural Australia, several of which recur in *Silver City*. In their turn, the continental European migrants, who had struggled to become millionaires, or who had sent their children to law and medical schools, have themselves become mythic figures. This mythologising has been achieved, in part, through a discourse which includes *The Far Country* and *Silver City*, in both their novel and film/television versions, and many other popular books, films, newspapers and television programmes.

European, if not yet Asian, migrants have become 'Aussie battlers', and the bigoted Australian officials who abused and exploited them in the early postwar years are now depicted in a manner similar to that of the arrogant British staff officers who despised the 'Diggers' in the First World War. This is certainly not to claim that the struggles of the migrants, any more than those of the Diggers or the farmers in the Depression, were unreal. Rather, it has taken time and, perhaps, popular fiction and other forms of popular culture to transform historical experience into popular myth.

Bromley[54] contends that 'the colonising of memory by popularised imagery is a complex process which is not simply a matter of personal recall.' (Autobiographies, oral histories and first-hand accounts are certainly of value but not, necessarily, face value.) 'Even the most prominent modes of remembering are subject to implicit social direction based on the power exercised by existing dominant/popular cultural forms.' The shifts, between

1952 and 1986, in what is deemed to be acceptable, appropriate or even credible in popular film, fiction and television would suggest that, as continental European outsiders have become Australian insiders, they have made a home not only in the nation but inside its memory as well.

NOTES

1 H.C. Darby, 'The regional geography of Thomas Hardy's Wessex', *Geographical Review*, 1948, vol. 69, pp. 43–62; J.H. Paterson and E. Paterson, 'Shropshire: reality and symbol in the work of Mary Webb', in D.C.D. Pocock (ed.), *Humanistic Geography and Literature*, London, Croom Helm, 1981, pp. 209–20.

2 J.D. Porteous, 'Literature and humanist geography', *Area*, 1985, vol. 17, pp. 117–22.

3 J. Sonnenfeld, 'Environmental perception and adaptation level in the Arctic', in D. Lowenthal (ed.), *Environmental Perception and Behavior*, Chicago, Chicago University Press, 1967, pp. 42–59.

4 C.A. Middleton, 'Roots and rootlessness: an exploration of the concept in the life and novels of George Eliot', in Pocock, op. cit., pp. 101–20.

5 P. White, 'On the use of creative literature in migration studies', *Area*, 1985, vol. 17, pp. 277–83.

6 E.K. Teather, 'Expatriate experience: becoming an outsider', *Area*, 1989, vol. 21, pp. 401–5.

7 B.S. Osborne, 'Fact, symbol and message: three approaches to literary landscapes', *Canadian Geographer*, 1988, vol. 32, pp. 267–9.

8 N. Shute, *The Far Country*, London, Heinemann, 1952.

9 Interview with Betty Lee of the *Toronto Globe* magazine, 21 February 1959, p. 12.

10 J. Smith, *Nevil Shute (Nevil Shute Norway)*, Boston, Twayne, 1976, p. 103.

11 S. Dowse, *Silver City*, Ringwood, Penguin, 1984.

12 Smith, op. cit., p. 109.

13 Shute, op. cit., p. 51 (Heinemann edition of 1953: other quotations are from the same edition).

14 ibid., p. 163.

15 ibid.

16 J.M. Powell, *Mirrors of the New World: Images and Image Makers in the Settlement Process*, Canberra, ANU Press, 1978; C. Lansbury, *Arcady in Australia: the Evocation of Australia in Nineteenth Century English Literature*, Melbourne, Melbourne University Press, 1970.

17 Quotations from Shute, op. cit., pp. 93, 98–9, 103.

18 Author's notes. National Library of Australia Manuscript Collection. Series 2.

19 Shute, op. cit., p. 108.

20 ibid., p. 77.

21 ibid., p. 13.

22 ibid., p. 153.

23 ibid., p. 67.

24 ibid., p. 91.

25 ibid., p. 179.

26 In the context of reverence for the monarchy, Shute approved of Australians, in 1952, being more British than the British. This was a major preoccupation of the author at the time. In his next novel, *In The Wet* (London, Heinemann, 1953) the hero was a part-Aboriginal pilot who rescued a grateful Queen and Duke of

Edinburgh from a Socialist, pro-republican Britain of the mid-1980s and transported them to a new home in an equally grateful Australia. This overtly royalist viewpoint did not survive when *The Far Country* was recast for television in 1986.

27 J. Wright, *Preoccupations in Australian Poetry*, Melbourne, Oxford University Press, 1965; G. Turner, *National Fictions: Literature, Film and the Construction of Australian Narrative*, Sydney, Allen and Unwin, 1986.

28 Dowse, op. cit., p. 11.

29 ibid., p. 36.

30 The housing crisis in Sydney, and indeed in many parts of Australia, was extremely severe in the early postwar years. It was also an issue upon which the Australian government was highly sensitive. The bleak picture of slum conditions in Sydney painted by Ruth Park in the novel *The Harp in the South* (Sydney, Angus and Robertson, 1948) generated both communal protests and official concern, and contrasted markedly with the view of Australia proffered by the government migration literature of the day. See E.K. Teather, 'Early postwar Sydney: a comparison of its portrayal in fiction and in official documents', *Australian Geographical Studies*, 1990, vol. 28, pp. 204–23; E.K. Teather, 'Visions and realities: images of early postwar Australia', *Transactions, Institute of British Geographers*, 1991, vol. 16, pp. 470–83.

31 Dowse, op. cit., p. 100.

32 ibid., p. 145.

33 ibid., p. 103.

34 ibid., p. 22.

35 ibid., p. 20.

36 ibid., p. 80.

37 ibid., p. 72.

38 Turner, op. cit., p. 22.

39 Shute, *The Far Country*, op. cit., pp. 101, 195.

40 Dowse, op. cit., pp. 73, 84.

41 ibid., p. 39.

42 ibid., p. 156.

43 Teather, 'Expatriate experience', op. cit., p. 405.

44 Dowse, op. cit., p. 104.

45 Joan Long, the producer of the film version of *Silver City*, had envisaged in 1971 a feature film on continental European migrants with the provisional title *A Boatload of Hope*. The title was initially retained by Turkiewicz.

46 One suspects that many expatriate academics will empathise with Porteous, who took fifteen years to 'feel Canadian' – see J.D. Porteous, 'Personally speaking', *Area*, 1989, vol. 21, pp. 419–21.

47 Dowse, op. cit., p. 194.

48 ibid., p. 111.

49 Teather, 'Expatriate experience', op. cit.

50 Also of interest is one character who exists in Shute's manuscript notes (National Library of Australia, MS 2199/2/13) but did not appear in the novel. Tamara Pesediak was a 24-year-old Eastern European refugee. Her family had all been killed in Europe. She was a hospital domestic in a small country town in Victoria and aspired to higher things, in this case a career in surgery. The similarities between this character *manqué* and the heroine of *Silver City* are considerable.

51 C. Cockburn, *Bestseller: The Books that Everyone Read 1900–1939*, London, Sidgwick and Jackson, 1972, p. 3. Today the importance of the bestseller list has been lessened by the existence of television ratings and videocassette hire figures, to name but two alternative criteria. In 1952 television was non-existent in

Australia and very limited in its coverage in Britain: the initial sales of *The Far Country* were therefore a far more reliable indicator of contemporary tastes than they would be today.

52 J. Walter, 'Necessary myths', *Australian Studies*, 1990, vol. 26, pp. 26–36. For a discussion of the role of art in the creation and modification of the Canadian national identity see B.S. Osborne, 'The iconography of nationhood in Canadian art', in D. Cosgrove and S. Daniels (eds), *The Iconography of Landscape*, Cambridge, Cambridge University Press, 1988, pp. 162–78; B. S. Osborne, 'The kindling touch of imagination: Charles William Jefferys and Canadian identity', in P. Simpson-Housley and G. Norcliffe (eds), *A Few Acres of Snow: Literary and Artistic Images of Canada*, Toronto, Dundurn, 1992, pp. 28–47.

53 A. Facey, *A Fortunate Life*, Fremantle, Fremantle Arts Centre, 1981.

54 R. Bromley, *Lost Narratives: Popular Fictions, Politics and Recent History*, London, Routledge, 1988.

17

IN SAMOAN WORLDS
Culture, migration, identity and Albert Wendt

John Connell

'Many of us want our lives to unfold like a novel.'
(Albert Wendt, *Ola*)[1]

INTRODUCTION

Few novels are more autobiographical than those of Albert Wendt, the major figure in South Pacific literature, who was born in the Polynesian island state of Western Samoa in 1939. He has written of his own cultural and migrant identity: 'I am Samoan with a dash of German' and 'New Zealand is a second home which I treasure.' He grew up in the Apia suburb of Vaipe, went to primary school in Samoa, secondary school, teachers' college and university in New Zealand, where he later married a New Zealander, and returned to Samoa to be principal of Samoa College. At the Victoria University of Wellington he was the first Samoan to write an MA thesis, which examined the interwar years when Samoa, then alone of Pacific nations, struggled against the yoke of colonialism. Some years later Wendt took up a lectureship in English at the University of the South Pacific in Suva, Fiji, before moving to the Chair of English at the University of Auckland. His migration history parallels yet amplifies that of many Samoan migrants. Migration – and the movement between and within cultures – are crucial elements in his life and throughout his work but particularly in his first novel, *Sons for the Return Home* (1973). Wendt has said about this novel 'it is an attempt . . . to show what it is like being Samoan and being Samoan in another culture . . . It is about every migrant's dream of the grand return home . . . To sound really grand, I think *Sons* is about Polynesia – what it was, what it is, what it is becoming.'[2] So is all of Wendt's work.

Wendt's subsequent novels and most of his poetry develop the themes of displacement and identity established in *Sons*, though there is a diversity of genre and style (including languages) in Wendt's work: from timeless fables, through *Leaves of the Banyan Tree* (1979), which incorporates much of what had gone before in a trilogy involving several generations of change in a

Samoan family in and away from the village of Sapepe and, finally, *Ola* (1991), a more experimental postmodernist novel that takes hitherto primarily Samoan themes beyond the Pacific into a global realm. For all the changes in style and genre, Wendt has remained focused on the past and its diverse and intangible heritage. That past primarily concerns *pouliuli* (the time of darkness) and the changing role of *fa'a Samoa* (the Samoan culture and way of life) as Samoa became influenced by European intrusions, and later by extensive migration to and from New Zealand. These themes are combined in 'a single oeuvre whose author is progressively mapping a fictional Samoa, rather as one of his acknowledged masters, William Faulkner, constructed his elaborate survey of Yoknapatawha county'[3] to demarcate changes and continuities in social organisation, history and landscape. For Wendt, 'Novels present the most complex histories that have been written.'[4] At the core of all Wendt's work are questions of identity and meaning, the diffusion and debasement of culture, the quest for modernity – in its various guises – and the changes in village institutions; neither culture nor village life are timeless. Migration emphasises change, yet change occurs without migration; very little is as it seems.

AN ISLAND MICRO-STATE

Western Samoa is one of several Polynesian island micro-states in the South Pacific. Initially a German colony, it became a New Zealand colony after the First World War; during the interwar years there was a violent struggle against New Zealand, and in 1962 Western Samoa became the first Pacific island state to achieve independence. About three-quarters of the population of 162,000 live on the island of Upolu – where the capital, Apia, has a population of around 35,000 – and almost all the remainder live on the larger island of Savai'i. Western Samoa experiences all the constraints that accompany smallness, isolation and insularity, including vulnerability to natural hazards (especially cyclones), but literacy rates and life expectancy are higher than in many small island states. The domestic economy is characterised by agricultural production, but agricultural exports have grown only slowly in the 1980s, and during the 1960s and 1970s the economy became increasingly dominated by the rise in importance of overseas aid and migrant remittances. Samoa and other nearby micro-states have been characterised as MIRAB societies, where Migration leads to Remittances, and the other principal income source, Aid, has contributed to the establishment of a government Bureaucracy.[5] In barely a quarter of a century migration has become of enormous social, economic and political importance, and despite some negative consequences is now widely perceived in a positive manner, not least by individual Samoans.[6] For all its vast social, economic and political significance, the extent and ramifications of migration are largely invisible in the landscape. Outside the small capital and port of Apia, Western Samoa is a country of coconuts, taro and bananas, with a population scattered in small villages, dominated by the towers of churches. Religion remains a powerful element in Samoan life.

Over time Samoa has become more urban, and the locus of contemporary power is increasingly in the sole urban centre. It is here that the new bourgeoisie – constant targets for Wendt – have emerged and become consolidated: the Pacific mimic men, fruitlessly and incessantly pretending and seeking to be different from what they are. Many are *afakasi* (half-caste) part-Europeans engaged in both business and politics – activities which are hopelessly and depressingly intertwined. The *afakasi* indigenous elite has become increasingly indistinguishable from the *papalagi* colonisers. The elite migration of foreigners into Samoa merely emphasises this situation. As Wendt has written, 'No party is complete without diplomats and UNO "experts and advisers". Samoa has one of the highest ratios of foreign experts in the world. Every type of expert you need to claim you're "developing". Development is the new gospel.'[7] The new experts are consequently the latter-day descendants of the missionaries, who came 'to help civilise these people and thereby make them worthy of independence and the modern world'.[8] That social presence has taken spatial form: 'Many of our rich (and aspiring rich), foreign diplomats and business people live in airy houses on the cool slopes of Vailima, above the heat, dust, stench and shabbiness of the town where they make their money, and pay no rates or accept any responsibility for the town's maintenance.' Though Samoa may still present, at times, 'a world suffused with light and hushly still as in a technicolour photograph in a travel book about the mythical, romantic South Seas', more generally 'the Hollywood/Margaret Mead myth' is exactly that.[9] Samoa has changed and, for Wendt, 'the rot has gotten worse ... but that doesn't make the country less spiritual for me ... I'm so disillusioned by politicians [though] some of my writing has changed the perceptions of some people.'[10] Pessimism and disappointment inform Wendt's vision.

A SAMOAN DIASPORA

In this once remote island state, now increasingly tied to the outside world, migration has rapidly grown in extent and significance. A major influence on migration has been the radical change in expectations over what constitutes a satisfactory standard of living, a desirable occupation and a suitable mix of accessible services and amenities. Different values following educational growth, reflecting and influencing the expansion of bureaucratic (largely urban) unemployment, and youthful disdain for agriculture, have further oriented migration streams away from Samoa, as new employment opportunities have not kept pace with population growth.

At the start of the 1990s there were about 78,000 Western Samoans – a third of all Western Samoans – living overseas, mainly in the United States,

American Samoa (a migratory bridge to the USA) and New Zealand. Though Samoans are extremely broadly distributed in global terms,[11] they are primarily concentrated in a few large New Zealand and US cities, where their presence is a magnet for future migration: Pacific chain migration exemplified.

> The second oldest brother went to live in Apia; he worked as a bulldozer driver and rarely visited the village. He married, and a few years later returned to tell them [his family] he was migrating to New Zealand. They dried copra and cocoa and sold it to help pay his fare. Alone he went to New Zealand. Some months later he sent for his wife and three children. He wrote letters to his family in Samoa, telling them how easy life was in New Zealand – good pay, good schools for the children and the Samoan church was the fastest growing church in the new land. Sometimes, usually just before Christmas, he sent them large sums of money. He bought a house and then asked the rest of the family to shift to New Zealand. He suggested that one of the older men should come over first, work, and send money back for the next one to come and so on.[12]

Samoans have used kinship ties in American Samoa to move there and, often, onwards to the USA.[13] Old colonial ties gave them special rights of entry into New Zealand. Since the mid-1970s, however, these rights have been more difficult to exercise because of more restrictive New Zealand legislation following economic recession, with substantial social consequences in Samoa, including a rise in the level of suicide.[14] From initially being recruited to help solve temporary labour shortages, prospective migrants are now more likely to be rejected unless they are skilled. Restrictions on migration have emphasised the economic vulnerability of an island micro-state, and the necessity for continuous migration, once dependence on remittances has been established. Remittances tend to decline over time, despite increased pressure on migrants from kin at home, sometimes involving specific trips to New Zealand to encourage, sponsor and collect remittances. Such wide-ranging uncertainties shape the nature of links with home. The rationale for migration has been broadly economic, a response to the extent of uneven development between small island and metropolitan states, and accentuated by land shortages and social change. In Samoa 'migrants seek in the West access to material goods, jobs in the industrial sector, better education for their young, and social mobility in a society they have believed free of the traditional barriers of rank and family status that made such mobility difficult at home.'[15] The ramifications of the wider world, different values, material rewards and frustrations, underlie the migratory experience.

Migration decisions are shaped within a family context, with migrants usually leaving to meet extended household (*'aiga*) expectations, such as the supply of remittances, rather than their being individualistic. Migration occurs under the auspices of kin already abroad who provide sponsorship, accommodation and even employment. Families in many respects operate on a world stage and, as in other Polynesian micro-states, households are characterised by remittance transfers among various component parts of the 'transnational corporations of kin'.[16] Social obligations within Samoa are rarely far from migrants' minds, and remittances are sustained for very long periods, especially when, as is usually the case, migrants intend to or believe they will return. Remittances are primarily directed into immediate consumption needs, including house construction and social obligations, whilst, through education and fares, they sustain the migratory system for subsequent generations. Remittances are rarely used for investment since, other than in small stores and transport businesses, opportunities are few. Hence 'migration was a far more lucrative investment than anything available in the village.'[17] Though high incomes enable the maintenance of social status, cash increasingly influences the votes for traditional chiefs (*matai*), though cash can also buy chainsaws, pesticides, fertiliser and other forms of modern technology which increase the productivity of village agriculture.[18]

Economic goals are paramount yet intricately related to prestige and social status, and hence to the discharge of the personal and family obligations (*faalavelave*) central to *fa'a Samoa*. Education, and especially tertiary education, is an integral component of social and family mobility yet largely impossible without access to overseas economic resources. Acquiring education and income may be a slow process and migrants' hopes and expectations are usually deferred to the next generation. In *Sons* the father 'had observed the changes taking place and had concluded that only people educated the papalagi [European] way would have a good future in Samoa. That was why he had brought them to New Zealand.' Despite the significance of New Zealand, and its growing Samoan population, migrants do not intend to move permanently: 'our whole life here is only a preparation for the grand return to our homeland. Their hopes and dreams all revolve round our return.'[19]

ON DISTANT SHORES

The capital city, Apia, once played a crucial role, not only in rural–urban migration but as the source of innovations, discoveries and new ways of thinking about and imagining the world.[20] Now Apia is by-passed by migrants en route for distant cities in more alien worlds. Samoan migrants live in an urban world, in a cold climate, ensuring that they are often enclosed within their houses, rather than in the unwalled houses of Samoa that are obviously part of a wider social and economic community. In the cities, the locus of migration, New Zealand is at its worst; without kin, or for the hero of *Sons*, without his partner,

> this city, this country, would be a barren place of exile. And, as you
> walk further into the maze of this city, these grey walls and floors of

concrete and steel and stone, you know that without her, the labyrinth would eventually turn you into stone, for modern cities are the new man-made deserts in which man traps himself and bleeds himself of all his rich warm fertile humanity and goodness.[21]

But to this distant, unfamiliar and hostile world thousands of Samoans have freely chosen to migrate, settle and achieve new forms of success.

Most migrants, though well educated by island standards, are not well educated by the standards of the countries to which they migrate, and few speak English fluently. They tend to enter the urban workforce, often with difficulty, in low-paying, unskilled, non-unionised blue-collar jobs at the bottom of the employment hierarchy.[22] Employment, especially in factories, takes new forms:

> He was afraid of the factory; he understood little of what went on in it. Caught in the noise, the overwhelming size of the building, the intricate system of machines and conveyor belts and cable, the large number of workers whose language he didn't understand, he felt small, lost . . . He told his wife that his job was not fit for a man; girls and old men could do it. He yearned to work with crops again . . . or to go out to sea to fish . . . He was not used to the monotonous routine of getting up early in the morning, catching a crowded bus and filling in a certain number of hours with humiliating work. But he enjoyed the big pay at the end of the day . . . He saved most of it so he could pay for his elder son to come from Samoa.[23]

Migrants often work long hours, because of a high level of financial commitments to their immediate family, the church and the relatives in their extended families in Samoa. These financial commitments influence perspectives, especially for the second or New Zealand-born generation, distanced from their kin in Samoa. One New Zealand-born Samoan woman has stated: 'Fa'a Samoa has its place in Samoa but here I think the family has to come first. Lots of Samoans are in debt up to their eyeballs because their extended family and the church come first.'[24] This declining interest in and acceptance of *fa'a Samoa* also accompanies socio-economic mobility, especially in the second generation. However, upward mobility for women out of the factories and cleaning jobs into the 'white-blouse' employment of offices is hampered by discrimination and the effect of economic restructuring. The second generation is also characterised by high levels of non-communicable diseases such as diabetes, stress and obesity. Most migrants are distanced from the mainstream of New Zealand economy and society.

Social tensions are recurrent, above all for those who are relatively assimilated. 'Most papalagi New Zealanders talked of racial integration, but what they wanted was assimilation, the conversion of Polynesians into middle-class papalagi. The process was one of castration, the creation of Uncle

Toms.'[25] First-generation migrants, working in groups, are often less conscious of this than the second generation, familiar with relative deprivation, uncertain of Samoan alternatives and incomplete within *fa'a Samoa*. But New Zealand is a country of diverse ethnicity, including of course the indigenous Polynesian Maori population. In *Sons for the Return Home*, Wendt writes:

> Like many other Samoans, he thought himself superior to Maoris ...
> He admitted that most Samoans believed the same racist myths about
> Maoris as pakehas [white New Zealanders] did: Maoris were dirty, lazy,
> irresponsible; they were intellectually inferior; they lacked initiative,
> drive, courage; they drank too much, were sexually immoral, treated
> their children cruelly; all that most of them were good at was rugby,
> bulldozer-driving, dancing and playing in bands; the only Maoris who
> had made the grade and were now teachers, doctors, lawyers and
> politicians were the descendants of Maori royalty; the quicker the
> Maoris adjusted to the pakeha way of life, which was based on thrift
> and cleanliness and godliness, sobriety and honesty and hard work, the
> quicker they would become worthy New Zealand citizens ... Most
> Samoans also believed that Maoris lacked pride. They had given in too
> easily to the pakeha and were no longer Polynesians.[26]

Not surprisingly, there have been many violent incidents in New Zealand between Samoans and Maoris. The incidents have become, according to Wendt, the very myths that *pakehas* believed about Samoans (and about islanders in general). Consequently

> the majority sometimes think of Samoans as part of a Polynesian brown
> peril that they fear will overrun their country. Tabloids in New Zealand
> portray Samoans as dangerous, as rapists, as drunks and as noisy and
> violent, even though many in the majority have Samoan neighbours
> who are every bit as quiet and law-abiding as they are.[27]

Samoan identity and ethnicity are thus created and recreated in the face of opposition. There are other Polynesian migrants in New Zealand, particularly Tongans, but also those from New Zealand's three Pacific territories, who are New Zealand citizens. However, 'we don't get on with Niueans, Tokelauans or Cook Islanders. You would think, because Maoris and Islanders are at the bottom of the social ladder, we would be brothers. But the sad fact is that we're not.'[28] Difference and distinctiveness are constantly emphasised in these lower echelons of society and the labour market.

In this alien environment new migrants are conscious of their intent to return home, despite the need to achieve enough success to accumulate appropriate material rewards and secure their children's education, and despite the success stories of other earlier migrants. While migrants intend to return home with the trappings of a successful period of time away, the actual return is often deferred, sometimes indefinitely. Though the mother in *Sons* observes that they are all going back home when the youngest son finishes university, that can take many years. Over time new commitments and new ties are established. Some of these are in the church; in *Sons* the father becomes a deacon in the local Pacific Islanders Church, the wife a social committee member, while the younger son teaches in Sunday School.[29] These examples are a further indication of the centrality of the church in Samoan life, both at home and abroad. Despite change there is a public ideology of return, yet a private recognition that return will not always occur; a variety of influences – local kin, employment, income levels, education, financial commitment and the emergence of new lifestyles – all constrain return migration.[30] Deliberate attempts are made by many households to maintain Samoan connections; children and grandchildren may be sent home 'to learn the real fa'a Samoa', 'to provide labour' and so on, enabling the migrants to work long hours of overtime and simultaneously meet their Samoan obligations and secure a place in New Zealand.[31]

Children educated in New Zealand, even if discriminated against as 'dirty coconut islanders',[32] are better able to cope with New Zealand society and correspondingly less well able to fit into a Samoa that they may barely know. In doing so, the centrality of *fa'a Samoa* declines, especially where linguistic competence may have withered, and connections in New Zealand have greater precedence and significance. Some connections are particularly significant. Marriage to a non-Samoan may mean the end of plans to return to Samoa both because the marriage itself marks a distinct move away from *fa'a Samoa* and because spouses are unlikely to want to move to and stay in Samoa, where it is widely assumed 'she won't fit into Samoa. She doesn't know our customs, our ways of doing things. And our people won't accept her. Our way of life, our people, may destroy her.'[33] It is no different for husbands.

Samoa is constructed and reconstructed in New Zealand. Parents stress to their children that, in the words of the mother in *Sons*, 'In Samoa villages are clean and tidy and widely scattered around the coast – one has a lot of room to live in. New Zealand is crowded, noisy and unhealthy. Families are crowded together.' (She had never left the confines of the city.) Samoa is lush green with tropical forests. New Zealand is made up of overcrowded cities, rife with crime. *Papalagi* were untrustworthy, most had false teeth – since *papalagi* food was too soft – and children 'were rude, destructive, spoilt and had no respect for their elders. In Samoa the children were the exact opposite.' Samoans were good Christians;

> the papalagi brought Christianity to Samoa but then, as they gained in atheistic knowledge and wordly wealth and power, they forgot God and became Pagans again . . . [whereas] the few misguided Samoans who have forsaken the Church are paying for it: they lead sinful lives . . . These are people who are giving us a bad name. What papalagi don't know is that these people are not real Samoans – they're half castes.

The extended Polynesian family remains paramount: 'There are no orphans or poor people in Samoa . . . No one starves either. We care for one another.' Wendt concludes the mother's dialogue:

And so she continued throughout the years, until a new mythology woven out of her romantic memories, her legends, her illusions and her prejudices, was born in her sons: a new fabulous Samoa to be attained by her sons when they returned home after surviving the winters of a pagan country.[34]

Similar themes are taken up by the father. 'Without self respect, life would have little meaning', but in the confines of *fa'a Samoa*, this is implicit: the way to leadership is through service. The father goes on:

Honour all your obligations to your family, church and village. Without your family you are lost, a bird without a nest to give you identity. You must sacrifice your personal ambitions if they clash with the wishes of your family. This is how it is in Samoa, how it is today, and how it will always be. God meant it to be that way . . . Our culture is wider than papalagi culture . . . Our culture is also based on sacred laws sanctioned by God and handed down to us by our forefathers. That is why in Samoa we have been able to remain Samoan, safe from the changes brought about by rapacious papalagi. In Samoa the magnificent tropical forest is never far away . . . And you are aware constantly of your own impermanence. Or the sound of children, like a mountain stream, . . . is always part of your day . . . you are reminded of your own mortality [yet] in Samoa it is difficult to accept death – everything invites you to live.[35]

Again Wendt sums up: 'Years later the youngest son would admit to himself that almost all he knew of Samoa was a creation of his parents and other Samoans he admired' in which the role of migration and the perceived necessity to leave Samoa were wholly absent.

Samoa must be constantly constructed since migrants necessarily inhabit a world of images shaped and distorted by memories that are inevitably selective and, for new generations, uninformed by personal experience. In *Sons* Samoan society is held by the hero's family to approximate to an 'ideal society' that somehow combines *fa'a Samoa* and the scriptures, whilst non-Samoan society has diverged from this in some important ways; these assertions and arguments depend not only on highly idealised pictures of Samoan society but also on fairly generalised pictures of non-Samoan society 'which draw heavily on images such as old people's homes and evidence from television'.[36] But cultures may not only be represented in quite different ways, they may be represented in different ways in the same household at different times. Culture is diverse, situational, intangible and constantly in flux. On distant shores very diverse images of Samoa, and of New Zealand, are created and provide the context in which return migration occurs or fails to occur.

THE RETURN HOME

Migrants almost always return home, both temporarily and much less frequently on the assumption of permanence. Duration of absence and reason for return are of enormous importance in the new accommodation to Apia and Samoa. Once again, from *Sons*, the palm trees, volcanic mountains, the smells of sweat and coconut oil, night-time darkness, rural noises and legs pitted with sores are initially familiar yet now unfamiliar. 'It was hard to believe that he had spent nearly twenty years preparing and waiting for this return. So many years and now nothing more than an uncomfortable seat, as a stranger, in a bus packed with the mythical characters of the legends his parents had nourished him on for so long.' Yet the myths and images were incomplete; 'he had returned unprepared for the flies and mosquitoes', the lack of solitude and the rudimentary standards of sanitation and hygiene, and had been made helpless 'by the comforts of electricity, instant food and over-efficient women'.[37] The reality is one of difference and distance.

Beyond the quickly apparent physical phenomena of the tropics are the cultural distinctions of *fa'a Samoa*, some of which are distasteful and different from the myths nurtured in New Zealand.

> His people – and it was difficult for him to refer to them as his people – measured life in proportion to their physical beauty – they measured themselves in terms of how much punishment and pleasure the flesh could consume and endure. Quantity of consumption was the measure. So every feast was an orgy of food ... they seemed to invite obesity, diabetes and heart attacks.

It was clear that 'loyalty to the family came before everything else, even one's life' and that the more than nascent individualism, partly constructed in foreign education systems, was still somewhat alien in Samoa.

> Their pretensions – like their exaggerated faith in their physical courage, were of gigantic proportions. Samoa was the navel of the universe: the world ended within the visible horizons and reefs ... What was real were their islands – magnified in their hearts into an emotional and spiritual heaven larger than the planet itself.[38]

Samoa achieved a centrality that was extraordinary to those who had spent time beyond the reefs.

Politics was ever-present, in personal and national form: a new or forgotten phenomenon for those marginalised in a primarily white New Zealand political system. In 'village and national politics ... the acquisition of titles, whether real or imaginary, was an endless battle, a dynamic force in village life'[39] and one that was wholly bound up with the hierarchical Samoan social structure. On distant shores heredity was of limited significance. The church too was vital and uncompromising. Religion was a social custom, a major strand of the social, economic and political web. 'If they found out he was an atheist ... [he] might as well return to New Zealand. There was no place in Samoa for atheists.'[40] Thus successful return migration meant substantial accommodation to the power of the church. *Fa'a Samoa*, political and religious hierarchies, and the uncertainties of the unknown, conspire against those who have been away for long.

Return migrants may face resentment from those who feel they have avoided social and economic obligations at home, or because migrants feel superior for having lived in New Zealand:

> their relatives at home believe that they made the migrants' sojourns possible by their sacrifices; hence, in expecting money on the migrants' return, they are receiving only what is rightfully theirs. Returnees quickly learn that they are judged by what they can provide when they come back as well as by what they have accomplished abroad.[41]

Pressures may be considerable: on the one hand return migrants may not wish to dissipate *'aiga* illusions of vast overseas wealth; whilst on the other hand they fear that their relative 'failure' may create disillusion in the face of great expectations. Both returnees and non-migrants may become victims of their own expectations. For the more educated or the overseas-born the task of becoming absorbed into Samoa and *fa'a Samoa* is particularly difficult. The hero of *Sons*, with two university degrees, can 'see too clearly [and] will never be happy with things as they are ... you will always be in permanent exile. You will never belong anywhere.' Hence his parents observe 'we shouldn't have taken you away from here. We shouldn't have tried to live our hopes, dreams, pretensions and lives through you and your brothers.'[42] Because of the significance of land, simple unfamiliarity with the unknown, mis-understandings and their own expectations, return migrants have

> found it difficult to return to village agriculture. Most often, returnees hoped to set up some kind of small business or to be retained in government service, the private sector or the churches. Some saw the potential for making money in commercial agricultural ventures, but not in village agriculture. Many bought land in Apia, far from their home villages

where there might be some degree of anonymity and some release from social obligations.[43]

A triumphant return is possible. The parents, in *Sons*, construct a large, new modern home, and nearby a store for selling food, whilst the elder brother, relatively uneducated, buys a bus to run between village and town, and is planning to buy two more.[44] The income brought back after a long period in New Zealand enables ready entry into the commercial world. Economic success, transferred to the village, leads to social rewards; the older brother is now the most eligible male in the village and the mother is a pillar of the Women's Committee and the first woman to be elected to the village school committee, 'envied by all the other women for her spectacular clothes and shoes and her cosmetics'. The new house with its water, electricity and flush toilet is 'a symbol of high status and sophistication and civilisation; [a] badge of honour ... a New Zealand oasis in the middle of wilderness, earned by over twenty years in exile'.[45] Beyond the display of the physical symbols of difference are cultural changes, such as speaking English more frequently. In *Sons* the mother 'had rarely spoken English in their home, but now in Samoa she used every available opportunity to impress the villagers with her English'.[46] Only the younger brother, university-educated and in New Zealand for much of his life, is dissatisfied; while the parents and older brother achieve permanence, for him

> the village seemed unreal and impermanent. He couldn't persuade himself that he could live and marry and die in it. Associated with this knowledge was a sense of guilt: if he left he would be betraying his parents and the twenty years they had spent in exile so that he could get a good education. Perhaps he could live in Apia.[47]

Though for him this fails, those who succeed are also often those who are relatively highly educated but who have spent their youth in Samoa, and for whom the return is an opportunity to use new and specialised skills for rapid advancement – if not always legitimately. Some return as lechers, others as unscrupulous businessmen or politicians: 'he is rumoured to have spent about a hundred thousand dollars to buy his three-hundred vote electorate in the last general elections; much of the money is said to have come from the sale of smuggled goods.'[48] One way or another, as in other Pacific states, many return migrants have assumed critical positions of power.

Successful return migration is partly a modern urban phenomenon. This may be true in terms of place of residence and new employment; or, in remote rural areas, a successful return is associated with the commerce and material rewards of once distant and different places. While return migration may prove successful and permanent there are still things about New Zealand to be missed, even the more unexpected:

> I miss that factory and that machine I worked with and lived with for nearly twenty years ... I miss that ugly, cruel city, with its insatiable roots stabbed into the earth, choking it; breathing all its poisons into the sky; its blood contaminating the people and turning them against one another in perpetual combat.'[49]

The city, with all its certainties and uncertainties, has ensured that migrants are changed yet remain Samoan, intent on the ideology of return. Yet return can be an alienating experience and success limited. Ultimately few definitely choose to return, and fewer succeed. Even those who visit Samoa are often glad to return to New Zealand. 'It seems that the closer the contact with the reality of home, the stronger is the migrant's resolve to consolidate his new life in New Zealand. Thus if the dream remains it seems likely that it is nostalgia for the past rather than a plan for the future.'[50] For the hero of *Sons* return also proves ultimately impossible; he cannot adjust and returns to New Zealand as a lonely exile. Others too re-emigrate to an established community of New Zealand Samoans, to their place in a growing diaspora. Dreams may be tarnished but they will remain, as the choice of return is postponed indefinitely.

THE AMBIGUITY OF ARRIVALS

Wendt depicts life as a constant struggle for survival where opposing forces measure their strength through subtle and shifting balances. Most pervasive is the conflict of light against darkness (*pouliuli*) but this duality is emphasised in other dichotomies – tradition versus modernity, the city and the village, Samoa and New Zealand, responsibility to the community versus individual ambition, self-enrichment or fulfilment, *fa'a Samoa* and European/New Zealand materialism – in a complex world where individuals shape and are buffeted by constant shifts in modernity. Even self-assertion and self-denigration proceed together, in arenas where cultural confrontation is both creative and destructive in the construction of new forms and meanings of Samoan identity. All are variations on darkness and light. For Wendt, modernity is corruption, evident in its early phase in *Pouliuli* and much later in *Ola*. Though migration accentuates change and focuses attention upon difference, change occurs without migration; hence an accusation to a *matai* (chief) who has changed: 'the individual freedom you have discovered and now want to maintain is contrary to the very basis of our way of life.'[51]

Inside and outside Samoa there have been substantial gains from migration and modernity – both of which, in one guise or another, have contributed to changes in *fa'a Samoa*. Samoans have migrated overseas for more than a quarter of a century and have been mobile for much longer than that. There are now four generations of Samoans in New Zealand exhibiting a diversity of lifestyles which, although more or less related to *fa'a Samoa*, are quantitatively and qualitatively different from those in Samoa. Samoan culture aggregates rather than integrates; ethnicity disguises diversity. For those who have been most successful, according to one Samoan, 'the choice of action is often a compromise between alternatives offered by both the Samoan and New Zealand cultures – choosing the best of both worlds.'[52] Migration is not escapism, nor the movement of the particularly disadvantaged and dissatisfied; it is 'well-adjusted' individuals who are most likely to migrate from Samoa and consequently to achieve some degree of success elsewhere.[53]

Wendt has observed 'I am of two worlds, but I do belong to the South Pacific. As a person I'm Samoan and I write about Samoa.'[54] In *Ola* the novel goes beyond two worlds – not only into Japanese and Jewish worlds, but into more detailed reflections on the other Polynesian world of New Zealand

Maoris. Dichotomies are no longer useful, as the world is revealed to be far more complex, involving 'movements in specific colonial, neo-colonial and postcolonial circuits, different diasporas, borderlands, exiles, detours and returns ... that generate *discrepant cosmopolitanisms*'.[55] There are many variants within and between Samoans, *papalagis*, *pakehas* and others. Two worlds are no longer enough.

Overseas is a world of diversity; for all those Samoans who experience culture clashes, conflicts, discrimination and disappointment, there are others who move easily between islands, oblivious to difference, sheltered in the confines of a transnational *aiga* or coping with change. Individuals may as easily manipulate and conquer different worlds as be fooled and frustrated by them. In Samoa responses to migration 'continue to involve the careful balancing of choices between traditional values and individualist tendencies, food security and cash hopes, personal control and dependence on an uncertain world market, and village autonomy and reliance on national institutions'.[56] Balancing multiple oppositions and dichotomies, and welding them into new situations and structures, are integral elements of migration; as elsewhere many Samoans, in and outside Samoa, become successful cosmopolitans 'at ease in multiple worlds, rather than natives of place torn by new and multiple allegiances'.[57] The bleakness of *Sons*, pervaded by images of abortion and crucifixion, the central theme of the tragic love affair between Samoan boy and white girl, and the apparent impossibility of moving from the colonial past into a partly neo-colonial present, has now been transcended by the sheer volume and ubiquity of migration but, even more so, by the extent that in-betweenness is a part of all Samoan lives.

Any notion of a condition of perpetual exile has been replaced by more positive responses to the diversity of lifeworlds, and a variety of means of finding and transforming cultural identities, and establishing new connections in unfamiliar settings a long way away from what might once have been home. Changes in identity and challenges to authenticity are endless and rarely other than subjective: 'We are what we remember: the actions we lived through or should have lived out and which we have chosen to remember' just as 'there is no difference between an imagined act and one actually committed.'[58] As migration continues, as new Samoan generations grow up in many different locations overseas, and as movement between them persists, the enigmas will increase. An already diverse population will merely become more so, and 'when you don't belong completely to any culture ... you will always be an outsider and suffer from a sense of unreality.'[59] Ambiguity and achievement are not incompatible but identity will constantly be challenged. Few people will confidently assert that they know where home is, or even what it is. 'In the final analysis our countries, cultures, nations and planets are what we imagine them to be',[60] because 'to have multiple roots is to have "no root". It is the modernity of a plural society whose sources of authenticity always lie elsewhere – in other continents or other times.'[61] Identities are necessary fictions and few spare time to assess or agonise over them.

In the era of modernisation that characterised the 1960s and 1970s, and the genesis and production of *Sons*, change was disruptive but there was a clear value attached to the re-establishment through whatever process of the coherence and stability of identity.[62] Two decades later the notions of stability and coherence have disappeared. In *Ola* the flexibility of metaphor, culture and geography has become apparent, reflecting a more complex Polynesian and other world, and the diversity of the lives of now middle-aged Samoans, as they overcome both distance and difference. In an uncertain global political economy, even the most cosmopolitan Samoan must ensure that Samoa is not merely a nostalgic fantasy, but a potentially real destination. The persistence of the ideology of return is just one means of bridging and welding together many different lifestyles and opportunities. Migration is rarely absolute, unambivalent or final; it is not a cause and consequence of a definite break with a cultural life that is part of history, but a partial and conditional state, characterised by ambiguity and indeterminacy.[63] A fixed status presupposes that the future can be foretold. Uncertainty defines the experience of migration, even in second generations. Ambivalence is the norm for Wendt too: 'I know I can't live away from Samoa for too long. I need a sense of roots, of home – a place where you live and die. I would die as a writer without roots; but when I go home I'm always reminded that I'm an outsider, palagified.'[64] The last words too must go to Wendt. 'All is real, whether borrowed or created or dreamed, or mixed together with facts, fictions, strange sauces and herbs and condiments',[65] and, through a combination of resistance and accommodation, 'we are what we remember.'[66] The identity of person and place is always continuously being produced.

NOTES

1 A. Wendt, *Ola*, Auckland, Penguin, 1991, p. 9.
2 The dust-jacket of A. Wendt, *Sons for the Return Home*, Auckland, Longman, 1973.
3 M. Neill, 'A cultivated innocence', *Landfall*, 1987, vol. 41, p. 214.
4 V. Hereniko and D. Hanlon, 'An interview with Albert Wendt', *The Contemporary Pacific*, 1992, vol. 5, p. 118.
5 I.E. Bertram and R. Watters, 'The MIRAB economy in South Pacific microstates', *Pacific Viewpoint*, 1985, vol. 27, pp. 497–520.
6 L.F. Va'a, 'The future of Western Samoan migration to New Zealand', *Asian and Pacific Migration Journal*, 1993, vol. 1, pp. 313–32; L.F. Va'a, 'Effects of migration on Western Samoa. An island viewpoint', in G. McCall and J. Connell (eds), *A World Perspective on Pacific Islander Migration*, Sydney, University of New South Wales, Centre for South Pacific Studies Monograph No. 6, 1993, pp. 343–57; J. Connell, 'Island microstates: the mirage of development', *The Contemporary Pacific*, 1991, vol. 3, pp. 251–87.
7 Wendt, *Ola*, op. cit., p. 258.
8 A. Wendt, *Flying Fox in a Freedom Tree*, Auckland, Longman Paul, 1974, p. 95.
9 Wendt, *Ola*, op. cit., p. 283, 321.

10 Hereniko and Hanlon, op. cit., pp. 113, 124, 129.
11 F.K. Sutter, *The Samoans: a Global Family*, Honolulu, University of Hawaii Press, 1989; J. Connell, *Migration Employment and Development in the South Pacific. Country Report No. 22 Western Samoa*, Noumea, South Pacific Commission, 1983; J. Connell, 'Paradise left? Pacific Island voyagers in the modern world', in J. Fawcett and B. Carino (eds), *Pacific Bridges. The New Immigration from Asia and The Pacific Islands*, New York, Center for Migration Studies, 1987, pp. 375–404.
12 Wendt, *Sons*, op. cit., p. 42.
13 E. Kallen, *The Western Samoan Kinship Bridge. A Study in Migration, Social Change and the New Ethnicity*, Leiden, Brill, 1982; D. Ahlburg and M. Levin, *The Northeast Passage. A Study of Pacific Islander Migration to American Samoa and the United States*, Canberra, National Centre for Development Studies, Pacific Research Monograph No. 23, 1990.
14 C. Macpherson, 'Stolen dreams: some consequences of dependency for Western Samoan youth', in J. Connell (ed.), *Migration and Development in the South Pacific*, Canberra, National Centre for Development Studies, Pacific Research Monograph No. 24, 1990, pp. 107–19.
15 B. Shore, 'Introduction', in C. Macpherson, B. Shore and R. Franco (eds), *New Neighbors, Islanders in Adaptation*, Santa Cruz, Center for Pacific Studies, 1978, p. xiii.
16 I. Bertram, 'Sustainable development in South Pacific micro-economies', *World Development*, 1986, vol. 14, pp. 809–22.
17 P. Shankman, *Migration and Underdevelopment. The Case of Western Samoa*, Boulder, Colo., Westview, 1976, p. 71.
18 P. Shankman, 'The Samoan exodus', in V. Lockwood, T.G. Harding and B. Wallace (eds), *Contemporary Pacific Societies*, Englewood Cliffs, NJ, Prentice Hall, 1993, p. 158; P. Fairbairn-Dunlop, 'A positive response to migration constraints', in McCall and Connell (eds), op. cit., pp. 327–42.
19 Wendt, *Sons*, op. cit., pp. 119–20, 140.
20 The role and significance of Apia, especially in the interwar years, is discussed in J. Connell, *The Return Home? Albert Wendt, Migration and Identity in Polynesia*, University of Leeds, School of Geography Working Paper 93/23, pp. 10–13.
21 Wendt, *Sons*, op. cit., p. 129.
22 K. Gibson, 'Political economy and international labour migration: the case of Polynesians in New Zealand', *New Zealand Geographer*, 1983, vol. 39, pp. 29–42; T. Loomis, 'The world economy, New Zealand restructuring and Pacific migrant labour', *Journal of International Studies*, 1991, vol. 12, pp. 59–70.
23 Wendt, *Sons*, op. cit., p. 53.
24 Quoted in W. Larner, 'Labour migration and female labour: Samoan women in New Zealand', *Australian and New Zealand Journal of Sociology*, 1991, vol. 27, p. 24.
25 Wendt, *Sons*, op. cit., p. 152.
26 ibid., p. 97.
27 Shankman, 'The Samoan exodus', op. cit., p. 161.
28 Wendt, *Sons*, op. cit., p. 98.
29 ibid., pp. 14, 30.
30 C. Macpherson, 'Public and private views of home: will Western Samoan migrants return?', *Pacific Viewpoint*, 1985, vol. 26, pp. 242–62.
31 Fairbairn-Dunlop, op. cit., p. 333.
32 Wendt, *Sons*, op. cit., p. 13.
33 ibid., p. 137.
34 ibid., pp. 74–6.

35 ibid., pp. 76–7.
36 C. Macpherson, 'On the future of Samoan ethnicity in New Zealand', in P. Spoonley, C. Macpherson, D. Pearson and C. Sedgwick (eds), *Tauiwi*, Palmerston North, Dunmore Press, 1984, p. 124.
37 Wendt, *Sons*, op. cit., pp. 171–5.
38 ibid., pp. 178–9.
39 ibid., p. 179.
40 ibid., pp. 180, 182.
41 ibid., p. 167.
42 ibid., p. 204; see also p. 120.
43 Shankman, 'The Samoan exodus', op. cit., p. 164.
44 Wendt, *Sons*, op. cit., pp. 190–1.
45 ibid., p. 202.
46 ibid., p. 212.
47 ibid., p. 191.
48 Wendt, *Ola*, op. cit., p. 184.
49 Wendt, *Sons*, op. cit., p. 209.
50 D. Pitt and C. Macpherson, *Emerging Pluralism. The Samoan Community in New Zealand*, Auckland, Longman Paul, 1974, p. 16.
51 A. Wendt, *Pouliuli*, Auckland, Longman Paul, 1977, p. 17.
52 L.M. Ioane, 'The lotus-eaters of the South Pacific', in Institute for Polynesian Studies, *Evolving Political Cultures in the Pacific Islands*, Laie, Brigham Young University, p. 319.
53 J.M. Hanna, M.H. Fitzgerald, J.D. Pearson, A. Howard and J.M. Hanna, 'Selective migration from Samoa: a longitudinal study of pre-migration differences in social and psychological attitudes', *Social Biology*, 1990, vol. 37, pp. 204–14.
54 J. Beston and R.M. Beston, 'An interview with Albert Wendt', *World Literature Written in English*, 1977, vol. 16, p. 153.
55 J. Clifford, 'Traveling cultures', in L. Grossberg, C. Nelson and P. Treichler (eds), *Cultural Studies*, New York, Routledge, 1992, p. 108.
56 Fairbairn-Dunlop, op. cit., pp. 340–1.
57 Yi-Fu Tuan, in J. Western, *A Passage to England. Barbadian Londoners Speak of Home*, London, UCL Press, 1992, p. 269.
58 Wendt, *Flying Fox*, op. cit., pp. 67, 128.
59 Wendt, in Beston and Beston, op. cit., p. 157.
60 A. Wendt, 'Towards a New Oceania', in G. Amirthanayagam (ed.), *Writers in East-West Encounter: New Cultural Bearings*, London, Macmillan, 1982, p. 203.
61 M. Strathern, 'Or, rather, on not collecting Clifford', *Social Analysis*, 1991, vol. 29, p. 90.
62 G. Marcus, 'Past, present and emergent identities: requirements for ethnographies of late twentieth-century modernity worldwide', in S. Lash and J. Friedman (eds), *Modernity and Identity*, Oxford, Blackwell, 1992, p. 312.
63 Ioane, op. cit., p. 315.
64 Wendt, in Beston and Beston, op. cit., p. 153.
65 Wendt, *Ola*, op. cit., p. 347.
66 A. Wendt, 'Novelists and historians and art of remembering', in A. Hooper *et al.* (eds), *Class and Culture in the South Pacific*, Suva, Institute of Pacific Studies, 1987, p. 8.

INDEX